AF801753

Interdisciplinary Cancer Research

Volume 10

Series Editor
Nima Rezaei
Department of Clinical Immunology, Karolinska Institutet, Stockholm, Sweden

Cancer Immunology Project (CIP), Universal Scientific Education and Research Network (USERN), Stockholm, Sweden

Editorial Board Members
Rodrigo Aguiar, Universidade Federal de São Paulo, São Paulo, Brazil
Atif A. Ahmed, University of Missouri–Kansas City, Kansas City, MO, USA
Maria R. Ambrosio, University of Siena, Siena, Italy
Mehmet Artac, Department of Medical Oncology, Necmettin Erbakan University, Konya, Türkiye
Tanya N. Augustine, Faculty of Health Sciences, University of the Witwatersrand, Johannesburg, South Africa
Rolf Bambauer, Institute for Blood Purification, Homburg, Germany
Ajaz Ahmad Bhat, Division of Translational Medicine, Sidra Medical and Research Center, Doha, Qatar
Luca Bertolaccini, European Institute of Oncology, Milan, Italy
Chiara Bianchini, University Hospital of Ferrara, Ferrara, Italy
Milena Cavic, Institute of Oncology and Radiology of Serbia, Belgrade, Serbia
Sakti Chakrabarti, Medical College of Wisconsin, Milwaukee, USA
William C. S. Cho, Department of Clinical Oncology, Queen Elizabeth Hospital, Kowloon, Hong Kong
Anna M. Czarnecka, Maria Sklodowska-Curie National Research Institute of Oncology, Warsaw, Poland
Cátia Domingues, University of Coimbra, Coimbra, Portugal
A. Emre Eşkazan, Istanbul University-Cerrahpaşa, Istanbul, Türkiye
Jawad Fares, Northwestern University, Chicago, IL, USA
Carlos E. Fonseca Alves, São Paulo State University, São Paulo, Brazil
Pascaline Fru, University of the Witwatersrand, Johannesburg, South Africa
Jessica Da Gama Duarte, Olivia Newton-John Cancer Research Institute, Heidelberg, Australia
Mónica C. García, Universidad Nacional de Córdoba, Córdoba, Argentina
Melissa A. H. Gener, Children's Mercy Hospital, Kansas City, MO, USA

José Antonio Estrada Guadarrama, Universidad Autónoma del Estado de México, Toluca, Mexico
Kristian M. Hargadon, Gilmer Hall, Hargadon Laboratory, Hampden–Sydney College, Hampden Sydney, VA, USA
Paul Holvoet, Catholic University of Leuven, Leuven, Belgium
Vladimir Jurisic, Faculty of Medical Sciences, University of Kragujevac, Kragujevac, Serbia
Yearul Kabir, University of Dhaka, Dhaka, Bangladesh
Theodora Katsila, National Hellenic Research Foundation, Athens, Greece
Jorg Kleeff, Martin-Luther-University Halle-Wittenberg, Halle, Germany
Chao Liang, Hong Kong Baptist University, Hong Kong, Hong Kong
Mei Lan Tan, Universiti Sains Malaysia, Pulau Pinang, Malaysia
Weijie Li, Children's Mercy Hospital, Kansas City, MO, USA
Sonia Prado López, Institute of Solid State Electronics, Technische Universität Wien, Vienna, Austria
Muzafar A. Macha, Islamic University of Science and Technology, Awantipora, India
Natalia Malara, Magna Graecia University, Catanzaro, Italy
Adile Orhan, University of Copenhagen, Copenhagen, Denmark
Heriberto Prado-Garcia, National Institute of Respiratory Diseases "Ismael Cosío Villegas", Mexico City, Mexico
Judith Pérez-Velázquez, Helmholtz Zentrum München, Munich, Germany
Wafaa M. Rashed, Children's Cancer Hospital, Cairo, Egypt
Francesca Sanguedolce, University of Foggia, Foggia, Italy
Rosalinda Sorrentino, University of Salerno, Fisciano, Italy
Irina Zh. Shubina, N.N. Blokhin National Medical Research Center of Oncology, Moscow, Russia
Heloisa Sobreiro Selistre de Araujo, Universidade Federal de São Carlos, Sao Carlos, Brazil
Ana Isabel Torres-Suárez, Universidad Complutense de Madrid, Madrid, Spain
Jakub Włodarczyk, Medical University of Lodz, Lodz, Poland
Joe Poh Sheng Yeong, Singapore General Hospital, Singapore, Singapore
Marta A. Toscano, Hospital de Endocrinología y Metabolismo Dr. Arturo Oñativia, Salta, Argentina
Tak-Wah Wong, National Cheng Kung University Medical Center, Tainan, Taiwan
Jun Yin, Central China Normal University, Wuhan, China
Bin Yu, Zhengzhou University, Zhengzhou, China
Nadia M. Hamdy, Biochemistry Department, Ain Shams University, Cairo, Egypt

The "Interdisciplinary Cancer Research" series publishes comprehensive volumes on different cancers and presents the most updated and peer-reviewed articles on human cancers. Over the past decade, increased cancer research has greatly improved our understanding of the nature of cancerous cells which has led to the development of more effective therapeutic strategies to treat cancers. This translational series is of special value to researchers and practitioners working on cell biology, immunology, hematology, biochemistry, genetics, oncology and related fields.

Nima Rezaei
Editor

Hematological Cancer Pathophysiology: An Interdisciplinary Approach

Springer

Editor
Nima Rezaei
Department of Clinical Immunology
Karolinska Institutet
Stockholm, Sweden

Research Center for Immunodeficiencies,
Children's Medical Center
Tehran University of Medical Sciences
Tehran, Iran

ISSN 2731-4561 ISSN 2731-457X (electronic)
Interdisciplinary Cancer Research
ISBN 978-3-031-75444-9 ISBN 978-3-031-75445-6 (eBook)
https://doi.org/10.1007/978-3-031-75445-6

© The Editor(s) (if applicable) and The Author(s), under exclusive license to Springer Nature Switzerland AG 2024

This work is subject to copyright. All rights are solely and exclusively licensed by the Publisher, whether the whole or part of the material is concerned, specifically the rights of translation, reprinting, reuse of illustrations, recitation, broadcasting, reproduction on microfilms or in any other physical way, and transmission or information storage and retrieval, electronic adaptation, computer software, or by similar or dissimilar methodology now known or hereafter developed.
The use of general descriptive names, registered names, trademarks, service marks, etc. in this publication does not imply, even in the absence of a specific statement, that such names are exempt from the relevant protective laws and regulations and therefore free for general use.
The publisher, the authors and the editors are safe to assume that the advice and information in this book are believed to be true and accurate at the date of publication. Neither the publisher nor the authors or the editors give a warranty, expressed or implied, with respect to the material contained herein or for any errors or omissions that may have been made. The publisher remains neutral with regard to jurisdictional claims in published maps and institutional affiliations.

This Springer imprint is published by the registered company Springer Nature Switzerland AG
The registered company address is: Gewerbestrasse 11, 6330 Cham, Switzerland

If disposing of this product, please recycle the paper.

Preface

Hematological cancers or blood cancers are a heterogeneous group of malignant neoplasms, which primarily affect the blood and bone marrow. Leukemia, lymphoma, and myeloma are some examples of hematological cancers, which affect more than a million individual annually worldwide.

The Interdisciplinary Cancer Research series publishes comprehensive volumes on different cancers. It plans to present the most updated and peer-reviewed interdisciplinary chapters on cancers. This interdisciplinary book series is of special value to researchers and practitioners working on cell biology, immunology, hematology, biochemistry, genetics, oncology, and related fields. This is the main concept of Cancer Immunology Project (CIP) and Network of Immunity in Infection, Malignancy and Autoimmunity (NIIMA), which are two active interest groups of the Universal Scientific Education and Research Network (USERN).

The tenth volume of the book entitled *Hematological Cancer Pathophysiology: An Interdisciplinary Approach* starts with a general title on interdisciplinary approach in hematological cancers. After presentation of the immune dysregulation in leukemia and lymphoma, the next chapter focuses on intestinal permeability in hematologic malignancies. Autophagy as a limiting factor for the effectiveness of tyrosine kinase inhibitors in blood cancers and the immunogenetic factors in the pathogenesis of myeloproliferative neoplasms are discussed in the following chapters. A few challenges on myeloproliferative neoplasms, T-cell Lymphoma, Hodgkin's lymphoma, non-Hodgkin's lymphoma, and intraocular lymphoma are explained in the next chapters. Finally, Waldenstrom macroglobulinemia and IgG4-related diseases are discussed in the last chapters.

I hope that this interdisciplinary book will be comprehensible, cogent, and of special value for hematologists and oncologists who wish to have an update on pathophysiology of hematological cancers.

Tehran, Iran　　　　Nima Rezaei

Contents

About the Editor

Nima Rezaei MD, PhD gained his medical degree (MD) from Tehran University of Medical Sciences and subsequently obtained an MSc in Molecular and Genetic Medicine and a PhD in Clinical Immunology and Human Genetics from the University of Sheffield, UK. He also spent a short-term fellowship of Pediatric Clinical Immunology and Bone Marrow Transplantation in the Newcastle General Hospital. Professor Rezaei is now the Full Professor of Immunology and Vice Dean of Research and Technologies, School of Medicine, Tehran University of Medical Sciences, and the co-founder and Head of the Research Center for Immunodeficiencies. He is also the Founder of Universal Scientific Education and Research Network (USERN). Prof. Rezaei has already been the Director of more than 200 research projects and has designed and participated in several international collaborative projects. Prof. Rezaei is the editor, editorial assistant, or editorial board member of more than 50 international journals. He has edited more than 100 international books, has presented more than 1000 lectures/posters in congresses/meetings, and has published more than 1500 scientific papers in the international journals.

Interdisciplinary Approach in Hematological Cancers

Arash Heidari, Sara Momtazmanesh, and Nima Rezaei

Abstract

Hematological cancers, namely leukemia, lymphoma, and multiple myeloma, are malignancies with high mortality and morbidity rates. Despite the advances in the treatment of these cancers, the survival rate of patients has not been much improved. These cancers also detrimentally affect patients' quality of life during the surviving years. The nature of hematological cancers and also the various treatment approaches, including chemotherapy, radiotherapy, and hematopoietic stem cell transplantation (HSCT), have the potential to involve other body systems, leading to neurological, psychological, renal, cardiovascular, metabolic and endocrine, gastrointestinal, musculoskeletal, reproductive, and ocular complications. This chapter reviews the involvement of different organ systems in hematologic cancers. These complications may further deteriorate the prognosis, impair the quality of life, reduce patients' compliance to treatment, and contribute to high morbidity and mortality. Moreover, a growing body of evidence supports the significant roles of the immune system in tumorigenesis

A. Heidari · S. Momtazmanesh
Research Center for Immunodeficiencies, Children's Medical Center, Tehran University of Medical Sciences, Tehran, Iran

Network of Immunity in Infection, Malignancy and Autoimmunity (NIIMA), Universal Scientific Education and Research Network (USERN), Tehran, Iran

N. Rezaei (✉)
Research Center for Immunodeficiencies, Children's Medical Center, Tehran University of Medical Sciences, Tehran, Iran

Department of Immunology, School of Medicine, Tehran University of Medical Sciences, Tehran, Iran

Cancer Immunology Project (CIP), Universal Scientific Education and Research Network (USERN), Stockholm, Sweden
e-mail: rezaei_nima@tums.ac.ir

© Springer Nature Switzerland AG 2022

Interdisciplinary Cancer Research, https://doi.org/10.1007/16833_2022_8
Published online: 20 August 2022

facilitating the progression of hematological cancers by mediating different mechanisms, including immune escape, immune suppression, and cancer-promoting cytokine balance in the tumor microenvironment. Therefore, immunotherapies have drawn considerable attention in addition to the conventional treatments targeting the tumor cells. To conclude, patients with hematologic cancers require a multidisciplinary healthcare team, including oncologists, cardiologists, internists, neurologists, psychiatrists, nephrologists, rheumatologists, and ophthalmologists. Interdisciplinary approaches are critical in improving quality of life, treatment adherence, and survival of these patients.

Keywords

Hematological cancers · Hematopoietic stem cell transplantation · Leukemia · Lymphoma · Multiple myeloma

1 Introduction

Hematological cancers are blood disorders causing excessive and uncontrolled production of different blood cell lineages with an abnormal function in the blood-forming tissues like bone marrow. The overproduction of these abnormal cells also interferes with the regular activity and development of other categories of blood cells and affects other body systems, which leads to various manifestations in affected patients (Jurlander 2011). Depending on which type of blood cell is involved, these malignancies are divided into three main categories: leukemia, lymphoma, and myeloma, each subcategorized based on the involved cell line. In this regard, acute myeloid leukemia (AML), acute lymphocytic leukemia (ALL), chronic lymphocytic leukemia (CLL), and chronic myeloid leukemia (CML) are the various types of leukemia, and Hodgkin and non-Hodgkin lymphoma are the main types of lymphoma (Miller et al. 2019).

In general, hematological cancers significantly impact the quality and quantity of life of affected individuals, thereby playing a substantial role in the global cancer burden (GBD 2017 DALYs and HALE Collaborators 2018). For example, multiple myeloma and non-Hodgkin lymphoma were estimated to be responsible for the 2.3 and 7 million disability-adjusted life-years (GBD 2017 DALYs and HALE Collaborators) in 2017, respectively (2018). Although the incidence and prognosis of leukemia depend greatly on its subtype, the incidence cases and mortality rate of leukemia generally in 2019 has been 643,000 and 334,000, respectively (2021). These statistics point to the major impact of hematological cancers on health status and their high burden on society and necessitates high quality of care, primarily associated with increased development status and gender equity in each nation (Keykhaei et al. 2021).

Despite advances in the treatment of blood cancers in recent years, the life expectancy of patients with most types of these cancers has not increased significantly. Furthermore, the quality of life (QoL) of most patients undergoing treatment is greatly affected. The main treatment approaches currently used for hematological cancers include chemotherapy, radiotherapy, targeted therapy, and stem cell transplantation. Studies have shown that most currently used treatments lead to the temporary remission of malignant cells, but the disease often recurs after a while. Also, the side effects of the current treatment approaches, including chemotherapy, are so significant that many patients cannot easily tolerate them (Devine and Larson 1994). The current therapies concentrate primarily on degenerating the malignant cells, and thereby they can cause severe damages to healthy cells sharing common receptors. On the other hand, they ignore seeing the patient as a whole, and they do not consider the individual differences between various patients using personalized medicine, which has shown its theoretical benefits in many disorders, including cancer and infectious diseases (Hosseini Bereshneh et al. 2017; Basiri et al. 2021).

The different effects of hematological cancers and their common treatments on various body systems, as well as the relative inadequacy of current tunnel view-based therapies in increasing life expectancy and improving the QoL of affected patients, remind the substantial need for an interdisciplinary approach to these cancers. Moreover, considering the involvement of the immune system in the pathogenesis of these cancers, novel approaches in the definitive treatment of hematological cancers may be introduced (Dufva et al. 2020). This chapter aims to more comprehensively investigate the involvement of the different body systems in hematological cancers and thereby provide a better understanding of the necessity for interdisciplinary approaches in treating these cancers.

2 Immune Modulations: The Core Component of Interdisciplinary Approaches in Hematological Cancers

The substantial role of the immune system in the pathogenesis and formation of many cancers has long been studied (Candeias and Gaipl 2016; Heidari et al. 2021). In this regard, hematological cancers are of particular importance due to their unique features regarding the application of immune modulation approaches compared to solid tumors. The main six different features of the hematological malignancies include:

1. The involvement of secondary lymphoid tissues in several forms of these cancers (Dhodapkar and Dhodapkar 2015)
2. The state of immune paresis in some hematological cancers like CLL, due to a shared niche between the tumor and healthy immune cells, leading to a reduction in normal B cell population and hypogammaglobulinemia (Dhodapkar and Dhodapkar 2015)
3. The bone marrow involvement, impacting the result of immune therapies, due to the substantial roles of the bone marrow in the production of immune cells (Tokoyoda et al. 2010)
4. The presence of fewer mutations and genetic changes in these cancers compared with solid tumors, making them suitable targets for therapies aiming at selected

driver mutations (However, fewer mutations result in less neo-antigens expression, so fewer options remain to be targeted by the adaptive immune response.) (Dhodapkar and Dhodapkar 2015)

5. The susceptibility of these cancers to be lysed by both adaptive and innate immune systems (Dhodapkar and Dhodapkar 2015)
6. The potential of these cancers to be served as novel targets for immunotherapies due to their physiologic interactions with immune cells because of the immune cell origins of most of these cancers, as shown by the interaction between T-follicular helper cells (TFH) with some hematological tumors by the expression of programmed-death-1 (PD-1) receptors, serving as a suitable immunotherapy target (Tangye et al. 2013)

The tumor microenvironment (TME), in which malignant cells are located, also plays a significant role in the pathogenesis and behaviors of these tumors (Heidari et al. 2021). TME in secondary lymph organs or bone marrow consists of various components like innate and adaptive immune cells, stromal cells, and vascular endothelial cells (Scott and Gascoyne 2014). Typically, at the first stages of tumor development, tumor-suppressing immune cells, such as T cells presenting in TME, prohibit tumor progression (Dhodapkar et al. 2002). As the tumor advances, due to several interactions between the tumor cells and various TME components, like the secretion of immunosuppressive cytokines by TME elements or the direct cell to cell connections between immune and tumor cells through multiple receptors and ligands like PD-1 and programmed death-ligand 1 (PDL-1), the performance of these immune cells is inhibited (Tarte 2017). Moreover, the secretion of immune inhibitory factors like IL-10 and transforming growth factor-beta (TGF-β) by stromal cells or other TME components provides an appropriate cytokine balance for the polarization of the immune cells toward an immunosuppressive direction (Tarte 2017). Therefore, different immunosuppressive cells are developed, including M2-type of tumor-associated macrophages (TAMs), TIMs (tumor-infiltrating myeloid cells), myeloid-derived suppressor cells (MDSCs), tumor-associated neutrophils (TANs), and tumor-associated dendritic cells (TADCs). These cells can suppress innate and adaptive anti-tumor responses, leading to immune escape and tumor resistance to the therapy (Höpken and Rehm 2019).

Due to the prominent role of the immune system in the pathogenesis and manifestations of hematological cancers, immune modulation methods are considered one of the core approaches for fighting against these cancers, and their benefits have long been recognized. Generally, immune modulation methods harness the host immune cells against the tumor or introduce new immune cells into the host body, like allogeneic hematopoietic stem cell transplantation. The latter could be considered the oldest immunotherapy method, which may be the only curative choice in some patients with end-stage leukemia (Dhodapkar and Dhodapkar 2015). Briefly, different methods of harnessing the immune system against blood cancers include the application of immune-modulating drugs (Haslett et al. 1998), monoclonal antibodies (Ab), vaccines for boosting immune response, immune checkpoint inhibitors using methods like inhibiting PD-PDL1 interaction, T-cell immunoglobulin domain, mucin domain 3 (Tim-3), or lymphocyte-activation gene 3 (Lag-3) (Bryan and Gordon 2015), targeting different TME components and inserting genetically manipulated immune cells like chimeric antigen receptor T (CAR-T) cells or chimeric antigen receptor natural killer (CAR-NK) cells (Maus et al. 2014).

Monoclonal antibodies can exert their anti-tumor effects by several mechanisms like direct anti-tumor invasion, triggering antibody-dependent cellular cytotoxicity (ADCC), triggering complement-mediated cytotoxicity, and increasing antigen-presentation by stimulating Fc receptors on dendritic cells, leading to more effective recognition and destruction of the tumor cells by the adaptive immune system (Weiner et al. 2009). In this regard, ADCC occurs when the Fc domains of the attached antibodies to tumor cells bind to its receptors in effector immune cells and thereby trigger the immune cell activity leading to tumor destruction.

Cancer vaccines are relatively novel approaches in hematological cancers immunomodulation, benefiting from the theoretical point of more specifically targeting the tumor cells. The clonal antibodies secreted by B-lymphocyte tumors are the most popular tumor antigens used by this method. Because of the immunosuppressive effects exerted by the tumor, most clinical trials using vaccine therapy in hematological cancers have failed until now (Nahas et al. 2018). However, it has been shown that vaccination with hybridoma-derived autologous tumor immunoglobulin in the first remission could enhance disease-free survival in follicular lymphoma (Schuster et al. 2011). In general, studies on vaccines are in their infancy, and more evidence is needed to identify vaccines' accurate effects on hematological cancers.

In hematological cancers, as in solid tumors, inhibition of cellular-immunity functions of T cells by engaging the co-inhibitory receptors with them is one of the recognized mechanisms for the immune escape of the tumor (Page et al. 2014). In this regard, several pathways, including PD-PDL1, Tim-3, Lag-3, and B- and T-lymphocyte attenuator (BTLA), play vital roles, having the potential to be suppressed by immune checkpoint inhibitors like pidilizumab (Bryan and Gordon 2015).

Furthermore, the growing data have shown the potentials and advantages of applying chimeric antigen-receptor T (CAR-T) cells to treat hematological cancers (Maus et al. 2014). In this approach, T cells are genetically manipulated to express receptors against some tumor-specific antigens, thereby prohibiting tumor progression. However, selecting the most appropriate receptor in this method can be challenging due to the potential of cross-reactivity with antigens on healthy cells (Linette et al. 2013). The other cell therapy approaches include the administration of engineered natural killer cells (NK), natural killer T cells (NKT), tumor-infiltrating lymphocytes, and T-cell receptors (Shimasaki et al. 2020).

3 Evidence of the Involvement of Different Organs in Hematological Cancers

Hematological malignancies, immune modulation methods and the adverse effects of the treatments can affect various organs, including neurological, psychological, renal, cardiovascular, metabolic and endocrine, gastrointestinal, musculoskeletal, reproductive, and ocular systems (Fig. 1). Therefore, a multidisciplinary approach is vital in managing hematological cancers and ensuring the optimal quality of care in these patients.

3.1 Neurologic Complications

Although treatment methods for hematological cancers are rapidly evolving, many of these methods, especially various chemotherapeutic drugs, can have side effects on both the central and peripheral nervous systems (CNS and PNS) (Magge and DeAngelis 2015). Unlike taste receptors and olfactory neurons, which can quickly regenerate when damaged by chemotherapy (Comeau et al. 2001), many neural cells

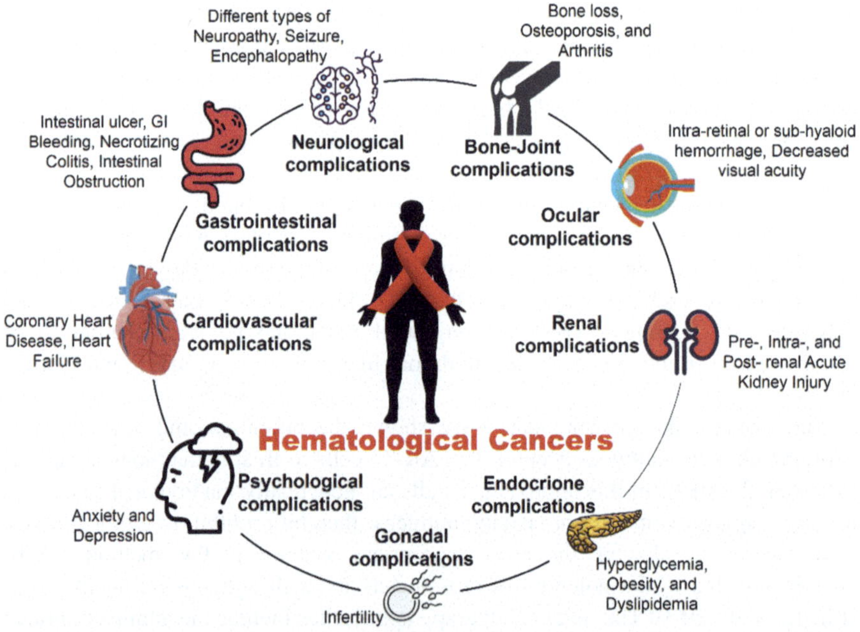

Fig. 1 Hematological malignancies and their related therapeutic approaches can interfere with the normal function of various organs, including neurological, psychological, renal, cardiovascular, metabolic and endocrine, gastrointestinal, musculoskeletal, reproductive, and ocular systems. This figure demonstrates the most remarkable manifestations of the involvement of the mentioned body systems

do not divide or divide very slowly. Therefore, the nervous system is susceptible to being irreversibly damaged by chemotherapy. Additionally, chemotherapy-induced neurotoxicity (CIPN) may make patients not tolerate the entire course of treatment, leading to treatment failure.

The most common CIPN observed in patients with hematological cancers is peripheral neuropathy, caused by the direct involvement of peripheral nerves (Magge and DeAngelis 2015). Moreover, the involvement of the central nervous system (CNS) may lead to more severe conditions, including seizure, vision loss, encephalopathy, speech abnormality, and headaches. In addition, chemotherapeutic drugs that cannot penetrate the blood–brain barrier (BBB) can cause a stroke by inducing coagulopathy, or they also can provoke cognitive changes in the setting of metabolic disturbances.

The two most common chemotherapy drugs widely used in hematological cancers are Vincristine and Bortezomib (Danish and Santomasso 2021). In addition to peripheral neuropathy, which can persist long after discontinuation, they have other significant effects on the nervous system. In this regard, Vincristine effects have been shown to mimic nerve cut changes by its direct toxic impacts on axons; on the other hand, Bortezomib induces several transcriptional pathways in the dorsal root ganglia activating various neuroinflammatory pathways and causing central sensitization (Danish and Santomasso 2021). The neurotoxicity effects of vincristine are dose dependent and can be cumulated with repeated dosages; therefore, it is suggested to stop the treatment after the cumulative dose reaches 30–50 mg (Legha 1986). Notably, patients receiving Bortezomib show the highest rate of peripheral neuropathy in the first five cycles of chemotherapy, and they reach a plateau in the following sessions. Hence, a higher cumulative dose of Bortezomib serves as a protective factor in peripheral neuropathy (Meregalli 2015). Along with chemotherapy, novel T-cell-based approaches such as CAR-T cell therapy in hematological cancers may lead to neurotoxic effects referred to as immune effector cell-associated neurotoxicity syndrome (ICANS) by various mechanisms, including endothelial cell damages, BBB degeneration, and increased CNS inflammatory markers (Kelly and Karcher 2005).

In addition to the drugs used to treat hematological cancers, the nature of the disease can also affect the nervous system. For example, it has been demonstrated that 20% of people with multiple myeloma had some degrees of paraneoplastic neuropathy even before starting treatment, and 50% of them showed small or large nerve fibers dysfunction (Mileshkin et al. 2006; Richardson et al. 2010). The most prevalent form of neuropathy in these patients is mild axonal sensorimotor neuropathy, making them susceptible to more dangerous forms of CIPN (5). Moreover, the initiating manifestations of non-Hodgkin lymphoma might be different types of neuropathies including, plexopathy, mononeuropathy, or polyneuropathy (Kelly and Karcher 2005). In contrast, Hodgkin lymphoma may trigger peripheral nerve inflammation and lead to Guillain-Barré syndrome or inflammatory plexopathy (Kelly and Karcher 2005).

The differentiation between tumor-caused neurological manifestations and therapy-induced neurotoxicity is essential in choosing the proper management in hematological cancers. In this regard, the timing of the onset of neurological symptoms and specialized neurological examinations can be beneficial. Early detection of neurological manifestations in hematological cancers may lead to drug change or dose adjustment of the medication regimen, leading to a better QoL and therapeutic compliance in these patients.

3.2 Psychiatric Complications

Psychological problems are one of the most prevalent comorbidities in patients with hematological cancers. The reasons for this high co-occurrence can be summarized into the severe nature of the disease, the general belief that most types of these cancers are incurable, and the side effects of different therapeutic approaches. In this regard, the most associated psychological disorders are anxiety disorders and depression (Gregurek et al. 2010). Patients with hematological cancers are reported in a study on 10,153 cancer patients to have the third-highest level of anxiety (with a prevalence rate of 22.8%) compared to other cancers following gynecological and lung cancers, and the second-highest level of depression (with a prevalence rate of 16.9%) following gynecological cancers, after the time point of diagnosis (Linden et al. 2012). Also, the prevalence of depression was approximately 40% higher in women (Linden et al. 2012). However, in another study on 76 hematological cancer patients using the Hospital Anxiety and Depression Scale in Greece, clinical depression and anxiety prevalence rates were reported even higher and were 53.9% and 63.7%, respectively (Bellali et al. 2020). Moreover, a study of 200 elderly hematological cancer patients aged 70 or more showed that their depressive symptoms were significantly higher than those in the control group (Meier et al. 2020). However, no significant difference was observed between these elderly patients and the control group in anxiety level (Meier et al. 2020). Besides, depression and anxiety are also significant problems in long-time survivors of hematological cancer, compared to the control group, which can significantly affect their QoL (Kuba et al. 2019). Overall, a meta-analysis of 94 interview-based studies, which included 10,071 hematological cancer patients, revealed that 38.2% had mood disorders, 20.7% had some forms of depression, and 19.4% had an adaptive disorder (El-Jawahri et al. 2016).

The hematological cancer patients' stress levels are reportedly maximal at the first visit by the physician and gradually decrease from the first visit to the middle time point and from there to the last stage of chemotherapy (Bergerot et al. 2015). In addition to anxiety and depression, other psychological disorders also may affect hematological cancer patients. In this regard, fatigue is frequently reported in these patients. A study on relapse-free hematological cancer patients with fatigue, who had not received chemotherapy for the last 3 months, showed that their fatigue level was strongly associated with depression and moderately associated with reduced performance status (Dimeo et al. 2004). Moreover, in patients with Hodgkin lymphoma, psychotic disorders including delirium, affective, and delusional as well as conversion and behavioral disorders have been reported (Olexenko and Vybornykh 2017).

Treatment approaches may also lead to psychological distress. For a successful bone marrow transplant (BMT), the patient needs to be isolated for several weeks, affecting mental well-being (Akaho et al. 2003). In this regard, BMT was shown to cause post-traumatic stress disorder in 28.4% of hematological cancer patients, measured at 6 months after HCT, and also it may increase the depression rate to 43.3% (El-Jawahri et al. 2016). Moreover, hospitalization for different treatments and HCT may significantly decrease the QoL of hematological cancer patients (El-Jawahri et al. 2016). Various chemotherapeutic drugs also have the potential to affect mental health. For example, Ifosfamide has been reported to cause isolated psychotic disorder and delusion in a hematological cancer patient (Alici-Evcimen and Breitbart 2007).

Psychological illnesses, in general, can reduce patients' compliance with continuing treatment courses, thereby increasing the rate of treatment failure. They also play a significant role in reducing the QoL of patients. Therefore, it is necessary to identify the psychological disorders in these patients as soon as possible and apply necessary modifications based on medication and behavioral therapies. This highlights the necessity of the presence of support groups and professional psychiatric support for these patients (Barata et al. 2016).

3.3 Renal Complications

Kidneys are essential organs that can be negatively affected in hematological cancers or during treatments. The kidney manifestations of hematological cancers are broad and include pre-renal, intra-renal, and post-renal acute kidney injury (AKI) (Luciano and Brewster 2014). A study showed that 43% of hematological cancer patients admitted to the Intensive Care Unit (ICU) had some form of AKI, and 29% of these AKI patients needed renal replacement therapy, of which 72% could not survive (Lanore et al. 1991). Also, in a study of 1,411 patients admitted to the ICU, the diagnosis of lymphoma or leukemia was the most significant risk factor for AKI compared to other factors such as infection, cirrhosis, and cardiovascular diseases (De Mendonça et al. 2000). Generally, AKI risk during the 5 years after diagnosing myeloma and leukemia has been reported to be 26% and 15.4%, respectively (Kitchlu et al. 2019). Pre-renal AKI is considered the most prevalent kidney injury in lymphoma and leukemia, resulting from depletion in circulating volume caused by factors like inadequate fluid intake, emesis, anorexia, and diarrhea (Luciano and Brewster 2014). On the other hand, post-renal AKI in hematological cancers is mainly caused by the compression effect of the tumor on urine outflow, retroperitoneal fibrosis, or nephrolithiasis due to tumor lysis syndrome (TLS) (Chong et al. 1981). The intra-renal etiologies of AKI in hematological cancer patients are broad and include ischemic and non-ischemic acute tubular necrosis (ATN), renovascular disorders, tubulointerstitial disorders, and various forms of glomerulopathies (Luciano and Brewster 2014). In this regard, ATN is considered the most prominent cause of intra-renal AKI, with a prevalence rate of 83% among hematological cancer patients with AKI (Harris et al. 1991). The most prevalent factors contributing to

ATN in hematological cancers are kidney ischemia, lysozymuria, TLS, and chemotherapy (Luciano and Brewster 2014). Hematological cancer patients are also more susceptible to developing renal vein and artery thrombosis due to different factors, including leukostasis, disease-associated nephrotic syndrome, chemotherapy, and comorbid conditions (Levin et al. 2003). Depending on the cancer type, the different glomerulopathies in patients with hematological cancers include focal segmental glomerulosclerosis, minimal change disease in ALL (Sathiapalan et al. 1998), and membranoproliferative glomerulonephritis in AML (Dosa et al. 1983).

Additionally, different therapeutic approaches used in hematological cancers might exert damage to the renal system. For example, chemotherapies, such as high dose methotrexate (Amitai et al. 2020), or Tocilizumab (Gupta et al. 2020), and several types of immunotherapies, including kinase inhibitors, and monoclonal antibodies, de-acetylase inhibitors and immune checkpoint inhibitors, and CAR-T cell therapy (Wanchoo et al. 2017; Perazella and Shirali 2018; Gupta et al. 2020), are reported to have nephrotoxic effects.

Renal involvement in hematological cancer may cause intolerance to the required doses of treatments, affecting the prognosis and increasing mortality. Moreover, in some cases, the nephrological manifestations appear as the earliest signs of the disease and can be used as early diagnostic clues in the suspected individuals (Luciano and Brewster 2014). Hence, a periodical consultation with a nephrologist should be considered at all stages, from diagnosis to follow-up visits in survivors.

3.4 Cardiovascular Complications

The cardiovascular system is highly prone to either complications from hematological cancers or the toxic effects of the therapies (Howden et al. 2018). It has been shown that cardiovascular diseases (CVD) are the second leading cause of mortality in hematological cancer patients after the deaths associated with direct tumor invasiveness (Gibson et al. 2016). CVD and hematological cancers share similar risk factors, including high BMI, smoking, and high blood pressure. Hematological cancers also increase the risk of cardiovascular diseases by inducing oxidative stress and chronic inflammation (Howden et al. 2018). Chronic inflammation and overproduced reactive oxygen synthetase (ROS) play major roles in the development and progression of atherosclerotic diseases (Libby et al. 2002) and insulin resistance (Di Meo et al. 2017). Thereby, hematological cancers could increase the risk of sclerotic plaque rupture and the development of acute coronary syndrome (ACS). Besides, ROS overproduction can reduce cardiac function by directly damaging myocardial cells (Powers et al. 2014). Notably, symptomatic cardiac failure is the most prevalent reported cardiac manifestation among non-Hodgkin lymphoma and acute leukemia patients (Gordon et al. 2016; Assuncao et al. 2017).

In addition, different therapeutic approaches may induce cardiac dysfunction, which may manifest during or even years after receiving the treatments. Anti-cancer therapies may affect both the heart and vascular system resulting in declining cardiac function and increasing the risk of thrombotic events. A study of 2285 cancer patients showed the highest incidence of treatment-related cardiac toxicity in hematological cancers than other cancers (Wang et al. 2015). The candidate therapeutic approaches responsible for the cardiotoxic effects in hematological cancers are chemotherapy, radiotherapy, immunotherapy, stem cell transplantation, and targeted therapies (Howden et al. 2018). For example, Anthracycline, one of the chemotherapeutic drug classes used in hematological cancers, can damage the DNA of cardiomyocytes by inhibiting topoisomerase 2B. It may reduce cardiac progenitor mesenchymal cells and impair the membrane transport system, leading to heart failure (Valcovici et al. 2016). These cardiac effects are reported to manifest on average 33 days after receiving Anthracycline; however, they may also present acutely (Valcovici et al. 2016). On the other hand, non-Anthracyclin-related chemotherapies like Cytarabine may also lead to several cardiac diseases, like pericarditis, congestive heart failure (CHF), and ischemia (Pai and Nahata 2000).

Furthermore, radiotherapy can damage the myocardium, pericardium, and coronary arteries through various mechanisms, including DNA damage, inflammation induction, fibrosis, and telomere erosion (Groarke et al. 2014). Radiotherapy may induce several cardiac disorders, including coronary artery disease (CAD), CHF, valvular and pericardial abnormalities, conduction disorders, and even sudden cardiac death (Stewart et al. 2013). On the other hand, several targeted therapeutic approaches like the inhibition of B-cell receptor pathways by targeting PI3K and/or Bruton's tyrosine kinase (BTK) can lead to heart dysfunction as these pathways are critical in the growth and function of cardiomyocytes with their cardioprotective effects (Maillet et al. 2013; McMullen et al. 2014). Immunotherapies can also induce cardiotoxic effects. For instance, proteasome inhibitors, checkpoint inhibitors, and some immunomodulatory drugs like pomalidomide and lenalidomide can increase the risk of cardiovascular events (Atrash et al. 2015), myocarditis (Läubli et al. 2015), and venous thromboembolism, respectively (Rajkumar et al. 2010).

Therefore, there is a great need for cardiovascular examination and consultation in patients with hematological cancers at different stages, i.e., at the time of diagnosis, before treatment, during treatment, and even for life in survivors.

3.5 Metabolic and Endocrine Complications

Metabolic disorders, including dysglycemia, obesity, and dyslipidemia, are common findings among patients with hematological cancer (Karmali et al. 2015). These disorders can be both the predisposing risk factors for hematological cancers and the subsequent effects of these malignancies (Majhail et al. 2009). High BMI, insulin resistance, and type 2 diabetes mellitus are associated with a higher incidence of hematologic cancers (Castillo et al. 2012; Nagel et al. 2012). The increased prevalence of metabolic disturbances has also been shown in survivors of these cancers, who had normal metabolic health before diagnosis. The course of cancer and the indicated therapies contribute to this increased prevalence. A study of 7287 hematological cancer survivors with a median age of 32 years reported a prevalence rate of 67% for at least one metabolic/endocrine disorder (Gebauer et al. 2019).

The cancer diagnosis is usually associated with reduced physical activity and increased smoking and dietary intake, promoting metabolic dysfunctions (Blanchard et al. 2003). For instance, 30% of hematological cancer survivors were less physically active than before diagnosis (Blanchard et al. 2003). On the other hand, a successful HSCT necessitates an extended hospitalization and inactivity to prevent infection (Sliwa and Marciniak 1999). Physical inactivity and cancer-associated cytokine release may lead to decreased muscle force and increased insulin resistance, contributing to metabolic syndrome (Chopard et al. 2009; Fearon et al. 2012). In this regard, 49% of adult survivors who received HSCT had reportedly metabolic syndrome (McMillen et al. 2014). In addition to long bed rest, Graft-versus-host disease (GVHD) and associated chemotherapies and radiotherapies play an essential role in this association (Majhail et al. 2009).

The various hematological cancer therapies, including chemotherapy, radiotherapy, immunotherapy, and HSCT, can cause irreversible damage to the endocrine organs, i.e., the thyroid, parathyroid, adrenal, hypothalamic-pituitary, and gonads (Gebauer et al. 2019). For example, treatment with the anti-PD-1 monoclonal antibody, an immune checkpoint inhibitor, can lead to thyroid dysfunction in patients with hematological cancer. In this regard, hypothyroidism was the most observed event in these patients (Elia et al. 2020). Moreover, hypothyroidism is reported as one of the most frequent endocrine abnormalities following HSCT (Paviglianiti 2020). Studies showed that the potential risk factors for HCT-related hypothyroidism include prolonged immunosuppressive therapies after HCT, chronic GVHD, HLA B35 in donors, and female to male transplantations (Savani et al. 2009).

Hyperglycemia is another prevalent endocrine abnormality in hematological cancer patients, frequently caused by glucocorticoid therapy as part of chemotherapy regimens. Glucocorticoids are also used for treating GVHD in these patients (Paviglianiti 2020). Moreover, calcineurin inhibitors like tacrolimus can cause hyperglycemia through their direct effects on insulin production and release (Vincenti et al. 2007) and their role in islet cell apoptosis after toxic doses (Clore and Thurby-Hay 2009). Another risk factor of hyperglycemia is total parenteral nutrition (TPN). In this regard, studies have shown significant hyperglycemia rates in HCT recipients treated with TPN than HCT recipients who were not treated with TPN (Sheean et al. 2006).

Given the better survival of hematological cancer patients following advances in cancer therapies, the prevalence of metabolic and endocrine disorders is high. Therefore, there is a need for regular follow-ups examining these disorders in hematological cancer survivors. On the other hand, the presence of metabolic and endocrine disorders during treatment may lead to decreased patient compliance and treatment failure. Hence, early diagnosis and treatment of metabolic endocrine disorders should be prioritized in hematological cancer patients.

3.6 Gastrointestinal Complications

The gastrointestinal tract can be affected during all three hematological cancers, including leukemia, lymphoma, and multiple myeloma. The prevalence of gastrointestinal involvement ranged from 14.7% to 25% in patients with leukemia at autopsy (Ebert and Hagspiel 2012). These involvements are more common in acute than chronic leukemia and occur mainly in the stomach, ileum, and proximal colon (Kirshbaum and Preuss 1943). The leukemia-related gastrointestinal involvement can be seen in the form of nodules, plaques, polyps, and diffuse infiltrates (Ebert and Hagspiel 2012). Notably, leukemic plaques may manifest with ulceration and intestinal perforation. In contrast, nodular lesions may be associated with intussusception and intestinal obstruction. Patients with diffuse infiltrate lesions may present unspecific manifestations, including abdominal pain, diarrhea, or gastrointestinal bleeding (Prolla and Kirsner 1964). The lesions in the esophagus and stomach are categorized into hemorrhagic, infiltrative, agranulocytic, and fungal and may imitate gastric carcinoma (Cavins et al. 1959; Prolla and Kirsner 1964). The intestinal lesions, in contrast, are primarily hemorrhagic and infiltrative (Ebert and Hagspiel 2012). Moreover, necrotizing colitis may also be seen in 11.6% to 32.5% of leukemia patients treated with chemotherapy, and it manifests with fever, abdominal pain, neutropenia, and thickening of the bowel wall (Zorrilla et al. 2006; Mullassery et al. 2009).

The gastrointestinal tract is also the most common site for extranodal non-Hodgkin lymphomas, involved in 30% to 40% of the patients (Thomas et al. 2019). Among the GI organs, the stomach and small intestine are the most common sites. Mucosa-associated lymphoid tissue (MALT) lymphoma, followed by diffuse large B-cell lymphoma (DLBL), is the most prevalent lymphoma involving the stomach (Fischbach 2014). The GI lymphoma presents very heterogeneously. Its non-specific manifestations range from mild symptoms, like dyspepsia, nausea, diarrhea, and abdominal pain, to more severe manifestations, including gastrointestinal bleeding, vomiting, weight loss, and intestinal obstruction. Celiac disease, inflammatory bowel disease, helicobacter pillory infection, and autoimmune disorders may predispose patients to develop GI lymphomas.

Furthermore, multiple myeloma is usually restricted to bone marrow, but its extraskeletal spread in the form of plasmacytomas is seen in a significant number of patients. Although these plasmacytomas may rarely cause GI manifestation, there are several case reports of overt GI bleeding due to these plasmacytomas (Telakis et al. 2009; Lin et al. 2013). Moreover, mucosal infiltration of amyloid protein has been reported in patients with multiple myeloma, which manifests with GI bleeding, malabsorption, or GI obstruction (Esfandyari et al. 2007). Treatment of hematological cancers may also cause several GI manifestations, including nausea, vomiting, dyspepsia, and even hepatitis B reactivation in patients receiving immunosuppressive drugs following HCT (Zannella et al. 2019). Therefore, the GI manifestations require special attention during treatment, as they can significantly affect the patient's compliance and QoL.

3.7 Musculoskeletal Complications

Bone involvements are common complications in hematological cancer patients. It is not yet clear to what extent the cancer behavior or the adverse effects of chemotherapy are responsible for bone complications in these patients. Bone loss is one of the most prevalent bone complications affecting lymphoma patients following chemotherapy, which may increase the likelihood of incident fractures, affecting the patients' QoL (Vishnevskia-Dai et al. 2020; Bodden et al. 2021). Osteoporosis and osteonecrosis are also prevalent manifestations in patients with non-Hodgkin lymphoma (NHL) before treatment (Thompson et al. 2007). A study on 181 adults with different types of hematological cancer excluding multiple myeloma reported a bone loss in 65% of patients, of whom 38% had osteopenia and 27% osteoporosis (Ruchlemer et al. 2018). This study found no association between the time of diagnosis and the extent of bone loss (Ruchlemer et al. 2018). However, dual-energy X-ray absorptiometry results varied depending on the diagnosis and showed a bone loss in 49% of CLL patients, compared with 67% in NHL and 88% in Hodgkin patients (Ruchlemer et al. 2018). The results of another study on NHL patients revealed a reduced bone mineral density and an increased bone turnover, bone resorption markers, osteoblast regulators, and bone formation markers following chemotherapy with rituximab, cyclophosphamide, hydroxydaunorubicin hydrochloride, vincristine, and prednisone with more severity in males and older participants (Anargyrou et al. 2019). The risk factors associated with bone loss following chemotherapy are female gender, lower creatinine clearance, and higher lactate dehydrogenase level (Paccou et al. 2014). In patients with lymphoma, treatment with zoledronic acid and calcium and vitamin D has effectively prevented further bone loss in multiple studies (Hari et al. 2013; Westin et al. 2013).

Joints are also highly vulnerable to involvement in hematological cancers. For example, in leukemia, the infiltration of the malignant cells into periosteal or capsular spaces may cause a synovial reaction in the form of synovitis (Ravindran and Anoop 2011). Moreover, peri-articular hemorrhages and arthritis may be seen due to synovial infiltration, pro-inflammatory cytokine release, and reactive joint inflammation secondary to the leukemic involvement of the adjacent bone (Ravindran and Anoop 2011). Moreover, 29.4% of patients with NHL-derived arthritis, caused by lymphoma cell-related synovial invasion and pro-inflammatory cytokine release by lymphoma cells, showed polyarthritis manifestations resembling rheumatoid arthritis, and 35.3% of them had monoarticular involvements (Demircay et al. 2013). The knee joints are considered the most prevalent site for arthritis in these patients (Demircay et al. 2013). In multiple myeloma, on the other hand, the joint involvements may be as symmetrical and asymmetrical polyarthritis due to several mechanisms including amyloid or immunoglobins infiltration or metabolic abnormalities (Jorgensen et al. 1996).

Therefore, bone and joint involvement in hematological cancers necessitate early and prophylactic treatments to improve the patient's QOL and tolerance to cancer treatments, leading to a high therapy success rate.

3.8 Fertility and Gonadal Complications

Given the increased survival of patients with hematological cancer due to the advanced therapeutic armamentarium, complications like infertility resulting from treatment or the disease are more pronounced in survivors. Infertility may arise from the gonadotoxic effects of various treatments used in hematological cancers, including chemotherapy, radiotherapy, and HSCT (Massenkeil et al. 2016). The disease course itself may also result in infertility (Ozdemir and Bozdag 2020). For instance, ovarian function is significantly affected by the effects of hemopathy on the ovarian follicular layer even before starting the treatment (Lawrenz et al. 2012). Lower anti-Mullerian hormone (AMH) levels, follicle count, and oocytes have been reported in newly diagnosed lymphoma patients compared to 365 healthy control and 128 patients with other types of cancers (Lekovich et al. 2016).

Chemotherapy-related infertility depends on the type of drug used, the intensity of the regimen, and the patient's age. Chemotherapeutic agents can damage ovaries by several mechanisms. They induce apoptosis of the oocytes in follicles as the main mechanism, leading to the destruction of germ cells and premature ovarian failure (Oktem and Oktay 2007a, b; Soleimani et al. 2011). Alkylating agents can also accelerate the reduction of oocytes by 10 years (Oktem and Oktay 2007a, b). Chemotherapeutic agents like doxorubicin may also be associated with vascular complications and stromal cells damage in the ovary (Bedoschi et al. 2016). In the male sex, some chemotherapeutic agents, including alkylating and platinum-based agents, can be associated with oligo- and azoospermia, and some cause testicular atrophy (Kinsella 1989; Meistrich 2013).

Radiotherapy may also damage the ovaries directly, leading to premature ovarian failure. The extent of damage depends on the irradiation dose and age of the patient. It has been reported that 20 grays (Gy) radiotherapy over 6 weeks can cause infertility in young women and children (Shapira et al. 2014). Moreover, testicles are significantly prone to be affected by radiotherapy due to the rapid division of their germ cells, resulting in oligospermia or azoospermia depending on the dose (Shalet 1993). In this regard, a fractionated radiation dose of 2 Gy may result in permanent infertility in males (Howell and Shalet 1998).

Infertility is one of the leading long-term complications of HSCT as well. The intensity of conditioning regimens, including chemoradiotherapy, given before HSCT significantly contributes to future infertility. A study on women treated with HSCT has shown the rate of recovery of gonadal function to be 10–14%, and the pregnancy rate in these patients was only 3% (Spinelli et al. 1994).

The potential treatment methods for preserving fertility include hormonal therapy, ovaries transposition, and cryopreservation of oocytes, embryo, ovarian tissue, or semen (Ozdemir and Bozdag 2020). In this regard, an in-depth conversation between the patients and their expert therapists before initiating the first therapies regarding the advantages and disadvantages of fertility-preservation methods is highly suggested (Lee et al. 2017). Besides, fertility counseling should be considered after the diagnosis of hematological cancers, during the treatments, and in survivors intending to have children.

3.9 Ocular Complications

Although ocular manifestations are considered rare in hematological cancers, they might give significant clues about different stages of the disease. In this regard, ocular symptoms may manifest as presenting signs of the disease, they could be recognized during the disease, or they could be the first signs of cancer relapse following remission (Vishnevskia-Dai et al. 2020). There is growing evidence of ocular manifestations of hematological cancers. A cross-sectional study on 133 hematological cancer patients has reported the prevalence rate of ophthalmopathy among ALL, AML, CLL, and CML patients to be 42%, 63%, 33%, and 13%, respectively. In this regard, vision-threatening disorders like subhyaloid hemorrhages were observed only in patients with acute leukemias (Soman et al. 2018). Another study on 102 patients with leukemia showed ocular manifestations in 39.3% of them. The most-reported presentation was intraretinal hemorrhage causing Roth spot. These manifestations were more prevalent in acute leukemia than chronic leukemia. The study also showed that ocular complications are significantly associated with low hemoglobin, high leukocyte count, and low platelet count (Dhasmana et al. 2016).

In addition to cancer, its treatments may also trigger ocular complications. A study on hematological cancer patients has shown the prevalence rate of dry eye to be 53% before HSCT. This rate has increased to 77%, with incidence cases of 51% following HSCT (Giannaccare et al. 2017). Besides, there is a growing literature reporting ocular complications as the first manifestations in hematological cancer patients. In this regard, ptosis, eyelid swelling (Alsalem et al. 2021), loss of intermittent color vision (Saleh et al. 2020), minor visual disturbance like seeing black dot (Hasan and Gomes 2021), and leukostasis retinopathy (Vicini et al. 2020) have been reported as first manifestations of hematological cancer in several case reports. A study has shown that intravitreal injection of methotrexate could resolve intraocular tumor cell infiltration and inflammatory reaction in the eyes (Vishnevskia-Dai et al. 2020).

Hence, the ocular examination should be considered in the baseline and then periodically in all hematological cancer patients, as they may be the first signs of the disease, and their early diagnosis may prevent visual morbidity and its effects on the QoL of these patients. Moreover, intravitreal injection of methotrexate may be an effective treatment approach in managing patients with hematological cancer suffering from ocular involvement. However, there is still a need for more elaboration regarding the prevention and management of ocular manifestations in hematological cancers in future studies.

4 Conclusion

Hematological cancers are a heterogeneous group of malignancies with a significant morbidity and mortality rate, affecting patients' quality of life detrimentally. Not only does the nature of these cancers make them a potential threat to other organ systems, but also the different therapeutic approaches may increase the likelihood of multi-system involvement. This chapter discussed the different neurological, psychological, renal, cardiovascular, metabolic and endocrine, gastrointestinal, musculoskeletal, reproductive, and ocular complications of these cancers and their management.

The most observed neurological complication in hematological cancer patients is neuropathy, resulting from chemotherapy or the nature of the disease. Anxiety and depression are highlighted as the two most prevalent psychological complications, significantly affecting the quality of life and compliance to treatment. Renal involvement can be observed as pre-, intra-, and post-renal acute kidney injury; each has significant effects on therapy results. Cardiovascular disorders and hematological cancers share relatively common risk factors, so they usually co-occur. The cardiac disorders in hematological cancer patients include coronary artery disease, congestive heart failure, and valvular and pericardial abnormalities. Metabolic disorders including hyperglycemia, obesity, and dyslipidemia are other common findings in these patients, caused by reduced physical activity, adverse effects of treatments, and the nature of the disease. On the other hand, gastrointestinal complications due to hematological cancers may include broad manifestations, such as intestinal ulcer, necrotizing colitis, intestinal obstruction, and non-specific manifestations like diarrhea, gastrointestinal bleeding, or abdominal pain. Furthermore, bone loss, osteoporosis, and arthritis are considered the most common bone and joint manifestations in these patients. Cancer, chemotherapy, radiotherapy, and hematopoietic stem cell transplantation may also have detrimental impacts on the gonads and lead to infertility. Ocular manifestations and ophthalmopathy have also been reported in these patients in several forms, including intraretinal or subhyaloid hemorrhage, loss of intermittent color vision, and ptosis.

The involvement of mentioned body systems in hematological cancers has significant effects on the patient's compliance with tolerating severe forms of therapies required for appropriate treatment, thereby contributing to high morbidity and mortality rate in these patients. Moreover, these manifestations may significantly influence the patients' quality of life during the surviving years. Besides, given the better survival rate of hematological cancer patients following advancements in treatments, the prevalence of these complications is increasing. Therefore, there is a substantial need for collaboration between different specialties, including cardiologists, neurologists, psychiatrists, nephrologists, orthopedists, ophthalmologists, internists, and oncologists, to manage these patients effectively. The significance of the cooperation between these specialties in the prognosis of hematological cancer patients necessitates considering interdisciplinary approaches in treating these cancers.

Acknowledgments None.

Compliance with Ethical Standards The authors declare that there is no conflict of interest.

References

Akaho R, Sasaki T, Yoshino M, Hagiya K, Akiyama H, Sakamaki H (2003) Bone marrow transplantation in subjects with mental disorders. Psychiatry Clin Neurosci 57:311–315

Alici-Evcimen Y, Breitbart WS (2007) Ifosfamide neuropsychiatric toxicity in patients with cancer. Psychooncology 16:956–960

Alsalem A, Almasoudi B, Alzahrani G, Sindi L, Alwan J (2021) Orbital swelling and ptosis as an initial presentation of childhood acute lymphoblastic leukaemia: a case report and review of the literature. Case Rep Pediatr 2021:5587767

Amitai I, Rozovski U, El-Saleh R, Shimony S, Shepshelovich D, Rozen-Zvi B, Raanani P, Gafter-Gvili A, Gurion R (2020) Risk factors for high-dose methotrexate associated acute kidney injury in patients with hematological malignancies. Hematol Oncol 38:584–588

Anargyrou K, Fotiou D, Vassilakopoulos TP, Christoulas D, Makras P, Dimou M, Ntanasis-Stathopoulos I, Masouridou S, Angelopoulou MK, Papatheodorou A (2019) Low bone mineral density and high bone turnover in patients with non-Hodgkin's lymphoma (NHL) who receive frontline therapy: results of a multicenter prospective study. Hemasphere 3:e303

Assuncao BMBL, Handschumacher MD, Brunner AM, Yucel E, Bartko PE, Cheng K-H, Campos O, Fathi AT, Tan TC, Scherrer-Crosbie M (2017) Acute leukemia is associated with cardiac alterations before chemotherapy. J Am Soc Echocardiogr 30:1111–1118

Atrash S, Tullos A, Panozzo S, Bhutani M, Van Rhee F, Barlogie B, Usmani S (2015) Cardiac complications in relapsed and refractory multiple myeloma patients treated with carfilzomib. Blood Cancer J 5:e272–e272

Barata A, Wood WA, Choi SW, Jim HS (2016) Unmet needs for psychosocial care in hematologic malignancies and hematopoietic cell transplant. Curr Hematol Malig Rep 11:280–287

Basiri A, Mansouri F, Azari A, Ranjbarvan P, Zarein F, Heidari A, Golchin A (2021) Stem cell therapy potency in personalizing severe COVID-19 treatment. Stem Cell Rev Rep 17:193–213

Bedoschi G, Navarro PA, Oktay K (2016) Chemotherapy-induced damage to ovary: mechanisms and clinical impact. Future Oncol 12:2333–2344

Bellali T, Manomenidis G, Meramveliotaki E, Minasidou E, Galanis P (2020) The impact of anxiety and depression in the quality of life and psychological well-being of Greek hematological cancer patients on chemotherapy. Psychol Health Med 25:201–213

Bergerot CD, Clark KL, Nonino A, Waliany S, Buso MM, Loscalzo M (2015) Course of distress, anxiety, and depression in hematological cancer patients: association between gender and grade of neoplasm. Palliat Support Care 13:115–123

Blanchard CM, Denniston MM, Baker F, Ainsworth SR, Courneya KS, Hann DM, Gesme DH, Reding D, Flynn T, Kennedy JS (2003) Do adults change their lifestyle behaviors after a cancer diagnosis? Am J Health Behav 27:246–256

Bodden J, Sun D, Joseph G, Huang L-W, Andreadis C, Hughes-Fulford M, Lang T, Link T (2021) Identification of non-Hodgkin lymphoma patients at risk for treatment-related vertebral density loss and fractures. Osteoporos Int 32:281–291

Bryan LJ, Gordon LI (2015) Blocking tumor escape in hematologic malignancies: the anti-PD-1 strategy. Blood Rev 29:25–32

Candeias SM, Gaipl US (2016) The immune system in cancer prevention, development and therapy. Anti Cancer Agents Med Chem 16:101–107

Castillo JJ, Mull N, Reagan JL, Nemr S, Mitri J (2012) Increased incidence of non-Hodgkin lymphoma, leukemia, and myeloma in patients with diabetes mellitus type 2: a meta-analysis of observational studies. Blood 119:4845–4850

Cavins JA, Levin HS, Day HJ (1959) Chronic myelogenous leukemia with gastric infiltration: report of a case simulating carcinoma of the stomach. N Engl J Med 260:1111–1114

Chong B, Trew P, Meng L, Pitney W (1981) Anuric renal failure due to encasement of the ureters by lymphoma—ureteric obstruction without dilatation. Aust NZ J Med 11:542–544

Chopard A, Hillock S, Jasmin BJ (2009) Molecular events and signalling pathways involved in skeletal muscle disuse-induced atrophy and the impact of countermeasures. J Cell Mol Med 13: 3032–3050
Clore JN, Thurby-Hay L (2009) Glucocorticoid-induced hyperglycemia. Endocr Pract 15:469–474
Comeau TB, Epstein JB, Migas C (2001) Taste and smell dysfunction in patients receiving chemotherapy: a review of current knowledge. Support Care Cancer 9:575–580
Danish H, Santomasso BD (2021) Neurotoxicity biology and management. Cancer J 27:126–133
De Mendonça A, Vincent J-L, Suter P, Moreno R, Dearden N, Antonelli M, Takala J, Sprung C, Cantraine F (2000) Acute renal failure in the ICU: risk factors and outcome evaluated by the SOFA score. Intensive Care Med 26:915–921
Demircay E, Hornicek FJ, Mankin HJ, Degroot H (2013) Malignant lymphoma of bone: a review of 119 patients. Clin Orthop Relat Res 471:2684–2690
Devine SM, Larson RA (1994) Acute leukemia in adults: recent developments in diagnosis and treatment. CA Cancer J Clin 44:326–352
Dhasmana R, Prakash A, Gupta N, Verma S (2016) Ocular manifestations in leukemia and myeloproliferative disorders and their association with hematological parameters. Ann Afr Med 15:97
Dhodapkar MV, Dhodapkar KM (2015) Immune modulation in hematologic malignancies. Semin Oncol 42:617–625
Dhodapkar MV, Krasovsky J, Olson K (2002) T cells from the tumor microenvironment of patients with progressive myeloma can generate strong, tumor-specific cytolytic responses to autologous, tumor-loaded dendritic cells. Proc Natl Acad Sci 99:13009–13013
Di Meo S, Iossa S, Venditti P (2017) Skeletal muscle insulin resistance: role of mitochondria and other ROS sources. J Endocrinol 233:R15–R42
Dimeo F, Schmittel A, Fietz T, Schwartz S, Köhler P, Böning D, Thiel E (2004) Physical performance, depression, immune status and fatigue in patients with hematological malignancies after treatment. Ann Oncol 15:1237–1242
Dosa S, Phillips T, Antonovych T, Segal A, Guha A, Thompson A (1983) Acute myelomonocytic leukemia associated with nephrotic syndrome. Nephron 34:125–129
Dufva O, Pölönen P, Brück O, Keränen MA, Klievink J, Mehtonen J, Huuhtanen J, Kumar A, Malani D, Siitonen S (2020) Immunogenomic landscape of hematological malignancies. Cancer Cell 38:380–399.e313
Ebert EC, Hagspiel KD (2012) Gastrointestinal manifestations of leukemia. J Gastroenterol Hepatol 27:458–463
El-Jawahri AR, Vandusen HB, Traeger LN, Fishbein JN, Keenan T, Gallagher ER, Greer JA, Pirl WF, Jackson VA, Spitzer TR (2016) Quality of life and mood predict posttraumatic stress disorder after hematopoietic stem cell transplantation. Cancer 122:806–812
Elia G, Ferrari SM, Galdiero MR, Ragusa F, Paparo SR, Ruffilli I, Varricchi G, Fallahi P, Antonelli A (2020) New insight in endocrine-related adverse events associated to immune checkpoint blockade. Best Pract Res Clin Endocrinol Metab 34:101370
Esfandyari T, Abraham SC, Arora AS (2007) Gastrointestinal plasmacytoma that caused anemia in a patient with multiple myeloma. Nat Clin Pract Gastroenterol Hepatol 4:111–115
Fearon KC, Glass DJ, Guttridge DC (2012) Cancer cachexia: mediators, signaling, and metabolic pathways. Cell Metab 16:153–166
Fischbach W (2014) Gastric MALT lymphoma–update on diagnosis and treatment. Best Pract Res Clin Gastroenterol 28:1069–1077
GBD 2017 DALYs and HALE Collaborators (2018, Nov 10) Global, regional, and national disability adjusted lifeyears (DALYs) for 359 diseases and injuries and healthy life expectancy (HALE) for 195 countries and territories, 1990–2017: a systematic analysis for the Global Burden of Disease Study 2017. Lancet 392(10159):1859–1922. https://doi.org/10.1016/S0140-6736(18)32335-3
Gebauer J, Higham C, Langer T, Denzer C, Brabant G (2019) Long-term endocrine and metabolic consequences of cancer treatment: a systematic review. Endocr Rev 40:711–767

Giannaccare G, Bonifazi F, Sessa M, Dan E, Arpinati M, Fresina M, Bandini G, Cavo M, Versura P, Campos E (2017) Ocular surface analysis in hematological patients before and after allogeneic hematopoietic stem cell transplantation: implication for daily clinical practice. Eye 31:1417–1426

Gibson TM, Ehrhardt MJ, Ness KK (2016) Obesity and metabolic syndrome among adult survivors of childhood leukemia. Curr Treat Options in Oncol 17:1–13

Global Burden of Disease Collaborative Network (2021) Global Burden of Disease Study 2019 (GBD 2019) reference life table. Institute for Health Metrics and Evaluation (IHME), Seattle

Gordon MJ, Danilova O, Spurgeon S, Danilov AV (2016) Cardiac non-Hodgkin's lymphoma: clinical characteristics and trends in survival. Eur J Haematol 97:445–452

Gregurek R, Bras M, Dordević V, Ratković AS, Brajković L (2010) Psychological problems of patients with cancer. Psychiatr Danub 22(2):227–230

Groarke JD, Nguyen PL, Nohria A, Ferrari R, Cheng S, Moslehi J (2014) Cardiovascular complications of radiation therapy for thoracic malignancies: the role for non-invasive imaging for detection of cardiovascular disease. Eur Heart J 35:612–623

Gupta S, Seethapathy H, Strohbehn IA, Frigault MJ, O'Donnell EK, Jacobson CA, Motwani SS, Parikh SM, Curhan GC, Reynolds KL (2020) Acute kidney injury and electrolyte abnormalities after chimeric antigen receptor T-cell (CAR-T) therapy for diffuse large B-cell lymphoma. Am J Kidney Dis 76:63–71

Hari P, DeFor TE, Vesole DH, Bredeson CN, Burns LJ (2013) Intermittent zoledronic acid prevents bone loss in adults after allogeneic hematopoietic cell transplantation. Biol Blood Marrow Transplant 19:1361–1367

Harris K, Hattersley J, Feehally J, Walls J (1991) Acute renal failure associated with haematological malignancies: a review of 10 years experience. Eur J Haematol 47:119–122

Hasan FM, Gomes RR (2021) Demographic profile & correlation between retinal exudates and hematological parameters in leukemic patients. J Hematol 2640:2823

Haslett PA, Corral LG, Albert M, Kaplan G (1998) Thalidomide costimulates primary human T lymphocytes, preferentially inducing proliferation, cytokine production, and cytotoxic responses in the CD8+ subset. J Exp Med 187:1885–1892

Heidari A, Sharif PM, Rezaei N (2021) The association between tumor-associated macrophages and glioblastoma: a potential target for therapy. Curr Pharm Des. https://doi.org/10.2174/1381612827666210816114003

Höpken UE, Rehm A (2019) Targeting the tumor microenvironment of leukemia and lymphoma. Trends Cancer 5:351–364

Hosseini Bereshneh A, Morshedi F, Hematyar M, Kaki A, Garshasbi M (2017) Pharmacogenetics and personalized medicine in pancreatic cancer. Acta Med Iran 55:194–199

Howden EJ, La Gerche A, Arthur JF, McMullen JR, Jennings GL, Dunstan DW, Owen N, Avery S, Kingwell BA (2018) Standing up to the cardiometabolic consequences of hematological cancers. Blood Rev 32:349–360

Howell S, Shalet S (1998) Gonadal damage from chemotherapy and radiotherapy. Endocrinol Metab Clin N Am 27:927–943

Jorgensen C, Guerin B, Ferrazzi V, Bologna C, Sany J (1996) Arthritis associated with monoclonal gammapathy: clinical characteristics. Rheumatology 35:241–243

Jurlander J (2011) Hematological malignancies, leukemias and lymphomas. In: Encyclopedia of cancer. Springer, Cham, pp 1640–1644

Karmali R, Dalovisio A, Borgia JA, Venugopal P, Kim BW, Grant-Szymanski K, Hari P, Lazarus H (2015) All in the family: clueing into the link between metabolic syndrome and hematologic malignancies. Blood Rev 29:71–80

Kelly JJ, Karcher DS (2005) Lymphoma and peripheral neuropathy: a clinical review. Muscle Nerve 31:301–313

Keykhaei M, Masinaei M, Mohammadi E, Azadnajafabad S, Rezaei N, Moghaddam SS, Rezaei N, Nasserinejad M, Abbasi-Kangevari M, Malekpour M-R (2021) A global, regional, and national

survey on burden and Quality of Care Index (QCI) of hematologic malignancies; global burden of disease systematic analysis 1990–2017. Exp Hematol Oncol 10:1–15
Kinsella TJ (1989) Effects of radiation therapy and chemotherapy on testicular function. Prog Clin Biol Res 302:157–171. Discussion 172–177
Kirshbaum JD, Preuss FS (1943) Leukemia: a clinical and pathologic study of one hundred and twenty-three fatal cases in a series of 14,400 necropsies. Arch Intern Med 71:777–792
Kitchlu A, McArthur E, Amir E, Booth CM, Sutradhar R, Majeed H, Nash DM, Silver SA, Garg AX, Chan CT (2019) Acute kidney injury in patients receiving systemic treatment for cancer: a population-based cohort study. J Natl Cancer Inst 111:727–736
Kuba K, Esser P, Mehnert A, Hinz A, Johansen C, Lordick F, Götze H (2019) Risk for depression and anxiety in long-term survivors of hematologic cancer. Health Psychol 38:187
Lanore JJ, Brunet F, Pochard F, Bellivier F, Dhainaut J-F, Vaxelaire J-F, Giraud T, Dreyfus F, Dreyfuss D, Chiche J-D (1991) Hemodialysis for acute renal failure in patients with hematologic malignancies. Crit Care Med 19:346–351
Läubli H, Balmelli C, Bossard M, Pfister O, Glatz K, Zippelius A (2015) Acute heart failure due to autoimmune myocarditis under pembrolizumab treatment for metastatic melanoma. J Immunother Cancer 3:1–6
Lawrenz B, Fehm T, von Wolff M, Soekler M, Huebner S, Henes J, Henes M (2012) Reduced pretreatment ovarian reserve in premenopausal female patients with Hodgkin lymphoma or non-Hodgkin-lymphoma—evaluation by using antimüllerian hormone and retrieved oocytes. Fertil Steril 98:141–144
Lee DY, Kim SK, Kim M, Hwang KJ, Kim SH (2017) Fertility preservation for patients with hematologic malignancies: the Korean Society for Fertility Preservation clinical guidelines. Clin Exp Reprod Med 44:187–192
Legha SS (1986) Vincristine neurotoxicity. Pathophysiology and management. Med Toxicol 1: 421–427
Lekovich J, Lobel AL, Stewart JD, Pereira N, Kligman I, Rosenwaks Z (2016) Female patients with lymphoma demonstrate diminished ovarian reserve even before initiation of chemotherapy when compared with healthy controls and patients with other malignancies. J Assist Reprod Genet 33:657–662
Levin M, Betjes M, Kwast Th H V d , Wenberg B, Leebeek F (2003) Acute renal cortex necrosis caused by arterial thrombosis during treatment for acute promyelocytic leukemia. Haematologica 88:ECR21
Libby P, Ridker PM, Maseri A (2002) Inflammation and atherosclerosis. Circulation 105:1135–1143
Lin M, Zhu J, Shen H, Huang J (2013) Gastrointestinal bleeding as an initial manifestation in asymptomatic multiple myeloma: a case report and review of the literature. Oncol Lett 5:218–220
Linden W, Vodermaier A, Mackenzie R, Greig D (2012) Anxiety and depression after cancer diagnosis: prevalence rates by cancer type, gender, and age. J Affect Disord 141:343–351
Linette GP, Stadtmauer EA, Maus MV, Rapoport AP, Levine BL, Emery L, Litzky L, Bagg A, Carreno BM, Cimino PJ (2013) Cardiovascular toxicity and titin cross-reactivity of affinity-enhanced T cells in myeloma and melanoma. Blood 122:863–871
Luciano RL, Brewster UC (2014) Kidney involvement in leukemia and lymphoma. Adv Chronic Kidney Dis 21:27–35
Magge RS, DeAngelis LM (2015) The double-edged sword: neurotoxicity of chemotherapy. Blood Rev 29:93–100
Maillet M, Van Berlo JH, Molkentin JD (2013) Molecular basis of physiological heart growth: fundamental concepts and new players. Nat Rev Mol Cell Biol 14:38–48
Majhail NS, Flowers ME, Ness KK, Jagasia M, Carpenter PA, Arora M, Arai S, Johnston L, Martin PJ, Baker KS (2009) High prevalence of metabolic syndrome after allogeneic hematopoietic cell transplantation. Bone Marrow Transplant 43:49–54

Massenkeil G, Alexander T, Rosen O, Dörken B, Burmester G, Radbruch A, Hiepe F, Arnold R (2016) Long-term follow-up of fertility and pregnancy in autoimmune diseases after autologous haematopoietic stem cell transplantation. Rheumatol Int 36:1563–1568

Maus MV, Grupp SA, Porter DL, June CH (2014) Antibody-modified T cells: CARs take the front seat for hematologic malignancies. Blood 123:2625–2635

McMillen KK, Schmidt EM, Storer BE, Bar M (2014) Metabolic syndrome appears early after hematopoietic cell transplantation. Metab Syndr Relat Disord 12:367–371

McMullen JR, Boey EJ, Ooi JY, Seymour JF, Keating MJ, Tam CS (2014) Ibrutinib increases the risk of atrial fibrillation, potentially through inhibition of cardiac PI3K-Akt signaling. Blood 124:3829–3830

Meier C, Taubenheim S, Lordick F, Mehnert-Theuerkauf A, Götze H (2020) Depression and anxiety in older patients with hematological cancer (70+)–geriatric, social, cancer-and treatment-related associations. J Geriatr Oncol 11:828–835

Meistrich ML (2013) Effects of chemotherapy and radiotherapy on spermatogenesis in humans. Fertil Steril 100:1180–1186

Meregalli C (2015) An overview of bortezomib-induced neurotoxicity. Toxics 3:294–303

Mileshkin L, Stark R, Day B, Seymour JF, Zeldis JB, Prince HM (2006) Development of neuropathy in patients with myeloma treated with thalidomide: patterns of occurrence and the role of electrophysiologic monitoring. J Clin Oncol 24:4507–4514

Miller KD, Nogueira L, Mariotto AB, Rowland JH, Yabroff KR, Alfano CM, Jemal A, Kramer JL, Siegel RL (2019) Cancer treatment and survivorship statistics, 2019. CA Cancer J Clin 69:363–385

Mullassery D, Bader A, Battersby AJ, Mohammad Z, Jones ELL, Parmar C, Scott R, Pizer BL, Baillie CT (2009) Diagnosis, incidence, and outcomes of suspected typhlitis in oncology patients—experience in a tertiary pediatric surgical center in the United Kingdom. J Pediatr Surg 44:381–385

Nagel G, Stocks T, Späth D, Hjartåker A, Lindkvist B, Hallmans G, Jonsson H, Bjørge T, Manjer J, Häggström C (2012) Metabolic factors and blood cancers among 578,000 adults in the metabolic syndrome and cancer project (Me-Can). Ann Hematol 91:1519–1531

Nahas MR, Rosenblatt J, Lazarus HM, Avigan D (2018) Anti-cancer vaccine therapy for hematologic malignancies: an evolving era. Blood Rev 32:312–325

Oktem O, Oktay K (2007a) A novel ovarian xenografting model to characterize the impact of chemotherapy agents on human primordial follicle reserve. Cancer Res 67:10159–10162

Oktem O, Oktay K (2007b) Quantitative assessment of the impact of chemotherapy on ovarian follicle reserve and stromal function. Cancer 110:2222–2229

Olexenko L, Vybornykh D (2017) Mental disorders in Hodgkin's lymphoma. Zh Nevrol Psikhiatr Im S S Korsakova 117:124–131

Ozdemir ZN, Bozdag SC (2020) Hematological malignancies and fertility. In: Cell biology and translational medicine, volume 9: stem cell-based therapeutic approaches in disease. Springer, Cham, pp 103–115

Paccou J, Merlusca L, Henry-Desailly I, Parcelier A, Gruson B, Royer B, Charbonnier A, Ursu D, Desailloud R, Garidi R (2014) Alterations in bone mineral density and bone turnover markers in newly diagnosed adults with lymphoma receiving chemotherapy: a 1-year prospective pilot study. Ann Oncol 25:481–486

Page DB, Postow MA, Callahan MK, Allison JP, Wolchok JD (2014) Immune modulation in cancer with antibodies. Annu Rev Med 65:185–202

Pai VB, Nahata MC (2000) Cardiotoxicity of chemotherapeutic agents. Drug Saf 22:263–302

Paviglianiti A (2020) Endocrine and metabolic disorders after hematopoietic cell transplantation. Turk J Haematol 37:111–115

Perazella MA, Shirali AC (2018) Nephrotoxicity of cancer immunotherapies: past, present and future. J Am Soc Nephrol 29:2039–2052

Powers S, Sollanek K, Wiggs M, Demirel H, Smuder A (2014) Exercise-induced improvements in myocardial antioxidant capacity: the antioxidant players and cardioprotection. Free Radic Res 48:43–51

Prolla J, Kirsner JB (1964) The gastrointestinal lesions and complications of the leukemias. Ann Intern Med 60:723–723

Rajkumar SV, Jacobus S, Callander NS, Fonseca R, Vesole DH, Williams ME, Abonour R, Siegel DS, Katz M, Greipp PR (2010) Lenalidomide plus high-dose dexamethasone versus lenalidomide plus low-dose dexamethasone as initial therapy for newly diagnosed multiple myeloma: an open-label randomised controlled trial. Lancet Oncol 11:29–37

Ravindran V, Anoop P (2011) Rheumatologic manifestations of benign and malignant haematological disorders. Clin Rheumatol 30:1143–1149

Richardson PG, Laubach JP, Schlossman RL, Mitsiades C, Anderson K (2010) Complications of multiple myeloma therapy, part 1: risk reduction and management of peripheral neuropathy and asthenia. J Natl Compr Canc Netw 8:S-4–S-12

Ruchlemer R, Amit-Kohn M, Tvito A, Sindelovsky I, Zimran A, Raveh-Brawer D (2018) Bone loss and hematological malignancies in adults: a pilot study. Support Care Cancer 26:3013–3020

Saleh S, Esmail K, Albreiki D (2020) Subjective intermittent colour vision loss as the initial presentation of chronic myeloid leukemia. Am J Ophthalmol Case Rep 19:100817

Sathiapalan RK, Velez MC, McWhorter ME, Irwin K, Correa H, Baliga R, Warrier RP (1998) Focal segmental glomerulosclerosis in children with acute lymphocytic leukemia: case reports and review of literature. J Pediatr Hematol Oncol 20:482–485

Savani BN, Koklanaris EK, Le Q, Shenoy A, Goodman S, Barrett A (2009) Prolonged chronic graft-versus-host disease is a risk factor for thyroid failure in long-term survivors after matched sibling donor stem cell transplantation for hematologic malignancies. Biol Blood Marrow Transplant 15:377–381

Schuster SJ, Neelapu SS, Gause BL, Janik JE, Muggia FM, Gockerman JP, Winter JN, Flowers CR, Nikcevich DA, Sotomayor EM (2011) Vaccination with patient-specific tumor-derived antigen in first remission improves disease-free survival in follicular lymphoma. J Clin Oncol 29:2787

Scott DW, Gascoyne RD (2014) The tumour microenvironment in B cell lymphomas. Nat Rev Cancer 14:517–534

Shalet SM (1993) Effect of irradiation treatment on gonadal function in men treated for germ cell cancer. Eur Urol 23:148–152

Shapira M, Raanani H, Cohen Y, Meirow D (2014) Fertility preservation in young females with hematological malignancies. Acta Haematol 132:400–413

Sheean PM, Freels SA, Helton WS, Braunschweig CA (2006) Adverse clinical consequences of hyperglycemia from total parenteral nutrition exposure during hematopoietic stem cell transplantation. Biol Blood Marrow Transplant 12:656–664

Shimasaki N, Jain A, Campana D (2020) NK cells for cancer immunotherapy. Nat Rev Drug Discov 19:200–218

Sliwa JA, Marciniak C (1999) Physical rehabilitation of the cancer patient. Cancer Treat Res 100: 75–89

Soleimani R, Heytens E, Darzynkiewicz Z, Oktay K (2011) Mechanisms of chemotherapy-induced human ovarian aging: double strand DNA breaks and microvascular compromise. Aging (Albany NY) 3:782

Soman S, Kasturi N, Srinivasan R, Vinod K (2018) Ocular manifestations in leukemias and their correlation with hematologic parameters at a tertiary care setting in South India. Ophthalmol Retina 2:17–23

Spinelli S, Chiodi S, Bacigalupo A, Brasca A, Menada MV, Petti A, Ravera G, Gualandi F, VanLint M, Sessarego M (1994) Ovarian recovery after total body irradiation and allogeneic bone marrow transplantation: long-term follow up of 79 females. Bone Marrow Transplant 14: 373–380

Stewart F, Seemann I, Hoving S, Russell N (2013) Understanding radiation-induced cardiovascular damage and strategies for intervention. Clin Oncol 25:617–624

Tangye SG, Ma CS, Brink R, Deenick EK (2013) The good, the bad and the ugly—T FH cells in human health and disease. Nat Rev Immunol 13:412–426

Tarte K (2017) Role of the microenvironment across histological subtypes of NHL. Hematology Am Soc Hematol Educ Program Book 2017:610–617

Telakis E, Tsironi E, Tavoularis G, Papatheodorou K, Tzaida O, Nikolaou A (2009) Gastrointestinal involvement in a patient with multiple myeloma: a case report. Ann Gastroenterol 22:287–290

Thomas AS, Schwartz M, Quigley E (2019) Gastrointestinal lymphoma: the new mimic. BMJ Open Gastroenterol 6:e000320

Thompson M, Huen A, Toth B, Vassilopoulou-Sellin R, Hoff A, Murphy W, Sanjoro H, Bekele B, Arbuckle R, Hagemeister F (2007) Osteopenia and osteoporosis in untreated non-Hodgkin's lymphoma patients: an important and potentially treatable survivorship issue in lymphoma. J Clin Oncol 25:9055–9055

Tokoyoda K, Hauser AE, Nakayama T, Radbruch A (2010) Organization of immunological memory by bone marrow stroma. Nat Rev Immunol 10:193–200

Valcovici M, Andrica F, Serban C, Dragan S (2016) Cardiotoxicity of anthracycline therapy: current perspectives. Arch Med Sci 12:428

Vicini G, Nicolosi C, Malandrino D, Tozzetti C, Rizzo S, Sodi A (2020) Leukostasis retinopathy with leukemic infiltrates as onset manifestation of chronic myeloid leukemia: a case report. Eur J Ophthalmol. https://doi.org/10.1177/1120672120930679

Vincenti F, Friman S, Scheuermann E, Rostaing L, Jenssen T, Campistol J, Uchida K, Pescovitz M, Marchetti P, Tuncer M (2007) Results of an international, randomized trial comparing glucose metabolism disorders and outcome with cyclosporine versus tacrolimus. Am J Transplant 7:1506–1514

Vishnevskia-Dai V, Sella King S, Lekach R, Fabian ID, Zloto O (2020) Ocular manifestations of leukemia and results of treatment with intravitreal methotrexate. Sci Rep 10:1994

Wanchoo R, Abudayyeh A, Doshi M, Edeani A, Glezerman IG, Monga D, Rosner M, Jhaveri KD (2017) Renal toxicities of novel agents used for treatment of multiple myeloma. Clin J Am Soc Nephrol 12:176–189

Wang L, Tan TC, Halpern EF, Neilan TG, Francis SA, Picard MH, Fei H, Hochberg EP, Abramson JS, Weyman AE (2015) Major cardiac events and the value of echocardiographic evaluation in patients receiving anthracycline-based chemotherapy. Am J Cardiol 116:442–446

Weiner LM, Dhodapkar MV, Ferrone S (2009) Monoclonal antibodies for cancer immunotherapy. Lancet 373:1033–1040

Westin JR, Thompson MA, Cataldo VD, Fayad LE, Fowler N, Fanale MA, Neelapu S, Samaniego F, Romaguera J, Shah J (2013) Zoledronic acid for prevention of bone loss in patients receiving primary therapy for lymphomas: a prospective, randomized controlled phase III trial. Clin Lymphoma Myeloma Leuk 13:99–105

Zannella A, Marignani M, Begini P (2019) Hematological malignancies and HBV reactivation risk: suggestions for clinical management. Viruses 11:858

Zorrilla AFC, Herault LR, Casasbuenas A, Aponte DM, Ramos PL (2006) Systematic review of case reports concerning adults suffering from neutropenic enterocolitis. Clin Transl Oncol 8:31–38

Immunodeficiency, Leukemia, and Lymphoma

Anand Srinivasan, Arpan A. Sinha, and J. Kimble Frazer

Abstract

Leukemias and lymphomas are the dominant malignancies in childhood, comprising over one third of all pediatric cancer cases. They are also prevalent in adulthood, each being among the ten most common cancer types. Inherited immunodeficiencies and immune dysregulatory syndromes, on the other hand, are quite rare, with nearly all of these conditions affecting few patients. Due to their rarity, they are undoubtedly under-recognized, and many have only recently been described, further complicating their diagnosis. Nonetheless, it is increasingly evident that many of the same genetic lesions that cause immune dysfunction also carry predilection to developing specific, or in some cases several, types of hematologic cancer. This chapter briefly summarizes the molecular and cellular pathophysiology and clinical features for many of these entities, including the leukemias and lymphomas associated with them. A table consisting of several additional immunodeficiency syndromes linked to hematologic malignancies is provided at its conclusion.

A. Srinivasan
Pediatric Transplant/Cell Therapy, Jimmy Everest Section of Pediatric Hematology-Oncology, University of Oklahoma Health Sciences Center, Oklahoma City, OK, USA

A. A. Sinha
Comprehensive Pediatric Sickle Cell Program, Jimmy Everest Section of Pediatric Hematology-Oncology, University of Oklahoma Health Sciences Center, Oklahoma City, OK, USA

J. K. Frazer (✉)
Jimmy Everest Section of Pediatric Hematology-Oncology, Children's Health Foundation E.L. & Thelma Gaylord Chair in Pediatric Oncology, University of Oklahoma Health Sciences Center, Oklahoma City, OK, USA
e-mail: Kimble-Frazer@ouhsc.edu

© The Author(s), under exclusive license to Springer Nature Switzerland AG 2022

Interdisciplinary Cancer Research, https://doi.org/10.1007/16833_2022_82
Published online: 29 November 2022

Keywords

ALL · AML · Burkitt · Hodgkin · Immunodeficiency · Leukemia · Lymphoma · MDS · Myelodysplasia · NHL

1 Introduction

Many genetic syndromes that cause immunodeficiency (IMD) also predispose patients to cancer, including leukemias and lymphomas. Conceptually, several mechanisms can explain this: First, some IMD syndromes are caused by underlying problems in DNA double-strand break (DSB) repair or related genome integrity issues. Such defects impair V(D)J recombination during immunoglobulin (Ig) and T-cell receptor (TCR) rearrangement, impeding B and T lymphopoiesis. Certain repair defects also make patients prone to somatically acquired point mutations, deletions, translocations, aneuploidies, or other genetic lesions. These events not only predispose to lymphomas and leukemias; many IMD are also linked to non-lymphoid cancers (Bilousova et al. 2005). However, translocations are particularly prevalent in lymphoid cancers, which may make IMD patients particularly prone to these malignancies. A second mechanism concerns the many viruses linked to specific cancers (discussed elsewhere in this series). Examples include human papillomavirus and genitourinary malignancies, human herpesvirus 8 and Kaposi sarcoma, Epstein-Barr virus (EBV) and multiple lymphomas, and others. EBV is especially noteworthy, as several IMD in this chapter promote chronic and/or fulminant EBV infection (Shannon-Lowe et al. 2017; Dupin 2020; Hu and Ma 2018). EBV-driven oncogenesis pertains to many IMD detailed here, predisposing to both non-Hodgkin lymphomas (NHL) and Hodgkin lymphomas (HL).

Other IMD features are also linked to hematologic cancer predisposition, with a third mechanism being the concept that many inherited IMD mutations can simultaneously serve as genetic “first hits” toward malignant transformation. No single DNA lesion causes cancer in isolation, but some inherited IMD result from mutations that are “pre-oncogenic.” These germline mutations are present in every cell (hematopoietic and non-hematopoietic) in IMD patients, whereas in non-IMD patients, they arise somatically in only one cell (and are then passed to its cellular progeny). Such mutations can be either gain- or loss-of-function (GOF, LOF), in either oncogenes or tumor suppressors, respectively. In each case, their existence in every cell vastly expands the size of the pre-malignant cell population that is subject to accruing random somatic second, third, and subsequent oncogenic genetic lesions that confer full transformation.

A fourth, and perhaps more nebulous, mechanism pertains to abnormal leukocyte numbers. Many IMD cause lymphopenia, while others lack this or have limited cytopenias of only one cell lineage or subtype. Alternatively, in other IMD, one cell type may actually expand polyclonally (i.e., lymphoproliferative disease; LPD). This enlarges a cell population that can then accrue somatically acquired malignant driver mutations, as seen when EBV-driven post-transplant lymphoproliferative disease

(PTLD) progresses to Burkitt lymphoma (BL) or other PTLD-derived lymphomas. However, returning to leukopenias, even without LPD, cytopenias can create a lower oncogenic threshold. Paradoxically, in animal models with genetic lesions that impede the development of a specific lineage or lymphoid subtype, cancers of that same lineage/cell type often occur (Martins et al. 2014; Seton-Rogers 2014; Moreno 2014; Porter et al. 2011). This reveals that rare surviving cells can overcome the differentiation "blockade" imposed by the genetic lesion by acquiring new mutations that allow them to persist. Such mutations are often pre-malignant, superseding apoptosis or other selection processes that normally curtail their development. In effect, constricting a specific cell population can drive a selective process that allows rare survivors to expand and fill the void. With normal cells absent, mutant cells that would normally be out-competed by their healthy counterparts are able to thrive. Thus, remarkably, even cytopenias that restrict hematopoiesis can also promote oncogenesis.

These mechanisms are not mutually exclusive. Genetic instability, aberrant antiviral responses, proto-malignant DNA lesions, LPD, and leukopenias co-exist in many IMD, with additional oncogenic mechanisms also likely at play. Thus, for many IMD syndromes—some of which are only represented by a few patients in the literature—assigning definitive IMD and oncogenic mechanisms is challenging. Even so, to organize the syndromes in this chapter, we have grouped leukemia- and/or lymphoma-predisposing IMD into categories based on their proposed mechanisms. However, we advise readers to recognize that no formal criteria have validated this classification schema. Specifically, we have divided cancer-linked IMD into four categories based on defects in (1) leukocyte development, (2) leukocyte signaling, (3) genomic stability, and (4) "other," for disorders not fitting the prior categories.

It is worth noting that several IMD could easily be classified into more than one category (genetic instability during Ig and TCR rearrangement and defects in leukocyte signaling both impact leukocyte development, e.g.), so our groupings are imperfect. Moreover, we have been deliberately inclusive, listing some IMD for which only a handful of—or, in some instances, only single—leukemia/lymphoma cases are reported. We have chosen this approach because IMD are rare, some are just recently described, and it is fair to speculate that the preponderance of cancer cases may be unreported in the literature, particularly if malignancies occur years after the initial IMD diagnosis. Consequently, it is possible that some entities detailed herein are not genuinely cancer predisposing, and it is certainly accurate that not all have been definitively proven to do so. There are also, undoubtedly, other IMD and/or cancer associations still awaiting discovery. For now, this chapter briefly summarizes several IMD linked to hematologic malignancies, with a listing of additional leukemia–/lymphoma-associated IMD at its conclusion that our text could not include due to space constraints.

Besides mechanistic categories, we have also partitioned text sections into groups of IMD that are linked to leukemia alone, leukemia *and* lymphoma, or lymphoma-only. Like prior caveats, it is possible that some categorizations may warrant revision in the future, should new cancer cases be reported. Throughout the chapter, we employ IMD, gene abbreviation, and other naming criteria as utilized by the US National Library of Medicine, National Center for Biotechnology Information, and its Online Mendelian Inheritance in Man resource (OMIM; https://www.ncbi.nlm.nih.gov/omim/).

2 IMD Associated with Leukemia

Several IMD have been associated with leukemia, but not lymphoma. Curiously, nearly all of these leukemias occur in myeloid lineage cells, rather than lymphocytes. Table 1 lists the five IMD chosen for discussion, with separate sections for each entity immediately following the table.

2.1 Defects of Leukocyte Development

2.1.1 IMD21, Mutations in *GATA2*

Pathophysiology: GATA2 is a member of a conserved group of transcription factors, consisting of six genes, GATA1–6 (Chlon and Crispino 2012). GATA2 is critical to hematopoiesis (Tsai et al. 1994), regulating gene expression in hematopoietic stem cells (HSC) (Orkin 2000; Rodrigues et al. 2012). GATA2 is necessary for primitive and definitive erythropoiesis and for HSC renewal, survival, and maturation to progenitors (Wlodarski et al. 2017; Spinner et al. 2014; Bigley et al. 2019). GATA2 haploinsufficiency in mice causes defective hematopoiesis, with marked reductions in monocytes and T and B lymphocytes (Rodrigues et al. 2005).

Different *GATA2* mutations have distinct phenotypes. Mutations in the ZF2 domain are typically germline and detailed below (Bresnick et al. 2012; Zhang et al. 2008). ZF1 mutations are usually somatic and occur in association with biallelic CEBPA mutations in adult acute myeloid leukemia (AML) of the M1 French-American-British subtype (Tien et al. 2018).

Clinical features: Germline GATA2 mutations cause four clinical phenotypes.

1. MonoMAC syndrome: monocytopenia and predisposition to mycobacterial infection (Vinh et al. 2010; Hsu et al. 2011).
2. DCML syndrome: dendritic cell, monocyte, B-cell, and natural killer (NK) cell lymphoid deficiency (Bigley et al. 2019).

Table 1 Featured IMD associated with leukemias

Leukocyte development defects	Leukocyte signaling defects	Genome stability defects	Other defects
IMD21: *GATA2* mutations SCN1: *ELANE* mutations	Cowden syndrome: *PTEN* mutations	Dyskeratosis congenita: *DKC1* mutations	fHLH4: *STX11* mutations

3. Familial myelodysplastic syndrome: MDS/AML (Hahn et al. 2011).
4. Emberger syndrome: primary lymphedema with MDS (Mansour et al. 2010; Ostergaard et al. 2011).

Germline mutations show autosomal dominant inheritance (Vinh et al. 2010; A. P. Hsu et al. 2011), with clinical hallmarks including IMD with susceptibility to human papillomaviruses and non-tuberculous mycobacteria (NTM), predisposition to MDS/AML, pulmonary alveolar proteinosis (PAP), and congenital lymphedema.

Age of onset varies, consistent with the differing phenotypes listed above; the earliest report is a 5-month infant with lymphedema (Spinner et al. 2014). Due to IMD, infection is the most common initial manifestation, with severe viral or NTM infection occurring in a majority of patients (60%). A smaller fraction (21%) present with MDS/AML (Spinner et al. 2014). GATA2 mutations are linked to pediatric hypoplastic MDS, especially MDS with monosomy 7, where GATA2 mutations are seen in up to 50% of patients (Hsu et al. 2013).

Spinner et al. described the largest cohort with GATA2 mutations, reporting B-cell, NK-cell, or monocytopenias in >75% of patients (Spinner et al. 2014). Marrow biopsy revealed myelodysplasia in 84%, with hypocellularity and multilineage (especially megakaryocytic) dysplasias. In this same cohort, another 14% had AML, with CMML occurring in 8% (Spinner et al. 2014). West et al. further analyzed this cohort and found younger, predominantly female patients carried somatic *ASXL1* mutations in addition to *GATA2* (West et al. 2014) and were more likely to develop CMML.

Apart from IMD and marrow findings, myriad symptoms may occur, including sensorineural hearing loss, hypothyroidism, solid organ tumors, erythema nodosum, chronic arthralgias, bone infarcts, and recurrent miscarriages (A. P. Hsu et al. 2015).

2.1.2 Severe Congenital Neutropenia 1 (SCN1), Mutations in *ELANE*

Pathophysiology: *ELANE* encodes a serine protease, neutrophil elastase (Korkmaz et al. 2010). Mutations act in an autosomal dominant fashion (Dale and Makaryan 2018) and include missense changes, insertions and deletions that preserve the reading frame, and truncations (Horwitz et al. 2013; Germeshausen et al. 2013). Interestingly, humans (Horwitz et al. 2019) and mice (Martinod et al. 2016) with biallelic ELANE LOF have functional neutrophils with normal blood counts. Neutrophil dysfunction is thought to be due to protease malfunction, not inadequate elastase activity, which causes global immune dysfunction due to dysregulated vesicular sorting and membrane trafficking (Nayak et al. 2015).

Clinical features: Certain ELANE mutations cause benign cyclic neutropenia, which does not predispose to cancer (Horwitz et al. 2013). In contrast, the severe neutropenia of SCN1 is present at birth, causing early and recurrent infections (Dale and Makaryan 2018). In a large cohort, MDS/AML incidence was 36% after G-CSF treatment (the primary therapy for SCN1) (Rosenberg et al. 2008), with G214R mutations showing the highest MDS/AML risk (Bellanné-Chantelot et al. 2004; Makaryan et al. 2015), precipitated by additional mutations in CSF3R, RAS, and RUNX1 (Skokowa et al. 2014). Secondary MDS/AML in patients with SCN1 has distinct genetic features from de novo MDS/AML, including monosomy 7 and chromosome 21 abnormalities (Freedman et al. 2000; Kalra et al. 1995; Link 2019). In contrast, monosomy 7 and mutations in RUNX1 and CSF3R are relatively uncommon in de novo AML (Germeshausen et al. 2007). Whole genome/exome sequencing studies of SCN1-associated MDS/AML are limited, so the full spectrum of leukemic mutations in SCN1 is unknown (Beekman et al. 2012). Rarely, ALL has been associated with ELANE mutations, and early allogeneic SCT prior to leukemia onset may alleviate this risk (Rotulo et al. 2020).

2.2 Defects of Leukocyte Signaling

2.2.1 Cowden Syndrome, Mutations in *PTEN*

Pathophysiology: PTEN is a known tumor suppressor; thus, PTEN LOF confers cancer predisposition (Browning et al. 2015). PTEN phosphatase antagonizes phosphatidylinositol 3-kinase (PI3K) signaling, which promotes B- and T-lymphocyte development, activation, and differentiation (Eissing et al. 2018).

Clinical features: The best-described syndrome regarding PTEN LOF is Cowden syndrome, characterized by hamartomatous tissue overgrowth and predisposition to breast, gastrointestinal, and thyroid cancers. Benign breast, thyroid, uterine, and skin lesions also occur (Pilarski 2009). IMD caused by PTEN LOF features T lymphocytopenia and recurrent infections. One AML case is reported in a Cowden syndrome patient, who was diagnosed with AML at 47 years after presenting with recurrent abscesses and cellulitis (Ruschak et al. 1981).

2.3 Defects of Genome Stability

2.3.1 X-Linked Dyskeratosis Congenita, Mutations in *DKC1* (Alternate Names: Zinsser-Cole-Engman Syndrome and Hoyeraal-Hreidarsson Syndrome)

Pathophysiology: *DKC1* encodes dyskerin, a multifunctional protein involved in rRNA synthesis, ribosomal assembly, and centromere/microtubule binding (Hassock et al. 1999; Heiss et al. 1998). Dyskerin belongs to two ribonucleoprotein (RNP) complexes. Its H/ACA domain binds small nucleolar RNA (snoRNA) sequences, providing stability during biogenesis and snoRNP assembly. The H/ACA domain also recognizes the telomerase RNA component (TERC) of telomerase, an RNP complex crucial for telomere extension and maintenance. In this context, dyskerin fosters TERC stability and function and, thus, telomerase activity (Mitchell et al. 1999). *DKC1* missense mutations typically cause dyskeratosis congenita, but noncoding and truncating variants are also reported (Marrone et al. 2007; Dokal 1996).

Clinical features: Symptoms include skin, mucosal, and hematopoietic abnormalities, with a classic triad of mucocutaneous features (skin hyperpigmentation, nail dystrophy, and oral leukoplakia) described (Parry et al. 2011). Young males typically develop progressive marrow failure by 30 years, which is the main cause of death (Marrone et al. 2007; Dokal 1996). IMD results from marrow failure, with cytopenias causing frequent infections (Wiedemann et al. 1984). Myeloid leukemia predilection is clear in dyskeratosis congenita, with one case of AML and eight MDS cases (two with refractory anemia, two with refractory anemia with excess blasts, two with monosomy 7 MDS, and two unclassified MDS) reported (Alter et al. 2009; Marsh et al. 1992).

2.4 Defects in Other Pathways

2.4.1 Familial Hemophagocytic Lymphohistiocytosis 4 (fHLH4), Mutations in *STX11*

Pathophysiology: Syntaxin-11 promotes vesicular exocytosis, facilitating intracellular transport (Tang et al. 1998). STX11 LOF causes familial hemophagocytic lymphohistiocytosis type 4, more common in Kurdish and Pakistani ethnicities (Balta et al. 2008; Muller et al. 2014). Missense, deletions, and non-sense mutations are reported, all leading to defective NK- and cytotoxic T-cell (CTL) degranulation with impaired cytotoxicity, although the mechanistic details remain under investigation (Sepulveda et al. 2013; Rudd et al. 2006).

Clinical features: fHLH4 patients are typically older and have less severe HLH than other familial types (Horne et al. 2008), with standard HLH features: fever, lymphadenopathy, hepatosplenomegaly, high ferritin and triglycerides, cytopenias, depressed NK-cell function, elevated soluble IL-2 receptor, and increased liver transaminases (Rosée et al. 2019). Single fHLH4 patients with MDS and AML are reported; both responded to conventional therapies (Rudd et al. 2006; Henter et al. 1993). One received epipodophyllotoxin HLH treatment at 3 years and subsequently developed MDS (refractory anemia with excess blasts); allogeneic SCT resulted in long-term cure (Henter et al. 1993). The second developed frank AML (M3 subtype with t [15;17]) 2 years after HLH treatment and responded to chemotherapy (Rudd et al. 2006).

3 IMD Associated with Leukemia and Lymphoma

A large number of IMD are convincingly linked to both leukemia and lymphoma. Not surprisingly, since lymphomas derive from the lymphoid lineage, most reported leukemias are also of lymphocytic origin. However, in some of these IMD, myeloid leukemias (AML, CML, JMML) have also been described. Table 2 lists the entities detailed in this section.

Table 2 Featured IMD associated with leukemias and lymphomas

Leukocyte development defects	Leukocyte signaling defects	Genome stability defects	Other defects
Cartilage-hair hypoplasia syndrome: *RMRP* mutations CVID13: *IKZF1* mutations	ADMIO1: *STAT3* mutations IMD11A and IMD11B: *CARD11* mutations IMD37: *BCL10* mutations IMD12: *MALT1* mutations CVID10 and CVID12: *NF-κB* mutations IMD14A/B and IMD36: *PIK3CD* and *PIK3R1* mutations	Bloom syndrome: *BLM* mutations LIG4 syndrome: *LIG4* mutations Nijmegen breakage syndrome: *NBS1* mutations MMRCS2 and MMRCS4: *MSH2* and *PMS2* mutations Ataxia-telangiectasia: *ATM* mutations Li-Fraumeni syndrome: *TP53* mutations	Wiskott-Aldrich syndrome: *WAS* mutations

3.1 Defects of Leukocyte Development

3.1.1 Cartilage-Hair Hypoplasia Syndrome, Mutations in *RMRP*

Pathophysiology: Cartilage-hair hypoplasia (CHH) syndrome is autosomal recessive, caused by *RMRP* gene variants. *RMRP* encodes the RNA component of a mitochondrial RNA endonuclease (Roifman, Gu, and Cohen 2006; Ridanpää et al. 2001). This RNA/protein complex has roles in nucleolar ribosomal RNA processing, mitochondrial RNA cleavage during mitochondrial DNA synthesis, and cyclin B2 mRNA cleavage, which regulates the cell cycle (Vakkilainen et al. 2019).

Clinical features: CHH heterogeneity is high. Features include metaphyseal dysplasia, short limbs, silky hair, ligament laxity, and anemia (deficient erythropoiesis) (Mäkitie and Vakkilainen 2020). IMD includes lymphopenia, decreased T-cell function, and hypogammaglobulinemia, with resulting respiratory disease, chronic viral infections, and cutaneous and visceral granulomatous disease (Narra and Shearer 2009; Bailly-Botuha et al. 2008). The most frequent malignancies are HL, NHL, leukemias, and laryngeal carcinomas (R. E. Harris et al. 1981; Mäkitie et al. 1999). In a large Finnish CHH registry, 5/122 patients developed cancer, with 3 NHL, 1 testicular progonoma (melanotic neuroectodermal tumor of infancy), and 1 vocal cord squamous cell carcinoma (Mäkitie et al. 1999).

3.1.2 Common Variable Immunodeficiency Syndrome 13 (CVID13), Mutations in *IKZF1*

Pathophysiology: *IKZF1* encodes Ikaros, a zinc-finger transcription factor that regulates immune cell development, mainly of B and T lymphocytes (Sellars et al. 2011; L. Sun et al. 1996; Molnár et al. 1996). IKZF1 guides fate decisions by early uncommitted lymphocytes and controls key genetic programs later in B- and T-cell development also. IKZF1 is detectable in the earliest stages of hematopoietic development (fetal liver) to the adult bone marrow, and its expression continues in mature B and T cells (Klug et al. 1998; Kelley et al. 1998; Morgan et al. 1997; Georgopoulos et al. 1992). IKZF1 lesions are typically large deletions and missense mutations (Boast et al. 2021). Haploinsufficiency and dominant-negative mutations both cause defective IKZF1 function, resulting in IMD and leukemia/lymphoma predisposition.

Clinical features: Recurrent skin and sinopulmonary infections, due to viral and mycobacterial pathogens, occur, with autoimmunity including ITP, AIHA, SLE, and arthritis (Nunes-Santos et al. 2020; Kuehn et al. 2021a, b). Dominant-negative mutations often cause more severe phenotypes than haploinsufficiency, presenting with early-onset SCID (Nunes-Santos et al. 2020). IMD consists of B-cell leukopenia, variable hypogammaglobulinemia, normal/high T cells with reduced memory, and effector T cells (Boast et al. 2021). Malignancies include B- and T-ALL and BL (Nunes-Santos et al. 2020; Kuehn et al. 2021b). Two cases of B-ALL are reported in haploinsufficient patients, with the first dying of infection and the second of B-ALL relapse (Kuehn et al. 2016). T-ALL has developed in patients with dominant-negative mutations, which responded to standard chemotherapy, and was then consolidated with allogeneic SCT for long-term cure (Yoshida et al. 2017).

3.2 Defects of Leukocyte Signaling

3.2.1 ADMIO1 (Autoimmune Disease, Multisystem, Infantile-Onset 1), Mutations in *STAT3*

Pathophysiology: STAT3 is a transcription factor that, upon phosphorylation by cytokine and growth factor receptor-associated Janus kinases (JAK), regulates genes critical to immune cell proliferation, apoptosis, and differentiation (Forbes et al. 2016; O'Shea et al. 2013; Villarino et al. 2015). Collectively, JAK/STAT family proteins control >50 cytokines (Tsilifis et al. 2021), with STAT3 modulating innate and adaptive immunity via several cytokines, including interferons and interleukins 2, 6, 7, 10, 12, 15, 21, 23, and 27 (T. J. Harris et al. 2007). Mutations in STAT3 or related proteins cause diverse pathology, including severe combined immunodeficiency (SCID), malignancy, autoimmunity, and myeloproliferation.

Clinical features: Differing phenotypes occur with STAT3 LOF versus GOF.

1. Autosomal Dominant Hyper-IgE Syndrome (AD-HIES): caused by germline STAT3 LOF. Most patients have eczematoid dermatitis and *Staphylococcus aureus* skin infections (Grimbacher et al. 1999; Chandesris et al. 2012; Jiao et al. 2008; Schimke et al. 2010; Woellner et al. 2010). They also have recurrent pneumonias with *Staphylococcus aureus* or *Streptococcus pneumoniae*, driving chronic lung damage and risk of *Aspergillus* and *Pseudomonas* pneumonias. Other features include dysmorphic facies, coronary artery dilatation, scoliosis, and hyperextensibility. Lab studies show elevated IgE, hypereosinophilia, and decreased Th17 cells (Grimbacher et al. 1999; Chandesris et al. 2012; Jiao et al. 2008; Schimke et al. 2010; Woellner et al. 2010). Patients have increased

lymphoma incidence, especially NHL, with 6 of 23 reported NHL being T-cell lymphomas. Reported B-cell lymphomas include primarily diffuse large B-cell lymphoma (DLBCL), as well as classic HL (CAHYANUR and RAHMAWATI 2020; Kumánovics et al. 2010).

2. Autoimmune Disease, Multisystem, Infantile-Onset 1 (ADMIO1): caused by germline STAT3 GOF. Patients show symptoms at an average of 4.6 years (range 0–17 years), with hematologic autoimmunity (ITP, autoimmune hemolytic anemia, and/or autoimmune neutropenia) commonly seen and LPD. Additional autoimmune disorders include enteritis, type I diabetes, lymphocytic interstitial pneumonia, thyroid disease, arthritis, hepatitis, and skin disease (Flanagan et al. 2014; Milner et al. 2015; Haapaniemi et al. 2015). Immune dysregulation also predisposes to recurrent and severe infections. Two patients have been reported with lymphoid malignancies [one adult with HL and an adolescent with large granular lymphocytic (LGL) leukemia] (Milner et al. 2015; Haapaniemi et al. 2015).

3.2.2 CARD11-BCL10-MALT1 (CBM) Signalosome Complex and NF-κB Defects

Pathophysiology: Caspase recruitment domain 11 (CARD), B-cell lymphoma 10 (BCL10), and mucosa-associated lymphoid tissue lymphoma translocation protein 1 (MALT1) form the CARD11-BCL10-MALT1 (CBM) complex, which bridges T-/B-cell receptor signaling to the NF-κB pathway (Meininger and Krappmann 2016; Schlauderer et al. 2018). Once activated, the NF-κB transcription factor promotes lymphocyte activation and differentiation. NF-κB family members, RelA and c-Rel, are key to mediating TCR signaling and naive T-cell activation (Oh and Ghosh 2013). Thus, CBM complex gene mutations affect downstream NF-κB function, causing IMD. Aberrant NF-κB activation is also linked to leukemias and lymphomas (Turvey et al. 2014; DiDonato et al. 2012; Liu et al. 2017; Ben-Neriah and Karin 2011). The NF-κB family comprises five related proteins, c-Rel, p65 (RelA), RelB, p50 (NF-κB1), and p52 (NF-κB2), that form homodimers and heterodimers with distinct regulatory functions (Q. Zhang et al. 2017; Vallabhapurapu and Karin 2009). The canonical pathway includes NF-κB1 and mediates inflammatory responses. The noncanonical pathway includes NF-κB2 and governs B-cell maturation, peripheral lymphoid development (organization of secondary lymphoid organs such as lymph nodes, splenic white pulp, and Peyer's patches), bone metabolism, and thymic development (Bonizzi and Karin 2004; Hayden and Ghosh 2011; S.-C. Sun 2012).

Clinical features:

1. **IMD11A and IMD11B syndromes, mutations in *CARD11***:
   - **IMD11A syndrome, homozygous LOF mutations in *CARD11***: Germline null mutations result in SCID, cured only by allogeneic SCT (Greil et al. 2013; Stepensky et al. 2013; Fuchs et al. 2015).
   - **IMD11B syndrome, heterozygous GOF mutations in *CARD11*** [B-cell expansion with NF-κB and T-cell anergy (BENTA syndrome)]: Features

include splenomegaly, profound polyclonal B-cell lymphocytosis (with 60–80% of peripheral blood leucocytes being $CD19^+$ B cells), increased transitional and naive B cells, but diminished class-switched B cells. Patients have frequent ear and sinopulmonary infections early in life and poor vaccine responses. In addition, T-cell hyporesponsiveness is associated with a milder IMD in some patients who present with molluscum contagiosum or chronic EBV. One BENTA patient who succumbed to HLH is reported (Buchbinder et al. 2015; Gupta et al. 2018).

- **Germline *CARD11* LOF heterozygous mutations**: Presents with atopy (moderate-to-severe atopic dermatitis) and recurrent infections including pneumonia, molluscum, abscesses, and bacteremia (C. A. Ma et al. 2017; Jun et al. 2003). Most patients have elevated IgE and eosinophilia (Ma et al. 2017). One patient developed a peripheral T-cell lymphoma that was cured after autologous SCT (Ma et al. 2017).

2. **IMD37 syndrome, mutations in *BCL10***: There is one report of germline homozygous LOF of BCL10. That patient's course was marked by influenza, adenovirus, and respiratory syncytial virus respiratory infections, gastroenteritis, otitis, recurrent diarrhea caused by *Campylobacter jejuni*, chronic colitis, suspected encephalitis, and eventual death due to respiratory failure. IMD in that patient affected both B and T cells. Impaired viral clearance due to IMD can promote lymphoma, especially DLBCL (Leeksma et al. 2017; Torres et al. 2014). Somatically acquired BCL10 GOF mutations also often occur in DLBCL, particularly the activated B-cell-like subtype (ABC-DLBCL) (Xia et al. 2020; Schmitz et al. 2018; Chapuy et al. 2018).
3. **IMD12 syndrome, mutations in *MALT1***: Germline LOF mutations cause recurrent infections (especially sinopulmonary), failure to thrive, dysmorphic facies, periodontal disease, and enteropathy (Jabara et al. 2013; McKinnon et al. 2014; Punwani et al. 2015b; Charbit-Henrion et al. 2017). Consanguinity existed in most cases. Patients had normal-to-decreased B cells with defective class-switching and impaired T-cell responses (Lu et al. 2018). Germline *MALT1* variants predispose to gastric MALT lymphomas linked to chronic *Helicobacter pylori* infection (Hellmig et al. 2009).
4. **Mutations in *NF-κB***: NF-κB1 and NF-κB2 mutations both cause common variable IMD (CVID).
   - **CVID10, mutations in *NFKB1***: NFKB1 LOF causes a spectrum of massive lymphadenopathy, splenomegaly, and organ-specific (alopecia areata, Hashimoto's thyroiditis) and/or hematologic (autoimmune hemolytic anemia and ITP) autoimmunity (Tuijnenburg et al. 2018). CVID10 is linked to B-cell NHL, including DLBCL and follicular lymphoma (FL), and T-cell lymphoma (Tuijnenburg et al. 2018; Voce et al. 2015; Gathmann et al. 2014).
   - **CVID12, mutations in *NFKB2***: Presents with recurrent sinopulmonary, giardial, and herpetic infections and can occur with central adrenal insufficiency, developmental delay, and asthma. Patients have decreased Ig levels and require IVIG supplementation (Chen et al. 2013).

3.2.3 PIK3 Pathway Defects

Pathophysiology: PI3K represents a family of signal transducer enzymes (Knight et al. 2006). Class IA molecules have a catalytic subunit (p110α, p110β, or p110δ encoded by *PIK3CA*, *PIK3CB*, or *PIK3CD*, respectively) and a regulatory subunit p55α or p85α (isoforms encoded by *PIK3R1*), p50α or p85β (isoforms encoded by *PIK3R2*), or p55γ (encoded by *PIK3R3*) (Okkenhaug 2013). Catalytic subunits bind regulatory subunits in response to extracellular signals. Regulatory units are crucial for PI3K activity, since they regulate stability by preventing proteasomal degradation and govern cellular localization and kinase activity (Burke and Williams 2015). PI3K activation is downstream of T-/B-cell receptor, cytokine receptor, and co-stimulatory receptor engagement, leading to phosphorylation of proteins that foster immune responses (Thouenon et al. 2021). Thus, PI3K pathway alterations result in several IMD caused by mutations enacting either GOF or LOF in both PIK3CD and PIK3R1.

Clinical features:

1. **GOF mutations in *PIK3CD* and *PIK3R1***: GOF mutations in PIK3CD were first identified in an IMD patient found to have increased PI3Kδ activity. Thus, the disease was named activated PI3K delta syndrome (APDS) (Angulo et al. 2013). Subsequent reports identified a germline mutation in PIK3R1 that also increased PI3Kδ activity, with an identical phenotype. This led to the recognition of two types of APDS, both autosomal dominant. Rare sporadic de novo mutational cases are also reported (Segundo et al. 2017; Mensa-Vilaró et al. 2019).
   - **IMD14A, GOF mutations in *PIK3CD*, type 1 APDS:** Recurrent sinopulmonary infections and LPD in childhood are the typical presentation, with developmental delay also reported in some cases (Thouenon et al. 2021; Michalovich and Nejentsev 2018). Bronchiectasis may be seen due to recurrent infections (Singh et al. 2019). Increased IgM, IgG deficiency, and CD4 lymphopenia are IMD features. Many patients develop autoimmune cytopenias (Coombs-positive trilineage cytopenia), glomerulonephritis, exocrine pancreatic insufficiency, and autoantibody-positive thyroiditis (Coulter et al. 2017). Patients may also develop lymphoma, with reported lymphomas including DLBCL (one EBV-positive), nodular sclerosing HL, nodal marginal zone lymphoma, lymphoplasmacytic lymphoma, EBV-driven Hodgkin-type LPD, and primary cutaneous anaplastic large cell lymphoma (ALCL) (Coulter et al. 2017).
   - **IMD36, mutations in *PIK3R1*, type 2 APDS:** Recurrent upper respiratory infections, pneumonitis, and LPD with splenomegaly are the most common features of type 2 APDS (Lucas et al. 2016); growth retardation, chronic diarrhea, and mild developmental delay are also reported. IMD features include decreased serum IgA and IgG, elevated IgM, B-cell lymphopenia, and decreased naive CD4 and CD8 cells, with elevated CD8 effector/memory T cells (Thouenon et al. 2021; Singh et al. 2019). Autoimmunity includes cytopenias (ITP, hemolytic anemia, and Evans syndrome), arthritis, insulin-dependent diabetes, autoimmune hepatitis, and eczema (Elkaim et al. 2016).

Reported malignancies include DLBCL, marginal zone B-cell lymphoma, nodular sclerosing HL, EBV+ nodular sclerosing HL, and chronic lymphocytic leukemia (CLL). One study estimated the cumulative lymphoid malignancy risk by 40 years to be up to 78% (Elkaim et al. 2016).

2. **LOF mutations in *PIK3CD* and *PIK3R1***:
   - **IMD14B, LOF mutations in *PIK3CD***: Less common than GOF mutations, with only four cases reported. Features include recurrent sinopulmonary infection, candidiasis, facial dysmorphism, developmental delay, and decreased T- and B-cell function. Cancers have not been reported, but most patients died due to infection-related complications (Sharfe et al. 2018; Sogkas et al. 2018; Cohen et al. 2019).
   - **SHORT syndrome, LOF mutations in *PIK3R1***: This rare multisystem disease presents with short stature, anterior chamber eye anomalies, abnormal facies, lipodystrophy, hernias, insulin-resistant diabetes, hyperextensibility, and delayed dentition (Lee et al. 2021; Zhang et al. 2020; Dyment et al. 2013). There is one report of pre-T-ALL. Cytogenetic studies detected a *SIL-TAL1* fusion (TAL/LMO T-ALL subgroup) and biallelic *CDKN2AB* and monoallelic *PTEN* deletions. The patient responded to conventional chemotherapy and received allogeneic SCT (Marzollo et al. 2020).

3.3 Defects of Genome Stability

3.3.1 Bloom Syndrome, Mutations in *BLM*

Pathophysiology: *BLM* encodes a RecQ-type helicase that unwinds DNA. Patients lack functional helicase activity due to homozygous or compound heterozygous mutations (Karow et al. 1997). Absent BLM leads to chromosome instability, excess homologous recombination (HR), and increased sister chromatid exchange (Cunniff et al. 2017).

Clinical features: Symptoms include failure to thrive; feeding intolerance; a characteristic erythematous, sun-sensitive rash on the nose and cheeks; and occasionally the dorsal hands and forearms. IMD features consist of decreased Ig levels and variable T- and B-cell abnormalities (M. Flanagan and Cunniff 2019). Several leukemias (ALL, AML, and biphenotypic) and lymphomas (HL and NHL) are reported (German 1997), with median onset in the mid-third decade. AML is twice as common as ALL, although many AML are likely chemotherapy-related (after prior cancer treatment) and often preceded by MDS (Brock et al. 1991). Nonetheless, primary AML and ALL both occur as well. A mean age of 22 years was reported for lymphoma, with types comprising both B- (including BL) and T-lineage, but not HL (Cunniff et al. 2017). In ~75% of cases, lymphoma was the first cancer diagnosis (Brock et al. 1991). Other malignancies (sarcomas, germ cell tumors, medulloblastoma, retinoblastoma) are also described (Cunniff et al. 2017). Patients also have increased colorectal cancer incidence (12% risk), usually in the second decade (Martinez et al. 2016).

3.3.2 LIG4 Syndrome, Mutations in *LIG4*

Pathophysiology: LIG4 is a DNA ligase in the non-homologous end joining (NHEJ) double-strand break (DSB) repair pathway. It also rectifies DSB during Ig and TCR VDJ recombination (Altmann and Gennery 2016). Consequently, LIG4 LOF mutations cause B- and T-cell deficiency while also making all cells susceptible to DSB throughout the genome.

Clinical features: B- and T-cell cytopenias with variable hypogammaglobulinemia. DSB due to Ig isotype class switching are also affected, resulting in elevated IgM and lower levels of IgG, IgA, and IgE. Patients exhibit frequent bacterial, viral, and fungal infections, with related failure to thrive. SCID was reported in four patients (Buck et al. 2006; van der Burg et al. 2006; IJspeert et al. 2013; Murray et al. 2014). Predictably, defective DSB repair confers substantial malignancy risk, with lymphomas (some EBV-driven) described, including DLBCL and T-cell lymphomas (Enders et al. 2006; Toita et al. 2007; Bacon et al. 2013; Ben-Omran et al. 2005), and one patient with T-ALL (Bacon et al. 2013; Slatter and Gennery 2010). Other features include microcephaly, polydactyly, and developmental delay (Murray et al. 2014; O'Driscoll et al. 2001; Unal et al. 2009).

3.3.3 Nijmegen Breakage Syndrome, Mutations in *NBS1*

Pathophysiology: Nijmegen breakage syndrome (NBS; alternatively designated Berlin breakage syndrome) is an autosomal recessive DNA repair defect caused by *NBS1* mutations (Iijima et al. 2004). *NBS1* encodes the protein Nibrin (NBN). DNA DSB are repaired by both the HR and NHEJ pathways. In the first NHEJ step, the MRN complex recognizes DSB. NBN regulates this complex, fostering MRN enzymatic activities and recruiting other proteins to DSB during early NHEJ. NBN also enacts a cell cycle checkpoint arrest to promote genome stability (Wang et al. 2010; Lu et al. 2006). Hence, NBS patients are highly sensitive to radiation and other genotoxic stressors.

Clinical features: Microcephaly, dysmorphic features, growth retardation, developmental delay, recurrent infections, and impaired fertility are features of NBS. Both humoral and cellular IMD occur, with severe hypogammaglobulinemia, isolated IgA, IgG2 and/or IgG4 deficiencies, and CD3 T-, CD4 T-, and B-cell lymphopenias (Varon et al. 2017; Group 2000). By 20 years, >40% of NBS patients develop cancer, predominantly lymphoid. B- and T-NHL are most common, especially DLBCL and T lymphoblastic lymphoma (T-LBL), with BL and Burkitt-like lymphomas also seen (Dembowska-Baginska et al. 2009; El Hasbaoui et al. 2020). Solid tumors also occur, including medulloblastoma, glioma, and rhabdomyosarcoma (Hiel et al. 2001; Bakhshi et al. 2003; Distel et al. 2003).

3.3.4 Constitutional Mismatch Repair Deficiency (CMMR-D), Mutations in *MSH2* and *PMS2*

Pathophysiology: CMMR-D is a rare childhood cancer syndrome caused by germline mutations in two of the four mismatch repair (MMR) genes: *MLH1*, *MSH2*, *MSH6*, and *PMS2* (Wimmer and Etzler 2008). MMR proteins correct single base-pair mismatches and small insertions and deletions that arise during

DNA replication and are also involved in responses to DNA-damaging agents (Jiricny 2006).

Mutations in two CMMR-D genes cause leukemias and lymphomas: *MSH2* and *PMS2*. *MSH2* encodes the mutS homolog 2 (MSH2). MSH2 forms heterodimers with MSH6 (de Wind et al. 1995; Brown et al. 2003). These heterodimers comprise the MutSα MMR complex and, when coupled with MSH3, form the MutSβ MMR complex. MSH2 is critical to several pathways, including transcription-coupled repair, HR, and base-excision repair (de Wind et al. 1995; Pitsikas et al. 2007; Mellon et al. 1996). PMS2 and MLH1 (mutL homolog 1) form another heterodimer, an endonuclease that excises base-pair mismatches during DNA replication (Nicolaides et al. 1995; PMS2 PMS1 Homolog 2, Mismatch Repair System Component [Homo Sapiens (Human)] - Gene - NCBI n.d., 1). MSH2 and PMS2 mutations both lead to hematological cancers, with variable IMD.

1. **Mismatch Repair Cancer Syndrome-2 (MMRCS2), Compound Heterozygous/Homozygous Mutations in *MSH2*.**
   - Clinical features: Features include abdominal pain caused by gastrointestinal polyps, café au lait spots, sphenoid bone dysplasia, pseudarthrosis, and optic gliomas (Whiteside et al. 2002). One T-ALL case in a 2-year-old MMRCS2 patient was reported (Whiteside et al. 2002). Bougeard et al. reported two MMRCS2 siblings, one who died of mediastinal T-cell lymphoma and the second of glioblastoma (Durno et al. 2021; Bougeard et al. 2003; Müller et al. 2006). Most patients have family histories with various tumors, including hematologic cancers, endometrial carcinomas, brain tumors, and gastrointestinal malignancies.
2. **Mismatch Repair Cancer Syndrome-4 (MMRCS4), Compound Heterozygous Mutations in *PMS2*.**
   - Clinical features: Variable phenotypic features include café au lait spots and neurofibromas (which can result in MMRCS4 being mis-diagnosed as neurofibromatosis 1) (Kratz et al. 2008; Giunti et al. 2009; De Vos et al. 2006). One case of IMD with defective Ig class-switching and recurrent infections has been reported (Péron et al. 2008). Hematologic malignancies include ALL, atypical CML, and NHL (B and T cell). Turcot syndrome refers to colorectal cancer and CNS tumors with *PMS2* mutations (Kratz et al. 2008; De Vos et al. 2006; Baas et al. 2013; Trimbath et al. 2001).

3.3.5 Ataxia-Telangiectasia (AT), Mutations in *ATM*

Pathophysiology: AT results from recessive mutations in the ataxia-telangiectasia mutated (*ATM*) gene, which has critical roles in DSB repair, cell cycle regulation, and cell survival (Lavin 2008; McKinnon 2012). Most ATM mutations create truncations, although some missense mutations and splicing variants show milder phenotypes (Gilad et al. 1998).

Clinical features: Symptoms include eye, ear, and skin (especially sun-exposed) telangiectasias, ataxia, hypotonia, chorea, and dystonia (due to cerebellar degeneration) (Boder and Sedgwick 1970; Shaikh et al. 2013; Pearson 2016). IMD consists of T- and B-cell lymphopenias and hypogammaglobulinemia (Buckley 2004; Fiorilli et al. 1983). Prevalent cancers include leukemias (T-ALL and T-cell prolymphocytic leukemia) and lymphomas (NHL and HL, some EBV-driven) (Taylor et al. 1996; Kulinski et al. 2012; Amirifar et al. 2019). Due to their sensitivity, some chemotherapies and radiation are highly cytotoxic to ATM patients, requiring caution (Suarez et al. 2015).

3.3.6 Li-Fraumeni Syndrome (LFS), Mutations in *TP53*

Pathophysiology: Li-Fraumeni syndrome (LFS) is caused by germline *TP53* mutations (Isobe et al. 1986; McBride et al. 1986). *TP53* encodes p53, which is vital to tumor suppression and cell stress responses. Upon activation, p53 regulates many genes critical to DNA repair, cell cycle arrest, apoptosis, senescence, and other vital processes.

Clinical features: There are many diagnostic criteria for LFS. Patients often have multiple tumor types at an early age, with high incidence of hypodiploid ALL, solid tumors (adrenocortical carcinomas, CNS tumors), AML, HL, and NHL (McBride et al. 2014; Schneider et al. 2019). IMD in LFS is underappreciated, but includes impaired lymphocyte responses to phytohemagglutinin-P and polyclonal IgM elevation; in most cases, cancers overshadow IMD features (Fraumeni et al. 1975). LFS and p53 are intimately linked to oncogenesis and covered extensively elsewhere in this series, so we will not review the many well-known aspects of this syndrome.

3.4 Defects in Other Pathways

3.4.1 Wiskott-Aldrich Syndrome, Mutations in *WAS*

Pathophysiology: Caused by mutations in *WAS* on the X chromosome (Kwan et al. 1991). *WAS* encodes Wiskott-Aldrich syndrome protein (WASp), which is key to cytoskeletal polymerization (Rivero-Lezcano et al. 1995; Symons et al. 1996; Miki and Takenawa 1998) and expressed by all hematopoietic cells (Parolini et al. 1997). It has critical immune functions, including formation of immunological synapses (Cory et al. 1996; Gallego et al. 1997; Sasahara et al. 2002), granule exocytosis (Orange et al. 2002; Nikolov et al. 2010), phagocytosis (J. Zhang et al. 1999; Lorenzi et al. 2000), and cell migration/motility (Linder et al. 2000; Badolato et al. 1998).

Clinical features: Patients exhibit variable features, including chronic ITP, colitis, and atopic dermatitis (Suri et al. 2021). The classic triad of eczema, thrombocytopenia, and IMD is often absent initially, but evolves later (Sullivan et al. 1994). Thrombocytopenia is universal, typically manifest as petechiae and spontaneous or prolonged bleeding in infancy. WAS platelets are characteristically small; coupled with X-linked inheritance, this aids in diagnosis.

Autoimmunity occurs in up to 70% of WAS cases and can even develop after SCT (Sullivan et al. 1994; Lee et al. 2009; Jin et al. 2019; Imai et al. 2004; Dupuis-Girod et al. 2003; Burroughs et al. 2020; Massaad et al. 2013). Autoimmune conditions include cytopenias (hemolytic anemia, neutropenia, and thrombocytopenia), inflammatory bowel disease, IgA nephropathy, Takayasu arteritis, and Guillain-Barré syndrome (Lau et al. 1992; Schurman and Candotti 2003).

Children with WAS have increased risk of leukemia and lymphoma, with a cumulative incidence reported to be as high as 13% (Sullivan et al. 1994; Imai et al. 2004). NHL is most common (EBV+ B-NHL, including DLBCL) with HL (Cotelingam et al. 1985) and JMML also described (Sullivan et al. 1994; Perry et al. 1980; Buchbinder et al. 2014; Yoshimi et al. 2013).

4 IMD Associated with Lymphoma

Several IMD syndromes are associated with lymphomas, but not leukemias (or, in one instance, a single LGL leukemia case). Table 3 lists the entities described in this section.

4.1 Defects of Leukocyte Development

4.1.1 Combined Cellular and Humoral Immune Defects with Granulomas, Mutations in *RAG1/RAG2*

Pathophysiology: *RAG1* and *RAG2* (recombination-activating genes 1 and 2) are essential for Ig and TCR V(D)J recombination in developing lymphocytes (Haines et al. 2006). Without V(D)J rearrangement, mature B and T cells do not develop (Niehues et al. 2010). *RAG1* and *RAG2* mutations cause a spectrum of IMD, including classical SCID, Omenn syndrome (OS), and other idiosyncratic variants (Sobacchi et al. 2006; Niehues et al. 2010; Asai et al. 2011). Aberrant *RAG1* or *RAG2* activity can also promote lymphocyte cancers by causing translocations, deletions, and insertions that destabilize the genome (Rommel et al. 2017).

Table 3 Featured IMD associated with lymphomas

Leukocyte development defects	Leukocyte signaling defects	Genome stability defects	Other defects
Combined cellular and humoral immune defects with granulomas: *RAG1* and *RAG2* mutations IMD24: *CTPS1* mutations WHIM-1: *CXCR4* mutations	LPFS2 and LPFS3: *CD27* and *CD70* mutations ALPS: *FAS* and *FASLG* mutations IMD5, *SH2D1A* mutations CVID8 with autoimmunity: *LRBA* mutations XMEN syndrome: *MAGT1* mutations TIIAC: *STK4* mutations	Partial SCID: *DCLRE1C* mutations IMD96: *LIG1* mutations	Chediak-Higashi syndrome: *LYST* mutations

Clinical features: Null *RAG1/RAG2* mutations lead to classical SCID, while hypomorphic mutations cause a variable spectrum of infections, autoimmunity, granulomatous disease, and malignancy (Niehues et al. 2010; Catharina Schuetz et al. 2008). Defects in RAG1/RAG2 lead to a combined IMD characterized by decreased number and defective function of B and T cells (Niehues et al. 2010). Scheutz et al. described three unrelated patients with compound heterozygous *RAG1* or *RAG2* mutations whose IMD included hypogammaglobulinemia, diminished T and B cells, granulomas in skin, mucous membranes and internal organs, sparse thymic tissue, and severe complications after viral infections. One of these patients also developed EBV-associated DLBCL at 5 years, successfully treated with rituximab followed by SCT (Catharina Schuetz et al. 2008). In a separate case, Maas et al. reported an EBV-associated DLBCL in a 4-month infant with homozygous RAG1-deficient SCID (Maas et al. 2018). B-NHL is highly unusual in infants; SCID patients either receive early SCT or succumb to opportunistic infection (Maas et al. 2018). However, this rare case demonstrates that complete RAG1/2 LOF also predisposes to lymphoma.

4.1.2 IMD24, Mutations in *CTPS1*

Pathophysiology: Cytidine triphosphate synthetase 1 is required for de novo pyrimidine nucleotide production, which is necessary for DNA replication and also lymphocyte development (Martin et al. 2014, n.d.). Homozygous *CTPS1* LOF mutations impair antigen receptor-mediated B- and T-cell activation and proliferation (Martin et al. 2014).

Clinical features: *CTPS1* splice-site mutations enact recessive LOF (Martin et al. 2014, n.d.). IMD24 usually presents with early and severe life-threatening IMD, although milder presentations are also described (Martin et al. n.d.; Trück et al. 2016). A spectrum of chronic viral (EBV, VZV, HHV6) and encapsulated bacterial infections reveal a complex IMD picture. Lab findings show variable lymphopenia—exacerbated during infection—and inversed CD4:CD8 ratio (Martin et al. 2014, n.d.). IgG levels are normal-to-elevated with increased IgG, but low IgG2 levels and *Streptococcus pneumoniae* titers (Martin et al. n.d.). Immunophenotyping in seven patients showed absent or low mucosal-associated T, invariant NKT, memory B, and NK cells. Overall, blood counts were near normal, with impaired antigen-induced T-cell proliferation (Martin et al. 2014, n.d.; Trück et al. 2016; Nademi et al. 2019; Kucuk et al. 2016). T-cell proliferation is required to control EBV, so patients manifest with EBV-driven immune dysregulation and high risk of EBV-driven LPD and lymphomas (Lino et al. 2021; Hislop et al. 2005). Martin et al. and Nademi et al. reported CNS, lung, and liver EBV+ LPD and CNS lymphomas (Martin et al. 2014; Nademi et al. 2019). SCT is curative (Kucuk et al. 2016; Nademi et al. 2019).

4.1.3 WHIM Syndrome-1, Mutations in *CXCR4*

WHIM is an acronym for a rare IMD presenting with warts, hypogammaglobulinemia, recurrent bacterial infections, and myelokathexis (abnormal retention of mature neutrophils in marrow) due to heterozygous mutations in *CXCR4* (Hernandez et al. 2003).

Pathophysiology: The chemokine receptor CXCR4 and its ligand, CXCL12, are required for B- and myeloid cell development (Nagasawa et al. 1996; Q. Ma et al. 1998), and they also likely promote T-cell maturation (Hernandez et al. 2003). In WHIM-1, impaired CXCR4 internalization prevents desensitization to CXCL12 binding, causing amplified CXCR4 signaling (i.e., GOF), which impedes normal chemotaxis and leukocyte trafficking (Balabanian et al. 2005).

Clinical features: As noted, features include a tetrad of warts, hypogammaglobulinemia, recurrent bacterial infections, and myelokathexis. In an extensive study of 105 WHIM patients (most with WHIM-1, rather than WHIM-2, due to CXCR2 mutations), the classic tetrad was present in 38%, with myelokathexis the most prevalent finding. Myelokathexis is near-pathognomonic for WHIM, with leukopenia, neutropenia, mild thrombocytopenia, and hypogammaglobulinemia also common (Heusinkveld et al. 2019; Tassone et al. 2009). Frequent infections include HPV-induced lesions and recurrent oto-sino-pulmonary and skin infections with gram-positive bacteria (Heusinkveld et al. 2019). WHIM-1 lacks gastroenteropathy and IBD, distinguishing it from CVID (Heusinkveld et al. 2019). Non-immunologic WHIM features include heart (especially tetralogy of Fallot), urogenital, and cerebellar developmental defects. Rare EBV-associated T- and B-cell LPD and EBV+ B-cell lymphomas all occur (Heusinkveld et al. 2019; Tassone et al. 2009; Imashuku et al. 2002; Beaussant Cohen et al. 2012), in addition to HPV-driven cancers. EBV-negative cutaneous B- and T-cell lymphomas are also reported (Heusinkveld et al. 2019; Beaussant Cohen et al. 2012).

4.2 Defects in Leukocyte Signaling

4.2.1 Lymphoproliferative Syndromes 2 and 3, Mutations in *CD27* and *CD70*

Pathophysiology: CD27 is a tumor necrosis factor receptor (TNFR) superfamily protein expressed by naive and memory T cells, germinal center and memory B cells, plasma cells, and some NK cells (Nolte and van Olffen 2009). Its ligand, CD70 (a TNF family member), is expressed by primarily activated lymphoid and myeloid cells (Nolte and van Olffen 2009). CD27 engagement by CD70 provides costimulatory signals leading to T-cell activation, survival, proliferation, and differentiation (Borst et al. 2005). CD27 signaling also promotes the generation and maintenance of antigen-specific CD4 and CD8 T cells (Hendriks et al. 2000; Munitic et al. 2013), as well as memory B-cell differentiation and plasma cell survival (Avery et al. 2005). Thus, the CD27-CD70 axis is essential for memory T- and B-cell generation; dysregulating this axis impairs cytokine production, CTL activity, and other vital mature B- and T-cell functions (Ghosh et al. 2020; Abolhassani et al. 2017).

1. **Lymphoproliferative Syndrome 2 (LPFS2), Mutations in *CD27*.**
   - Clinical features: Occurs primarily in patients of Moroccan/Turkish/Lebanese descent, presenting with severe recurrent viral infections due to immune dysfunction. The most common features are EBV-related (infectious mononucleosis, LPD), with 36% of patients developing lymphoma by a median of 8.5 years. Lymphomas (in decreasing order of frequency) include HL, DLBCL, and other B-NHL (Ghosh et al. 2020).
2. **Lymphoproliferative Syndrome 3 (LPFS3), Mutations in *CD70*.**
   - Clinical features: Occurs primarily in patients of Iranian/Turkish descent, presenting with recurrent otitis, sinusitis, chronic diarrhea, and respiratory tract infections. These patients are also unable to clear EBV, resulting in predisposition to EBV-driven LPD. Over half of LPFS3 patients (56%) developed lymphoma by a median age of 3 years. Lymphomas (in decreasing order of frequency) include HL (nodular sclerosing and mixed-cellularity subtypes), BL, and other B-NHL (Ghosh et al. 2020).

Allogeneic SCT is curative for both LPFS2 and LPFS3.

4.2.2 Autoimmune Lymphoproliferative Syndrome (ALPS), Mutations in *FAS* and *FASLG* (Alternate Name: Canale-Smith Syndrome)

Pathophysiology: ALPS is caused by mutations in either *FAS* or *FASLG*, genes that encode another TNFR/TNF pair, analogous to CD27/CD70. Upon binding by FASLG (a TNF member), FAS (a TNFR member) transmits an intracellular apoptotic signal (Inazawa et al. 1992; Lichter et al. 1992). Expression of FAS and FASLG are both triggered by prolonged T-cell activation in order to dampen immune responses (Le Deist et al. 1996). Their interaction also controls peripheral T-cell lifespans, eliminating potentially autoreactive cells (Singer and Abbas 1994). Apoptosis of activated immune cells is a critical element of immune homeostasis; thus, FAS or FASLG LOF causes immune dysregulation and LPD, hence the term autoimmune lymphoproliferative syndrome (Sneller et al. 1992, 1997).

Clinical features: The earliest ALPS manifestations are lymphadenopathy and/or splenomegaly (at times, incidentally identified) (Shah et al. 2014). It can also present with refractory autoimmune multilineage cytopenias (most severe in childhood), and bacterial infections can occur due to neutropenia. Autoimmune problems of other organs, such as the liver, kidneys, and eyes, or LPD involving other organ systems can also arise (Neven et al. 2011). ALPS diagnostic criteria requires the following: (1) chronic, non-malignant, non-infectious lymphadenopathy and/or splenomegaly and (2) elevated $CD3^+$, $TCR\alpha\beta^+$, double-negative ($CD4^-$/$CD8^-$) T cells ($\geq$1.5% of total lymphocytes or 2.5% of $CD3^+$ lymphocytes), in the setting of a normal or elevated lymphocyte count (Oliveira et al. 2010). Accessory criteria also exist (Oliveira et al. 2010).

ALPS has established lymphoma predilection. In a 150-patient ALPS cohort, 18 developed lymphoma (Price et al. 2014), with median onset by 18 years. B-lineage lymphomas predominated (16/18, ~89%), with 10 HL and 6 NHL. Every lymphoma patient in this cohort had a dominant-negative FAS mutation.

However, 20% in this cohort were FAS haploinsufficient, none of whom developed lymphoma, suggesting risk is highest in (and perhaps exclusive to) patients with dominant-negative FAS mutations (Price et al. 2014). Two non-hematologic cancers (one tongue squamous cell carcinoma, one pleomorphic adenoma) also occurred in this cohort (Price et al. 2014), and another ALPS patient with HL developed a secondary histiocytic sarcoma in a separate report (Venkataraman et al. 2010).

4.2.3 IMD5, X-Linked Lymphoproliferative Disease, Mutations in *SH2D1A* (Alternate Name: Duncan's Disease)

Pathophysiology: *SH2D1A* encodes an SH2 domain protein known as SLAM-associated protein (SAP), with "SLAM" representing "signaling lymphocytic activation molecule" (Coffey et al. 1998; Sayos et al. 1998). SAP interacts with SLAM, inhibiting this transmembrane protein by blocking the recruitment of another SH2 domain-containing signal transduction molecule, SHP-2 (Detre et al. 2010). SAP regulates T-, NK-, and NKT-cell function directly and indirectly governs B-cell function (Sharifi et al. 2004; Okano and Gross 1996; Dupré et al. 2005). Defects in SAP impact these cells, with humoral immune defects in XLP thought to arise from impaired CD4 T:B-cell interaction, rather than an intrinsic B-cell problem (Ma et al. 2005; Nichols et al. 2005a, b; Chung et al. 2005).

Clinical features: XLP patients develop fulminant infectious mononucleosis, dysgammaglobulinemia, and lymphoma. Upon EBV infection, LPD, hepatosplenomegaly, cytopenias, and erythrophagocytosis all occur (Bar et al. 1974). Rarer characteristics include aplastic anemia, autoimmune colitis, psoriasis, and pulmonary lymphoid granulomatosis (Dutz et al. 2001; Purtilo and Grierson 1991; Schuster and Kreth 2000). Humoral immune defects range from Ig subclass deficiencies or elevated IgM to hypogammaglobulinemia (Grierson et al. 1991), making distinction between XLP and CVID challenging.

About 30% of XLP patients develop EBV-driven lymphomas. Not surprisingly, most are B-lineage, such as BL or other B-NHL, although immunoblastic and extranodal lymphomas are also reported (Rezaei et al. 2011a, b). Early diagnosis and management is vital to decreasing mortality, with SCT the definitive treatment (Marsh et al. 2014; Shamriz et al. 2014).

4.2.4 CVID8 with Autoimmunity, Mutations in *LRBA*

CVIDs are a heterogeneous group of primary IMD featuring decreased antibody production, infections, inflammatory and autoimmune diseases, LPD, and lymphoma predilection (Lo et al. 2015; Cunningham-Rundles 2012).

Pathophysiology: CVID8 with autoimmunity is caused by recessive mutations in *LRBA* (lipopolysaccharide-responsive vesicle trafficking, beach- and anchor-containing) (Lopez-Herrera et al. 2012), a protein homologous to other vesicle trafficking proteins. LRBA regulates the expression of and downstream signaling by CTLA4, a T-cell checkpoint inhibitor (Lo et al. 2015; Li et al. 2018; Cagdas et al. 2019). Ultimately, LRBA LOF results in diminished regulatory T-cell (T_{reg}) function (Cagdas et al. 2019).

Clinical features: CVID- or ALPS-like presentations are typical (Cagdas et al. 2019), with variable phenotypes, even within families (Lopez-Herrera et al. 2012; Alangari et al. 2012). Patients present with early recurrent infections (mostly respiratory), chronic diarrhea, and variable autoimmunity—including cytopenias (ITP, AIHA, Evans syndrome), IBD, thyroiditis, and others (Lopez-Herrera et al. 2012; Cagdas et al. 2019; Alangari et al. 2012). Lab findings include B-cell leukopenia, hypogammaglobulinemia, and $CD4^+$ T_{reg} deficiency. LPD is also common. A review of 15 CVID8 patients reported splenomegaly in >90%. Four of these patients (>25%) developed lymphoma, with three B-NHL and one HL (Cagdas et al. 2019).

4.2.5 X-Linked Immunodeficiency with Magnesium Defect, Epstein-Barr Virus Infection, and Neoplasia (XMEN) Syndrome, Mutations in *MAGT1*

Pathophysiology: In 2011, Li et al. identified LOF mutations in the magnesium transporter gene *MAGT1* and described a new inborn error of glycosylation and immunity, XMEN syndrome (Li et al. 2011).

MAGT1 is a conserved, ubiquitous protein that is a critical regulator of intracellular Mg^{2+} homeostasis (Li et al. 2014; Ravell et al. 2014). Loss of MAGT1 abolishes the transient Mg^{2+} influx induced by TCR ligation that is required for T-cell activation (Li et al. 2011, 2014). Lower intracellular Mg^{2+} also impairs glycosylation of the NK-activating receptor NKG2D by NK and CD8 T cells. This blunts their activation, resulting in decreased cytolysis of EBV-infected cells by these cells, as well as diminished anti-tumor immunity (Li et al. 2014; Chaigne-Delalande et al. 2013). Ultimately, MAGT1 mutations cause NKG2D instability (Matsuda-Lennikov et al. 2019; Ravell et al. 2020b). Since XMEN patients invariably have decreased NKG2D levels on NK and CD8 T cells, this finding is pathognomonic for the syndrome (Ravell et al. 2020a, b).

Clinical features: XMEN syndrome is phenotypically heterogeneous and primarily affects males, since it is X-linked. Females with heterozygous *MAGT1* mutations are healthy carriers, due to biased X-inactivation of the chromosome bearing the mutant allele in hematopoietic cells (Li et al. 2011; Ravell et al. 2020a; Blommaert et al. 2019). Key features include chronic EBV infection and LPD, splenomegaly, variable Ig deficiencies, and decreased CD4:CD8 ratio (<1.3). Other features include recurrent otic and sinopulmonary infections, at times causing bronchiectasis, susceptibility to viral infections like varicella/recurrent zoster and molluscum contagiosum, autoimmune cytopenias, and CD4 T-cell lymphopenia (Li et al. 2014; Ravell et al. 2020a). Ravell et al. reported 12 XMEN patients who developed lymphoma or EBV+ LPD (Ravell et al. 2020a). Classical HL was most common, with EBV+ LPD, BL, and DLBCL also occurring (Ravell et al. 2020a). Most were in the second or third decade, and five patients experienced more than one malignancy (Ravell et al. 2020a).

Magnesium supplementation was proposed as a possible therapy, but a recent trial proved its futility (Li et al. 2014; Chauvin et al. 2022). SCT is potentially curative, but associated with high mortality (Brault et al. 2021). CRISPR gene editing was recently used to restore MAGT1-mediated glycosylation of NKG2D in mice and could be a therapeutic option (Brault et al. 2021).

4.2.6 T-Cell Immunodeficiency, Recurrent Infections, and Autoimmunity with or without Cardiac Malformations (TIIAC), Mutations in *STK4* (Also Known as *MST1*)

Pathophysiology: Serine/threonine kinase 4 (STK4) is a ubiquitous kinase and key component of the Hippo signaling pathway that regulates biological processes including proliferation, differentiation, apoptosis, and immune responses (Abdollahpour et al. 2012; Zhao et al. 2010; Bata et al. 2022; Jørgensen et al. 2021). In 2012, 5 groups described 13 patients with STK4 LOF (Abdollahpour et al. 2012; Nehme et al. 2012; Dang et al. 2016; Crequer et al. 2012a; Halacli et al. 2015). Functions in innate and adaptive immunity, apoptosis, and autophagy have subsequently been ascribed to STK4, with abnormalities of adhesion, vesicle trafficking, and migration impairing T-cell function (Dang et al. 2016; Kurz et al. 2016). AKT goes unchecked in T cells, decreasing the activity of the FOXO1 transcription factor and its targets in naive T cells, including FOXO1-mediated *IL7R* transcription (Nehme et al. 2012; Risma 2012). Collectively, SKT4 deficiency leads to increased cell death of naive and proliferating T cells, accentuated by increased FAS expression on these cells (Nehme et al. 2012). It also acts as a tumor suppressor by activating the large tumor suppressor kinases (LATS1/2), the transcription factor Yes-Associated Protein (YAP), and the TAZ transcriptional co-activator (also known as WWTR1) (Jørgensen et al. 2021; Kurz et al. 2016).

Clinical features: Common features include recurrent bacterial, viral, and fungal infections, mucocutaneous candidiasis, dermatitis, cutaneous warts, molluscum contagiosum, recurrent skin abscesses, autoimmune cytopenias like AIHA and ITP, lymphadenopathy, EBV+ LPD, and cardiac defects (ASD, pulmonary and tricuspid valve insufficiency, and others) (Abdollahpour et al. 2012; Jørgensen et al. 2021; Nehme et al. 2012; Dang et al. 2016; Crequer et al. 2012a; Halacli et al. 2015; Guennoun et al. 2021). Lab findings show profound T- and B-cell lymphopenia, intermittent neutropenia, hypergammaglobulinemia with low IgG2, and reduced T-cell proliferation.

STK4-deficient patients are prone to various lymphomas, with HL, extra-nodal marginal zone lymphoma of MALT, BL, maxillary sinus DLBCL, and primary cardiac T-NHL all reported (Radwan et al. 2020; Schipp et al. 2018; Nehme et al. 2012; Ashrafi et al. 2020; Sherkat et al. 2017; Lum et al. 2016). Interestingly, many lymphomas were EBV negative (Radwan et al. 2020; Schipp et al. 2018), indicating the stand-alone role of STK4 as a tumor suppressor.

Since its discovery, a confusing disease spectrum has been recognized, with some cases showing overlap with other primary IMD like DOCK8 deficiency (Radwan et al. 2020) and ALPS (Schipp et al. 2018). Additional features of TIIAC patients include immune complex glomerulonephritis, salt-losing tubulopathy suggestive of acquired Gitelman syndrome, Castleman-like disease, juvenile idiopathic arthritis, human beta-papillomavirus-associated epidermodysplasia verruciformis, and short stature (Jørgensen et al. 2021; Guennoun et al. 2021; Sharafian et al. 2019; Crequer et al. 2012a).

4.3 Defects of Genome Stability

4.3.1 Partial SCID Syndrome, Mutations in *DCLRE1C*

Pathophysiology: *DCLRE1C* encodes ARTEMIS, a nuclease in NHEJ pathway (D. Moshous et al. 2000, 2001) that is essential for both hairpin opening during V(D)J recombination and general DSB repair. ARTEMIS directly interacts with LIG4 (discussed previously in this chapter) (Malu et al. 2012). NHEJ defects cause genomic radiation sensitivity and cancer susceptibility (Lee et al. 2013). *DCLRE1C* null mutations result in absent B and T cells manifesting as radiosensitive $T^{-}B^{-}$ SCID. Hypomorphic *DCLRE1C* mutations cause a spectrum that varies based on the degree of remaining ARTEMIS function (i.e., "leaky SCID") (Lee et al. 2013; Felgentreff et al. 2015; Fevang et al. 2018). Most of these patients have one null allele and one hypomorphic allele, allowing some residual V(D)J recombination and DSB repair activity (Felgentreff et al. 2015).

Clinical features: Complete Artemis LOF presents as SCID in early infancy with life-threatening infections, diarrhea, failure to thrive, and radiosensitivity (Fevang et al. 2018). Hypomorphic *DCLRE1C* mutations are more variable, with unpredictable clinical syndromes ranging from isolated Ig deficiency to partial/atypical SCID, Omenn syndrome, hyper-IgM syndrome, inflammatory bowel disease, and other phenotypes (Volk et al. 2015). Patients with hypomorphic ARTEMIS are also predisposed to B-cell lymphomas (Fevang et al. 2018; Despina Moshous et al. 2003; Jacobs et al. 2011; de Miranda et al. 2013). Presentations include recurrent respiratory infections, candidiasis, diarrhea, immune dysregulation, and malignancies that may even occur in adulthood (Volk et al. 2015). Moshous et al. reported two partial SCID patients who developed EBV-associated DLBCL. IMD in these patients included candidiasis, diarrhea, recurrent pulmonary infection, lymphopenia, hypogammaglobulinemia, and T- and B-cell leukopenia (Despina Moshous et al. 2003). Fevang et al. described a 23-year-old with hypomorphic *DCLRE1C* who developed lymphoproliferative skin lesions at 2 years that transformed into two EBV+ B-cell lymphomas: one DLBCL and one intermediate between DLBCL and classical HL (Fevang et al. 2018). In a unique case, Bajin et al. described a child with recurrent infections, chronic diarrhea, growth retardation, sclerosing cholangitis, and hyper-IgM syndrome with mutant ARTEMIS who subsequently developed LGL leukemia (Bajin et al. 2013).

4.3.2 IMD96, Mutations in *LIG1*

Pathophysiology: DNA Ligase I (LIG1) is a member of the conserved DNA ligase family that plays a key role in DNA replication by facilitating joining of Okazaki fragments (Harrison et al. 2002; Peng et al. 2012; Barnes et al. 1992). *LIG1* mutations result in IMD, increased sensitivity to DNA-damaging agents, and resulting cancers (Maffucci et al. n.d.).

Clinical features: IMD manifests as recurrent infections, with lab findings of hypogammaglobulinemia, lymphopenia, erythrocyte macrocytosis, and NK-cell deficiency (Maffucci et al. n.d.). Webster et al. reported a case with IMD characterized by recurrent infections, stunted growth, sun sensitivity, an "elf-like" appearance, absent secondary sexual characteristics in adolescence, delayed bone age, and telangiectasias as a teen, who subsequently developed hepatosplenomegaly and died of a possible lymphoma at 19 years (Barnes et al. 1992; Webster et al. 1992). Lab studies showed erythrocyte macrocytosis, hypogammaglobulinemia, and progressive lymphopenia and neutropenia (Webster et al. 1992). Due to phenotypic variability and overlap with other IMD, next-generation sequencing can aid diagnosis. Curative strategies include SCT, although poor tolerance of genotoxic conditioning regimens is a concern; gene therapy may represent an option (Punwani et al. 2017).

4.4 Defects in Other Pathways

4.4.1 Chediak-Higashi Syndrome (CHS), Mutations in *LYST*

Pathophysiology: CHS results from autosomal recessive lysosomal trafficking regulator (*LYST*) mutations (Kaplan et al. 2008; Barbosa et al. 1996; Perou et al. 1996). Most tissues express LYST, with higher levels in the bone marrow (Sharma et al. 2020). LYST protein is involved in vesicle trafficking and lysosome size and movement, and LYST mutations cause abnormal, enlarged lysosomes that affect immune cell lytic granules, platelet dense bodies, and melanosomes (Burkhardt et al. 1993; Introne et al. 1999).

Clinical features: IMD results in frequent, often severe, bacterial infections and periodontitis. Patients also have bleeding tendency (due to defective platelet dense bodies) and HLH predilection (Toro et al. 2018). Reported immune dysfunction includes abnormal NK-cell function, defective antibody-dependent cell-mediated cytolysis, and poor CTL killing (Introne et al. 1999; Baetz et al. 1995; Klein et al. 1980). In ~85% of patients, an accelerated phase occurs with fever, LPD, hepatosplenomegaly, and pancytopenia (Blume and Wolff 1972). One CHS patient with T-cell lymphoma is reported, who rapidly developed lymphadenopathy and initiated treatment, but died due to sepsis (Argyle et al. 1982).

5 Additional IMD Associated with Leukemia and/or Lymphoma

Due to space constraints, we were unable to provide text summaries of all IMD syndromes linked to leukemia and/or lymphoma. Key aspects of these other entities are listed in Table 4.

6 Conclusion

Immunodeficiency and immune dysregulatory syndromes are rare entities. Adult and pediatric oncologists encounter patients with these disorders infrequently, unless practicing at specialized centers. However, when present, these diseases may go

Table 4 Additional IMD associated with leukemias and lymphomas

Syndrome, mutated gene	IMD and other clinical features	Cancer(s) reported	References
IMD4, *IL2RG* mutations	X-linked SCID, agammaglobulinemia, thymic atrophy, recurrent infection	Two T-NHL cases (after gene therapy, implicating treatment, not IMD4)	Buckley et al. (1999), Woods et al. (2006) and Hacein-Bey-Abina et al. (2003)
IMD7, *TRAC* mutations	Decreased T cells and T-cell subsets, increased IgE and IgA, recurrent respiratory infection, chronic diarrhea	EBV+ B-NHL (1 case)	Rawat et al. (2021) and Morgan et al. (2011)
MIRAGE syndrome, *SAMD9* mutations	Myelodysplasia, bacterial infections, decreased B and NK cells, growth restriction, adrenal hypoplasia, genital phenotypes, enteropathy	MDS AML	Buonocore et al. (n.d.), Narumi et al. (2016) and Sahoo et al. (2021)
Ataxia-pancytopenia syndrome, *SAMD9L* mutations	Progressive neurologic symptoms, bacterial infection, decreased B and NK cells, multilineage cytopenia with marrow hypoplasia		
IMD27A, *IFNGR1* mutations	Complete deficiency has early (often fatal) mycobacterial infections; decreased T-cell proliferation and production of IFN-γ, atypical mycobacterial infections	CLL described with certain *IFNGR* gene polymorphisms, not reported in IMD27A	Toyoda et al. (2004) and Zhou et al. (2010)
MVA1 (mosaic variegated aneuploidy) syndrome, homozygous/ compound heterozygous *BUB1B* mutations	Innate and adaptive IMD, microcephaly, developmental delay, seizures, restricted growth	MDS (1 case) ALL (1 case)	Jacquemont et al. (2002) and Miller et al. (1990)
AGM6 (agammaglobulinemia 6) syndrome, *CD79b* mutations	Hypogammaglobulinemia recurrent infection	DLBCL	Leeksma et al. (2017) and Ferrari et al. (2007)
Lymphoproliferative syndrome 1 (LPFS1), *ITK* mutations	Immune dysregulation with lymphadenopathy, hepatosplenomegaly, AIHA, ITP, HLH, recurring fever, stomatitis	EBV+ HL (2 cases) EBV- HL (8 cases) DLBCL (2 cases) NHL (1 case) LPD (1 case)	Huck et al. (2009), Stepensky et al. (2011), Linka et al. (2012), Ghosh et al. (2018) and Fang et al. (2019)

(continued)

Table 4 (continued)

Syndrome, mutated gene	IMD and other clinical features	Cancer(s) reported	References
SCN3 (severe congenital neutropenia) syndrome, *HAX1* mutations	Agranulocytosis/ neutropenia, life-threatening infection, cognitive defects, developmental delay	MDS AML ALL	Klirlat et al. (1989), Germeshausen et al. (2008) and Boztug et al. (2010)
IMD23, *PGM3* mutations	Neutrophil chemotaxis defect, neutropenia, eosinophilia, poor ab vaccine responses, elevated serum IgA, recurrent bacterial infection, neurologic abnormalities	EBV+ HL	Yu Zhang et al. (2014), Hay et al. (2004) and Lundin et al. (2015)
AISIMD (auto-inflammatory syndrome with IMD), *SOCS1* mutations	Hypogammaglobulinemia, T-cell lymphopenia, Evans syndrome, AIHA, ITP, chronic granulomatous uveitis, recurrent infection, psoriasis, SLE with nephritis, spondyloarthropathies	HL (1 case)	Hadjadj et al. (2020), Lee et al. (2020) and Thaventhiran et al. (2020)
IMD75, *TET2* mutations	Hypogammaglobulinemia, chronic lymphadenopathy, autoimmune cytopenias, recurrent infection, hypothyroidism	EBV+ polymorphic B-NHL resembling PTLD, Primary Mediastinal B-cell Lymphoma, nodal Peripheral T-cell Lymphoma	Spegarova et al. (2020)
Epidermodysplasia verruciformis, *RHOH* mutations	HPV-associated flat warts and psoriatic-like lesions, rosacea, chronic bronchial disease with emphysema, gingivostomatitis; decreased naive T cells and increased "exhausted" memory and effector T cells	BL	Crequer et al. (2012b)
IMD76, *FCHO1* mutations	Recurrent infection, lymphopenia, hypogammaglobulinemia, lymphadenopathy, variable neurologic symptoms, HLH	EBV+ HL DLBCL	Łyszkiewicz et al. (2020) and Calzoni et al. (2019)

(continued)

Table 4 (continued)

Syndrome, mutated gene	IMD and other clinical features	Cancer(s) reported	References
IMD82 with systemic inflammation, *SYK* mutations	Recurrent infection, failure to thrive, systemic inflammation (colitis, arthritis, vasculitis, CNS), hypogammaglobulinemia, lymphadenopathy, elevated CRP	DLBCL (2 adult cases)	Wang et al. (2021)
IMD-centromeric instability-facial anomalies syndrome 1 (ICF1), *DNMT3B* mutations	Recurrent infection, abnormal facies, mental retardation, variable IMD with invariant IgA deficiency	HL (1 case)	Schuetz et al. (2007)
DiGeorge syndrome, *TBX1* mutations	Thymic aplasia and T-cell lymphopenia, recurrent infection, abnormal facies, and cardiac, abdominal, genitourinary, neurologic, and endocrine problems	EBV+ DLBCL (1 case)	Sato et al. (1999)
IMD73B with defective neutrophil chemotaxis and lymphopenia, *RAC2* mutations	Recurrent infection, lymphadenitis, lymphopenia, hypogammaglobulinemia, neutropenia, defective neutrophil chemotaxis, and enhanced oxidative burst	B-NHL	Lougaris et al. (2019)
IMD71 with inflammatory disease and congenital thrombocytopenia, *ARPC1B* mutations	Recurrent infection, platelet abnormalities, B-cell lymphocytosis, eosinophilia, vasculitis	ALCL (1 case)	Thaventhiran et al. (2020)
IMD105 (T-,B+,NK+ SCID), *PTPRC* mutations	Recurrent infection, failure to thrive, hypogammaglobulinemia, hepatosplenomegaly, low-to-absent CD45	B-cell lymphoma (1 case)	Kung et al. (2000)
Purine nucleoside phosphorylase deficiency (PNPD), *PNP* mutations	T-cell IMD, recurrent infection, failure to thrive, splenomegaly, neurologic impairment	B-cell immunoblastic lymphoma EBV+ DLBCL	Graham-Pole et al. (1975), Watson et al. (1981) and Schejter et al. (2020)
Familial hemophagocytic lymphohistiocytosis, type 1 (FHL1), *FHL1* mutations	Fever, hepatosplenomegaly, cytopenias, CNS involvement, low fibrinogen, high triglycerides	T-NHL (2 cases)	Mache et al. (1994)

(continued)

Table 4 (continued)

Syndrome, mutated gene	IMD and other clinical features	Cancer(s) reported	References
IMD84, *IKZF3* mutations	Recurrent infection, B-cell leukopenia, hypogammaglobulinemia, EBV susceptibility, splenomegaly, HLH	FL and EBV+ DLBCL (1 case) EBV+ DLBCL (1 case)	Yamashita et al. (2021)
IMD64, *RASGRP1* mutations	Recurrent infection, failure to thrive, EBV susceptibility, LPD, T-cell leukopenia, autoimmunity	EBV+ DLBCL, marginal zone lymphoma, HL, LPD	Platt et al. (2017), Winter et al. (2018), Somekh et al. (2018) and Salzer et al. (2016)
T-cell lymphoma, subcutaneous panniculitis-like (SPTCL), *HAVCR2* mutations	Elevated, variable HLH, variable autoimmunity, cytopenias, splenomegaly, subcutaneous nodules, panniculitis, atypical CD8 T cells rimming adipocytes	T-NHL: Subcutaneous panniculitis-like T-cell lymphoma (SPTCL)	Polprasert et al. (2019) and Gayden et al. (2018)
CVID2, *TNFRSF13B* mutations	Recurrent infection, lymphadenopathy, hypogammaglobulinemia, EBV LPD	EBV+ HL, DLBCL, BL	Rodriguez et al. (2019), Alosaimi et al. (2019), Somekh et al. (2019) and Castigli et al. (2005)
IMD8, *CORO1A* mutations	Recurrent infection, lymphopenia, hypogammaglobulinemia, EBV susceptibility, mucocutaneous IMD syndrome, CNS and behavioral /psychiatric symptoms	EBV+ DLBCL, HL, CNS B-NHL	Punwani et al. (2015a), Stray-Pedersen et al. (2014), Moshous et al. (2013) and Vignesh et al. (2020)
Immune deficiency, polyendocrinopathy, enteropathy, X-linked (IPEX) syndrome, *FOXP3* mutations	Enteropathic diarrhea, autoimmune endocrinopathies (type 1 DM, hypothyroidism), dermatitis, failure to thrive, AIHA, ITP, lymphadenopathy	DLBCL (1 case)	Lucas et al. (2008)

unrecognized due to their varied clinical presentations. Unless a patient already carries an IMD diagnosis, at initial presentation to an oncologist for evaluation of possible malignancy, only an experienced and savvy practitioner is likely to be sufficiently prescient to consider IMD, absent obvious clinical and/or family history to suggest it. In addition, many IMD require complex genetic testing for proper diagnosis, which may entail multidisciplinary consultations from genetic and immunology subspecialists. As exemplified by this chapter, these conditions are myriad, and their features are varied. Many are only recently described, and established clinical management guidelines are still evolving. It is near-certain that additional IMD with hematologic cancer predilection also remain undiscovered. With next-generation sequencing, other emerging technologies, and increasing knowledge of IMD, this list will continue to grow. Nonetheless, it is important to recognize and accurately diagnose such patients, because it impacts their cancer treatment, their risk of subsequent malignancy, and, in some cases, the healthcare of their family members. A high index of suspicion, thorough evaluation of past medical and family history, inquisitiveness by healthcare providers, and effort to seek expert counsel are all necessary to ensure optimal care for these unfortunate patients. Our hope is that this chapter can be a useful resource in pursuit of that goal.

Acknowledgments The authors would like to thank Rikin Shah, M.D., Director of Pediatric BMT-Cellular Therapy, Orlando Health-Arnold Palmer Hospital for Children, for his contributions in organizing this chapter's content.

References

Abdollahpour H, Appaswamy G, Kotlarz D, Diestelhorst J, Beier R, Schäffer AA, Michael Gertz E et al (2012) The phenotype of human STK4 deficiency. Blood 119(15):3450–3457. https://doi.org/10.1182/blood-2011-09-378158

Abolhassani H, Edwards ESJ, Ikinciogullari A, Jing H, Borte S, Buggert M, Du L et al (2017) Combined immunodeficiency and Epstein-Barr virus-induced B cell malignancy in humans with inherited CD70 deficiency. J Exp Med 214(1):91–106. https://doi.org/10.1084/jem.20160849

Alangari A, Alsultan A, Adly N, Massaad MJ, Kiani IS, Aljebreen A, Raddaoui E et al (2012) LPS-responsive beige-like anchor (LRBA) gene mutation in a family with inflammatory bowel disease and combined immunodeficiency. J Allergy Clin Immunol 130(2):481–488. https://doi.org/10.1016/j.jaci.2012.05.043

Alosaimi M, Hoenig M, Jaber F, Platt CD, Jones J, Wallace J, Debatin K-M et al (2019) Immunodeficiency and Epstein Barr virus induced lymphoproliferation caused by 4-1BB deficiency. J Allergy Clin Immunol 144(2):574–583. https://doi.org/10.1016/j.jaci.2019.03.002

Alter BP, Giri N, Savage SA, Rosenberg PS (2009) Cancer in dyskeratosis congenita. Blood 113(26):6549–6557. https://doi.org/10.1182/blood-2008-12-192880

Altmann T, Gennery AR (2016) DNA ligase IV syndrome; a review. Orphanet J Rare Dis 11(1): 137. https://doi.org/10.1186/s13023-016-0520-1

Amirifar P, Ranjouri MR, Yazdani R, Abolhassani H, Aghamohammadi A (2019) Ataxia-telangiectasia: a review of clinical features and molecular pathology. Pediatric Allergy Immunol Off Pub Eur Soc Pediatric Allergy Immunol 30(3):277–288. https://doi.org/10.1111/pai.13020

Angulo I, Vadas O, Garçon F, Banham-Hall E, Plagnol V, Leahy TR, Baxendale H et al (2013) Phosphoinositide 3-kinase δ gene mutation predisposes to respiratory infection and airway damage. Science (New York, N.Y.) 342(6160):866–871. https://doi.org/10.1126/science.1243292

Argyle JC, Kjeldsberg CR, Marty J, Shigeoka AO, Hill HR (1982) T-cell lymphoma and the Chediak-Higashi syndrome. Blood 60(3):672–676

Asai E, Wada T, Sakakibara Y, Toga A, Toma T, Shimizu T, Nampoothiri S et al (2011) Analysis of mutations and recombination activity in RAG-deficient patients. Clin Immunol (Orlando, Fla.) 138(2):172–177. https://doi.org/10.1016/j.clim.2010.11.005

Ashrafi F, Klein C, Poorpooneh M, Sherkat R, Khoshnevisan R (2020) A case report of sinusoidal diffuse large B-cell lymphoma in a STK4 deficient patient. Medicine 99(9):e18601. https://doi.org/10.1097/MD.0000000000018601

Avery DT, Ellyard JI, Mackay F, Corcoran LM, Hodgkin PD, Tangye SG (2005) Increased expression of CD27 on activated human memory B cells correlates with their commitment to the plasma cell lineage. J Immunol (Baltimore, Md.: 1950) 174(7):4034–4042. https://doi.org/10.4049/jimmunol.174.7.4034

Baas AF, Gabbett M, Rimac M, Kansikas M, Raphael M, Nievelstein RAJ, Nicholls W et al (2013) Agenesis of the corpus callosum and gray matter heterotopia in three patients with constitutional mismatch repair deficiency syndrome. Eur J Hum Genet 21(1):55–61. https://doi.org/10.1038/ejhg.2012.117

Bacon CM, Wilkinson SJ, Spickett GP, Barge D, Lucraft HH, Jackson G, Rand V, Gennery AR (2013) Epstein-Barr virus-independent diffuse large B-cell lymphoma in DNA ligase 4 deficiency. J Allergy Clin Immunol 131(4):1237–1239. https://doi.org/10.1016/j.jaci.2012.10.027

Badolato R, Sozzani S, Malacarne F, Bresciani S, Fiorini M, Borsatti A, Albertini A, Mantovani A, Ugazio AG, Notarangelo LD (1998) Monocytes from Wiskott-Aldrich patients display reduced chemotaxis and lack of cell polarization in response to monocyte chemoattractant protein-1 and formyl-methionyl-leucyl-phenylalanine. J Immunol (Baltimore, Md.: 1950) 161(2):1026–1033

Baetz K, Isaaz S, Griffiths GM (1995) Loss of cytotoxic T lymphocyte function in Chediak-Higashi syndrome arises from a secretory defect that prevents lytic granule exocytosis. J Immunol (Baltimore, Md.: 1950) 154(11):6122–6131

Bailly-Botuha C, Jaubert F, Taam RA, Galmiche L, Picard C, Bellon G, de Blic J (2008) Diffuse lymphoplasmacytic bronchiolitis in cartilage-hair hypoplasia. J Pediatr 152(3):429–433. https://doi.org/10.1016/j.jpeds.2007.10.010

Bajin İY, Ayvaz DÇ, Ünal Ş, Özgür TT, Çetin M, Gümrük F, Tezcan İ, de Villartay J-P, Sanal Ö (2013) Atypical combined immunodeficiency due to Artemis defect: a case presenting as hyperimmunoglobulin M syndrome and with LGLL. Mol Immunol 56(4):354–357. https://doi.org/10.1016/j.molimm.2013.05.004

Bakhshi S, Cerosaletti KM, Concannon P, Bawle EV, Fontanesi J, Gatti RA, Bhambhani K (2003) Medulloblastoma with adverse reaction to radiation therapy in Nijmegen breakage syndrome. J Pediatr Hematol Oncol 25(3):248–251. https://doi.org/10.1097/00043426-200303000-00013

Balabanian K, Lagane B, Pablos JL, Laurent L, Planchenault T, Verola O, Lebbe C et al (2005) WHIM syndromes with different genetic anomalies are accounted for by impaired CXCR4 desensitization to CXCL12. Blood 105(6):2449–2457. https://doi.org/10.1182/blood-2004-06-2289

Balta G, Caliskan U, Guler E, Patiroglu T, Oner AF, Gurgey A (2008) Familial hemophagocytic lymphohistiocytosis type IV: possible founder effect in syntaxin 11 gene common mutation endemic to Turkish population. Blood 112(11):1259. https://doi.org/10.1182/blood.V112.11.1259.1259

Bar RS, DeLor CJ, Clausen KP, Hurtubise P, Henle W, Hewetson JF (1974) Fatal infectious mononucleosis in a family. N Engl J Med 290(7):363–367. https://doi.org/10.1056/NEJM197402142900704

Barbosa MD, Nguyen QA, Tchernev VT, Ashley JA, Detter JC, Blaydes SM, Brandt SJ et al (1996) Identification of the homologous beige and Chediak-Higashi syndrome genes. Nature 382(6588):262–265. https://doi.org/10.1038/382262a0

Barnes DE, Tomkinson AE, Lehmann AR, David A, Webster B, Lindahl T (1992) Mutations in the DNA ligase I gene of an individual with immunodeficiencies and cellular hypersensitivity to DNA-damaging agents. Cell 69(3):495–503. https://doi.org/10.1016/0092-8674(92)90450-Q

Bata N, Chaikuad A, Bakas NA, Limpert AS, Lambert LJ, Sheffler DJ, Berger LM et al (2022) Inhibitors of the hippo pathway kinases STK3/MST2 and STK4/MST1 have utility for the treatment of acute myeloid leukemia. J Med Chem 65(2):1352–1369. https://doi.org/10.1021/acs.jmedchem.1c00804

Cohen B, Sarah OF, Plouvier E, Rohrlich P-S, Daltroff G, Plantier I, Dupuy A et al (2012) Description and outcome of a cohort of 8 patients with WHIM syndrome from the French severe chronic neutropenia registry. Orphanet J Rare Dis 7(1):71. https://doi.org/10.1186/1750-1172-7-71

Beekman R, Valkhof MG, Sanders MA, van Strien PMH, Haanstra JR, Broeders L, Geertsma-Kleinekoort WM et al (2012) Sequential gain of mutations in severe congenital neutropenia progressing to acute myeloid leukemia. Blood 119(22):5071–5077. https://doi.org/10.1182/blood-2012-01-406116

Bellanné-Chantelot C, Clauin S, Leblanc T, Cassinat B, Rodrigues-Lima F, Beaufils S, Vaury C et al (2004) Mutations in the ELA2 gene correlate with more severe expression of neutropenia: a study of 81 patients from the French neutropenia register. Blood 103(11):4119–4125. https://doi.org/10.1182/blood-2003-10-3518

Ben-Neriah Y, Karin M (2011) Inflammation meets cancer, with NF-KB as the matchmaker. Nat Immunol 12(8):715–723. https://doi.org/10.1038/ni.2060

Ben-Omran TI, Cerosaletti K, Concannon P, Weitzman S, Nezarati MM (2005) A patient with mutations in DNA Ligase IV: Clinical features and overlap with Nijmegen breakage syndrome. Am J Med Genet Part A 137A(3):283–287. https://doi.org/10.1002/ajmg.a.30869

Bigley V, Cytlak U, Collin M (2019) Human dendritic cell immunodeficiencies. Semin Cell Dev Biol 86(February):50–61. https://doi.org/10.1016/j.semcdb.2018.02.020

Bilousova G, Marusyk A, Porter CC, Cardiff RD, DeGregori J (2005) Impaired DNA replication within progenitor cell pools promotes leukemogenesis. PLoS Biol 3(12):e401. https://doi.org/10.1371/journal.pbio.0030401

Blommaert E, Péanne R, Cherepanova NA, Rymen D, Staels F, Jaeken J, Race V et al (2019) Mutations in MAGT1 lead to a glycosylation disorder with a variable phenotype. Proc Natl Acad Sci U S A 116(20):9865–9870. https://doi.org/10.1073/pnas.1817815116

Blume RS, Wolff SM (1972) The Chediak-Higashi syndrome: studies in four patients and a review of the literature. Medicine 51(4):247–280

Boast B, de Jesus Nunes-Santos C, Kuehn HS, Rosenzweig SD (2021) Ikaros-associated diseases: from mice to humans and back again. Front Pediatr 9. https://www.frontiersin.org/article/10.3389/fped.2021.705497

Boder E, Sedgwick RP (1970) Ataxia-Telangiectasia. (Clinical and immunological aspects). Psychiatrie Neurol Medizinische Psychol Beihefte 13–14:8–16

Bonizzi G, Karin M (2004) The two NF-KappaB activation pathways and their role in innate and adaptive immunity. Trends Immunol 25(6):280–288. https://doi.org/10.1016/j.it.2004.03.008

Borst J, Hendriks J, Xiao Y (2005) CD27 and CD70 in T cell and B cell activation. Curr Opin Immunol 17(3):275–281. https://doi.org/10.1016/j.coi.2005.04.004

Bougeard G, Charbonnier F, Moerman A, Martin C, Ruchoux MM, Drouot N, Frébourg T (2003) Early-onset brain tumor and lymphoma in MSH2-deficient children. Am J Hum Genet 72(1): 213–216. https://doi.org/10.1086/345297

Boztug K, Ding X-Q, Hartmann H, Ziesenitz L, Schäffer AA, Diestelhorst J, Pfeifer D et al (2010) HAX1 mutations causing severe congenital neuropenia and neurological disease lead to cerebral microstructural abnormalities documented by quantitative MRI. Am J Med Genet A 152A(12): 3157–3163. https://doi.org/10.1002/ajmg.a.33748

Brault J, Liu T, Bello E, Liu S, Sweeney CL, Meis RJ, Koontz S et al (2021) CRISPR-targeted MAGT1 insertion restores XMEN patient hematopoietic stem cells and lymphocytes. Blood 138(26):2768–2780. https://doi.org/10.1182/blood.2021011192

Bresnick EH, Katsumura KR, Lee H-Y, Johnson KD, Perkins AS (2012) Master regulatory GATA transcription factors: mechanistic principles and emerging links to hematologic malignancies. Nucleic Acids Res 40(13):5819–5831. https://doi.org/10.1093/nar/gks281

Brock PR, de Zegher F, Casteels-Van Daele M, Vanderschueren-Lodeweyckx M (1991) Malignant disease in Bloom's syndrome children treated with growth hormone. Lancet (London, England) 337(8753):1345–1346. https://doi.org/10.1016/0140-6736(91)93017-4

Brown KD, Rathi A, Kamath R, Beardsley DI, Zhan Q, Mannino JL, Baskaran R (2003) The mismatch repair system is required for S-phase checkpoint activation. Nat Genet 33(1):80–84. https://doi.org/10.1038/ng1052

Browning MJ, Chandra A, Carbonaro V, Okkenhaug K, Barwell J (2015) Cowden's syndrome with immunodeficiency. J Med Genet 52(12):856–859. https://doi.org/10.1136/jmedgenet-2015-103266

Buchbinder D, Nugent DJ, Fillipovich AH (2014) Wiskott–Aldrich syndrome: diagnosis, current management, and emerging treatments. Appl Clin Genet 7(April):55–66. https://doi.org/10.2147/TACG.S58444

Buchbinder D, Stinson JR, Nugent DJ, Heurtier L, Suarez F, Sukumar G, Dalgard CL et al (2015) Mild B-Cell lymphocytosis in patients with a CARD11 C49Y mutation. J Allergy Clin Immunol 136(3):819–821. https://doi.org/10.1016/j.jaci.2015.03.008

Buck D, Moshous D, de Chasseval R, Ma Y, le Deist F, Cavazzana-Calvo M, Fischer A, Casanova J-L, Lieber MR, de Villartay J-P (2006) Severe combined immunodeficiency and microcephaly in siblings with hypomorphic mutations in DNA ligase IV. Eur J Immunol 36(1):224–235. https://doi.org/10.1002/eji.200535401

Buckley RH (2004) Pulmonary complications of primary immunodeficiencies. Paediatric Respir Rev 5(Suppl A):S225–S233. https://doi.org/10.1016/s1526-0542(04)90043-7

Buckley RH, Schiff SE, Schiff RI, Louise Markert M, Williams LW, Roberts JL, Myers LA, Ward FE (1999) Hematopoietic stem-cell transplantation for the treatment of severe combined immunodeficiency. N Engl J Med 340(7):508–516. https://doi.org/10.1056/NEJM199902183400703

Buonocore F, Kühnen P, Suntharalingham JP, Del Valle I, Digweed M, Stachelscheid H, Khajavi N et al (n.d.) Somatic mutations and progressive monosomy modify SAMD9-related phenotypes in humans. J Clin Invest 127(5):1700–1713. https://doi.org/10.1172/JCI91913

van der Burg M, van Veelen LR, Verkaik NS, Wiegant WW, Hartwig NG, Barendregt BH, Brugmans L et al (2006) A new type of radiosensitive T-B-NK+ severe combined immunodeficiency caused by a LIG4 mutation. J Clin Invest 116(1):137–145. https://doi.org/10.1172/JCI26121

Burke JE, Williams RL (2015) Synergy in activating class I PI3Ks. Trends Biochem Sci 40(2):88–100. https://doi.org/10.1016/j.tibs.2014.12.003

Burkhardt JK, Wiebel FA, Hester S, Argon Y (1993) The giant organelles in beige and Chediak-Higashi fibroblasts are derived from late endosomes and mature lysosomes. J Exp Med 178(6):1845–1856. https://doi.org/10.1084/jem.178.6.1845

Burroughs LM, Petrovic A, Brazauskas R, Liu X, Griffith LM, Ochs HD, Bleesing JJ et al (2020) Excellent outcomes following hematopoietic cell transplantation for Wiskott-Aldrich syndrome: A PIDTC report. Blood 135(23):2094–2105. https://doi.org/10.1182/blood.2019002939

Cagdas D, Halaçlı SO, Tan Ç, Lo B, Çetinkaya PG, Esenboğa S, Karaatmaca B et al (2019) A spectrum of clinical findings from ALPS to CVID: several novel LRBA defects. J Clin Immunol 39(7):726–738. https://doi.org/10.1007/s10875-019-00677-6

Cahyanur R, Rahmawati S (2020) Diffuse large B-cell lymphoma and job's syndrome: a case report. Mædica 15(2):269–271. https://doi.org/10.26574/maedica.2020.15.2.269

Calzoni E, Platt CD, Keles S, Kuehn HS, Beaussant-Cohen S, Zhang Y, Pazmandi J et al (2019) F-BAR domain only protein 1 (FCHO1) deficiency is a novel cause of combined immune deficiency in humans. J Allergy Clin Immunol 143(6):2317–2321. https://doi.org/10.1016/j.jaci.2019.02.014

Castigli E, Wilson SA, Garibyan L, Rachid R, Bonilla F, Schneider L, Geha RS (2005) TACI is mutant in common variable immunodeficiency and IgA deficiency. Nat Genet 37(8):829–834. https://doi.org/10.1038/ng1601

Chaigne-Delalande B, Li F-Y, O'Connor GM, Lukacs MJ, Jiang P, Zheng L, Shatzer A et al (2013) Mg2+ regulates cytotoxic functions of NK and CD8 T cells in chronic EBV infection through NKG2D. Science 341(6142):186–191. https://doi.org/10.1126/science.1240094

Chandesris M-O, Melki I, Natividad A, Puel A, Fieschi C, Yun L, Thumerelle C et al (2012) Autosomal dominant STAT3 deficiency and hyper-IgE syndrome: molecular, cellular, and clinical features from a French national survey. Medicine 91(4):1–19. https://doi.org/10.1097/MD.0b013e31825f95b9

Chapuy B, Stewart C, Dunford AJ, Kim J, Kamburov A, Redd RA, Lawrence MS et al (2018) Molecular subtypes of diffuse large B cell lymphoma are associated with distinct pathogenic mechanisms and outcomes. Nat Med 24(5):679–690. https://doi.org/10.1038/s41591-018-0016-8

Charbit-Henrion F, Jeverica AK, Bègue B, Markelj G, Parlato M, Avcin SL, Callebaut I et al (2017) Deficiency in mucosa-associated lymphoid tissue lymphoma translocation 1: a novel cause of IPEX-like syndrome. J Pediatr Gastroenterol Nutr 64(3):378–384. https://doi.org/10.1097/MPG.0000000000001262

Chauvin SD, Price S, Zou J, Hunsberger S, Brofferio A, Matthews H, Similuk M et al (2022) A double-blind, placebo-controlled, crossover study of magnesium supplementation in patients with XMEN disease. J Clin Immunol 42(1):108–118. https://doi.org/10.1007/s10875-021-01137-w

Chen K, Coonrod EM, Kumánovics A, Franks ZF, Durtschi JD, Margraf RL, Wu W et al (2013) Germline mutations in NFKB2 implicate the noncanonical NF-KB pathway in the pathogenesis of common variable immunodeficiency. Am J Hum Genet 93(5):812–824. https://doi.org/10.1016/j.ajhg.2013.09.009

Chlon TM, Crispino JD (2012) Combinatorial regulation of tissue specification by GATA and FOG factors. Development 139(21):3905–3916. https://doi.org/10.1242/dev.080440

Chung B, Aoukaty A, Dutz J, Terhorst C, Tan R (2005) Signaling lymphocytic activation molecule-associated protein controls NKT cell functions. J Immunol (Baltimore, Md.: 1950) 174(6):3153–3157. https://doi.org/10.4049/jimmunol.174.6.3153

Coffey AJ, Brooksbank RA, Brandau O, Oohashi T, Howell GR, Bye JM, Cahn AP et al (1998) Host response to EBV infection in X-linked lymphoproliferative disease results from mutations in an SH2-domain encoding gene. Nat Genet 20(2):129–135. https://doi.org/10.1038/2424

Cohen SB, Bainter W, Johnson JL, Lin T-Y, Wong JCY, Wallace JG, Jones J et al (2019) Human primary immunodeficiency caused by expression of a kinase-dead P110δ mutant. J Allergy Clin Immunol 143(2):797–799. https://doi.org/10.1016/j.jaci.2018.10.005

Cory GO, MacCarthy-Morrogh L, Banin S, Gout I, Brickell PM, Levinsky RJ, Kinnon C, Lovering RC (1996) Evidence that the Wiskott-Aldrich syndrome protein may be involved in lymphoid cell signaling pathways. J Immunol (Baltimore, Md.: 1950) 157(9):3791–3795

Cotelingam JD, Witebsky FG, Hsu SM, Blaese RM, Jaffe ES (1985) Malignant lymphoma in patients with the Wiskott-Aldrich syndrome. Cancer Investig 3(6):515–522. https://doi.org/10.3109/07357908509039813

Coulter TI, Chandra A, Bacon CM, Babar J, Curtis J, Screaton N, Goodlad JR et al (2017) Clinical spectrum and features of activated phosphoinositide 3-kinase δ syndrome: a large patient cohort study. J Allergy Clin Immunol 139(2):597–606. https://doi.org/10.1016/j.jaci.2016.06.021

Crequer A, Picard C, Patin E, D'Amico A, Abhyankar A, Munzer M, Debré M et al (2012a) Inherited MST1 deficiency underlies susceptibility to EV-HPV infections. Edited by Mark M. Wurfel. PLoS ONE 7(8):e44010. https://doi.org/10.1371/journal.pone.0044010

Crequer A, Troeger A, Patin E, Ma CS, Picard C, Pedergnana V, Fieschi C et al (2012b) Human RHOH deficiency causes T cell defects and susceptibility to EV-HPV infections. J Clin Invest 122(9):3239–3247. https://doi.org/10.1172/JCI62949

Cunniff C, Bassetti JA, Ellis NA (2017) Bloom's syndrome: clinical Spectrum, molecular pathogenesis, and cancer predisposition. Mol Syndromol 8(1):4–23. https://doi.org/10.1159/000452082

Cunningham-Rundles C (2012) The many faces of common variable immunodeficiency. Hematol Am Soc Hematol Edu Prog 2012:301–305. https://doi.org/10.1182/asheducation-2012.1.301

Dale DC, Makaryan V (2018) ELANE-related neutropenia. GeneReviews® [Internet]. University of Washington, Seattle. https://www.ncbi.nlm.nih.gov/books/NBK1533/

Dang TS, Willet JDP, Griffin HR, Morgan NV, O'Boyle G, Arkwright PD, Hughes SM et al (2016) Defective leukocyte adhesion and chemotaxis contributes to combined immunodeficiency in humans with autosomal recessive MST1 deficiency. J Clin Immunol 36(2):117–122. https://doi.org/10.1007/s10875-016-0232-2

de Miranda NFCC, Peng R, Georgiou K, Wu C, Sörqvist EF, Berglund M, Chen L et al (2013) DNA repair genes are selectively mutated in diffuse large B cell lymphomas. J Exp Med 210(9): 1729–1742. https://doi.org/10.1084/jem.20122842

de Wind N, Dekker M, Berns A, Radman M, te Riele H (1995) Inactivation of the mouse Msh2 gene results in mismatch repair deficiency, methylation tolerance, Hyperrecombination, and predisposition to cancer. Cell 82(2):321–330. https://doi.org/10.1016/0092-8674(95)90319-4

De Vos M, Hayward BE, Charlton R, Taylor GR, Glaser AW, Picton S, Cole TR et al (2006) PMS2 mutations in childhood cancer. JNCI: J Nat Cancer Inst 98(5):358–361. https://doi.org/10.1093/jnci/djj073

Dembowska-Baginska B, Perek D, Brozyna A, Wakulinska A, Olczak-Kowalczyk D, Gladkowska-Dura M, Grajkowska W, Chrzanowska KH (2009) Non-Hodgkin lymphoma (NHL) in children with Nijmegen breakage syndrome (NBS). Pediatr Blood Cancer 52(2):186–190. https://doi.org/10.1002/pbc.21789

Detre C, Keszei M, Romero X, Tsokos GC, Terhorst C (2010) SLAM family receptors and the SLAM-associated protein (SAP) modulate T cell functions. Semin Immunopathol 32(2): 157–171. https://doi.org/10.1007/s00281-009-0193-0

DiDonato JA, Mercurio F, Karin M (2012) NF-KB and the link between inflammation and cancer. Immunol Rev 246(1):379–400. https://doi.org/10.1111/j.1600-065X.2012.01099.x

Distel L, Neubauer S, Varon R, Holter W, Grabenbauer G (2003) Fatal toxicity following radio- and chemotherapy of medulloblastoma in a child with unrecognized Nijmegen breakage syndrome. Med Pediatr Oncol 41(1):44–48. https://doi.org/10.1002/mpo.10275

Dokal I (1996) Dyskeratosis congenita: an inherited bone marrow failure syndrome. Br J Haematol 92(4):775–779. https://doi.org/10.1046/j.1365-2141.1996.355881.x

Dupin N (2020) Update on oncogenesis and therapy for Kaposi sarcoma. Curr Opin Oncol 32(2): 122–128. https://doi.org/10.1097/CCO.0000000000000601

Dupré L, Andolfi G, Tangye SG, Clementi R, Locatelli F, Aricò M, Aiuti A, Roncarolo M-G (2005) SAP controls the cytolytic activity of CD8+ T cells against EBV-infected cells. Blood 105(11): 4383–4389. https://doi.org/10.1182/blood-2004-08-3269

Dupuis-Girod S, Medioni J, Haddad E, Quartier P, Cavazzana-Calvo M, Le Deist F, de Saint Basile G et al (2003) Autoimmunity in Wiskott-Aldrich syndrome: risk factors, clinical features, and outcome in a single-center cohort of 55 patients. Pediatrics 111(5 Pt 1):e622–e627. https://doi.org/10.1542/peds.111.5.e622

Durno C, Ercan AB, Bianchi V, Edwards M, Aronson M, Galati M, Atenafu EG et al (2021) Survival benefit for individuals with constitutional mismatch repair deficiency undergoing surveillance. J Clin Oncol 39(25):2779–2790. https://doi.org/10.1200/JCO.20.02636

Dutz JP, Benoit L, Wang X, Demetrick DJ, Junker A, de Sa D, Tan R (2001) Lymphocytic vasculitis in X-linked lymphoproliferative disease. Blood 97(1):95–100. https://doi.org/10.1182/blood.V97.1.95

Dyment DA, Smith AC, Alcantara D, Schwartzentruber JA, Basel-Vanagaite L, Curry CJ, Karen Temple I et al (2013) Mutations in PIK3R1 cause SHORT syndrome. Am J Hum Genet 93(1): 158–166. https://doi.org/10.1016/j.ajhg.2013.06.005

Eissing M, Ripken L, Schreibelt G, Westdorp H, Ligtenberg M, Netea-Maier R, Netea MG, de Vries JM, Hoogerbrugge N (2018) PTEN hamartoma tumor syndrome and immune dysregulation. Transl Oncol 12(2):361–367. https://doi.org/10.1016/j.tranon.2018.11.003

El Hasbaoui B, Elyajouri A, Abilkassem R, Agadr A (2020) Nijmegen breakage syndrome: case report and review of literature. Pan Afr Med J 35, 85(March). https://doi.org/10.11604/pamj.2020.35.85.14746

Elkaim E, Neven B, Bruneau J, Mitsui-Sekinaka K, Stanislas A, Heurtier L, Lucas CL et al (2016) Clinical and immunologic phenotype associated with activated phosphoinositide 3-kinase δ syndrome 2: a cohort study. J Allergy Clin Immunol 138(1):210–218. https://doi.org/10.1016/j.jaci.2016.03.022

Enders A, Fisch P, Schwarz K, Duffner U, Pannicke U, Nikolopoulos E, Peters A et al (2006) A severe form of human combined immunodeficiency due to mutations in DNA ligase IV. J Immunol (Baltimore, Md.: 1950) 176(8):5060–5068. https://doi.org/10.4049/jimmunol.176.8.5060

Fang M, Abolhassani H, Pan-Hammarström Q, Sandholm E, Liu X, Hammarström L (2019) Compound heterozygous mutations of IL2-inducible T cell kinase in a Swedish patient: the importance of early genetic diagnosis. J Clin Immunol 39(2):131–134. https://doi.org/10.1007/s10875-019-00598-4

Felgentreff K, Lee YN, Frugoni F, Du L, van der Burg M, Giliani S, Tezcan I et al (2015) Functional analysis of naturally occurring DCLRE1C mutations and correlation with the clinical phenotype of ARTEMIS deficiency. J Allergy Clin Immunol 136(1):140–150. https://doi.org/10.1016/j.jaci.2015.03.005

Ferrari S, Lougaris V, Caraffi S, Zuntini R, Yang J, Soresina A, Meini A et al (2007) Mutations of the Igβ gene cause agammaglobulinemia in man. J Exp Med 204(9):2047–2051. https://doi.org/10.1084/jem.20070264

Fevang B, Fagerli UM, Sorte H, Aarset H, Hov H, Langmyr M, Keil TM et al (2018) Runaway train: a leaky radiosensitive SCID with skin lesions and multiple lymphomas. Case Rep Immunol 2018(May):2053716. https://doi.org/10.1155/2018/2053716

Fiorilli M, Businco L, Pandolfi F, Paganelli R, Russo G, Aiuti F (1983) Heterogeneity of immunological abnormalities in ataxia-telangiectasia. J Clin Immunol 3(2):135–141. https://doi.org/10.1007/BF00915484

Flanagan M, Cunniff CM (2019) Bloom syndrome. GeneReviews® [Internet]. University of Washington, Seattle. https://www.ncbi.nlm.nih.gov/books/NBK1398/

Flanagan SE, Haapaniemi E, Russell MA, Caswell R, Allen HL, De Franco E, McDonald TJ et al (2014) Activating germline mutations in STAT3 cause early-onset multi-organ autoimmune disease. Nat Genet 46(8):812–814. https://doi.org/10.1038/ng.3040

Forbes LR, Milner J, Haddad E (2016) STAT3: A year in review. Curr Opin Hematol 23(1):23–27. https://doi.org/10.1097/MOH.0000000000000206

Fraumeni JF, Wertelecki W, Blattner WA, Jensen RD, Leventhal BG (1975) Varied manifestations of a familial lymphoproliferative disorder. Am J Med 59(1):145–151. https://doi.org/10.1016/0002-9343(75)90333-2

Freedman MH, Bonilla MA, Fier C, Bolyard AA, Scarlata D, Boxer LA, Brown S et al (2000) Myelodysplasia syndrome and acute myeloid leukemia in patients with congenital neutropenia receiving G-CSF therapy. Blood 96(2):429–436

Fuchs S, Rensing-Ehl A, Pannicke U, Lorenz MR, Fisch P, Jeelall Y, Rohr J et al (2015) Omenn syndrome associated with a functional reversion due to a somatic second-site mutation in CARD11 deficiency. Blood 126(14):1658–1669. https://doi.org/10.1182/blood-2015-03-631374

Gallego MD, Santamaría M, Peña J, Molina IJ (1997) Defective actin reorganization and polymerization of Wiskott-Aldrich T cells in response to CD3-mediated stimulation. Blood 90(8):3089–3097

Gathmann B, Nizar Mahlaoui CEREDIH, Gérard L, Oksenhendler E, Warnatz K, Schulze I et al (2014) Clinical picture and treatment of 2212 patients with common variable immunodeficiency. J Allergy Clin Immunol 134(1):116–126. https://doi.org/10.1016/j.jaci.2013.12.1077

Gayden T, Sepulveda FE, Khuong-Quang D-A, Pratt J, Valera ET, Garrigue A, Kelso S et al (2018) Germline HAVCR2 mutations altering TIM-3 characterize subcutaneous panniculitis-like T cell lymphomas with hemophagocytic lymphohistiocytic syndrome. Nat Genet 50(12):1650–1657. https://doi.org/10.1038/s41588-018-0251-4

Georgopoulos K, Moore DD, Derfler B (1992) Ikaros, an early lymphoid-specific transcription factor and a putative mediator for T cell commitment. Science (New York, N.Y.) 258(5083): 808–812. https://doi.org/10.1126/science.1439790

German J (1997) Bloom's syndrome. XX. The first 100 cancers. Cancer Genet Cytogenet 93(1): 100–106. https://doi.org/10.1016/s0165-4608(96)00336-6

Germeshausen M, Ballmaier M, Welte K (2007) Incidence of CSF3R mutations in severe congenital neutropenia and relevance for leukemogenesis: results of a long-term survey. Blood 109(1): 93–99. https://doi.org/10.1182/blood-2006-02-004275

Germeshausen M, Deerberg S, Peter Y, Reimer C, Kratz CP, Ballmaier M (2013) The spectrum of ELANE mutations and their implications in severe congenital and cyclic neutropenia. Hum Mutat 34(6):905–914. https://doi.org/10.1002/humu.22308

Germeshausen M, Grudzien M, Zeidler C, Abdollahpour H, Yetgin S, Rezaei N, Ballmaier M, Grimbacher B, Welte K, Klein C (2008) Novel HAX1 mutations in patients with severe congenital neutropenia reveal isoform-dependent genotype-phenotype associations. Blood 111(10):4954–4957. https://doi.org/10.1182/blood-2007-11-120667

Ghosh S, Drexler I, Bhatia S, Adler H, Gennery AR, Borkhardt A (2018) Interleukin-2-inducible T-cell kinase deficiency-new patients, new insight? Front Immunol 9:979. https://doi.org/10.3389/fimmu.2018.00979

Ghosh S, Bal SK, Edwards ESJ, Pillay B, Heredia RJ, Cipe FE, Rao G et al (2020) Extended clinical and immunological phenotype and transplant outcome in CD27 and CD70 deficiency. Blood 136(23):2638–2655. https://doi.org/10.1182/blood.2020006738

Gilad S, Chessa L, Khosravi R, Russell P, Galanty Y, Piane M, Gatti RA, Jorgensen TJ, Shiloh Y, Bar-Shira A (1998) Genotype-phenotype relationships in ataxia-telangiectasia and variants. Am J Hum Genet 62(3):551–561. https://doi.org/10.1086/301755

Giunti L, Cetica V, Ricci U, Giglio S, Sardi I, Paglierani M, Andreucci E et al (2009) Type A microsatellite instability in pediatric gliomas as an indicator of Turcot syndrome. Eur J Hum Genet 17(7):919–927. https://doi.org/10.1038/ejhg.2008.271

Graham-Pole J, Ferguson A, Gibson AA, Stephenson JB (1975) Familial dysequilibrium-diplegia with T-lymphocyte deficiency. Arch Dis Child 50(12):927–932. https://doi.org/10.1136/adc.50.12.927

Greil J, Rausch T, Giese T, Bandapalli OR, Daniel V, Bekeredjian-Ding I, Stütz AM et al (2013) Whole-exome sequencing links caspase recruitment domain 11 (CARD11) inactivation to severe combined immunodeficiency. J Allergy Clin Immunol 131(5):1376–1383. https://doi.org/10.1016/j.jaci.2013.02.012

Grierson HL, Skare J, Hawk J, Pauza M, Purtilo DT (1991) Immunoglobulin class and subclass deficiencies prior to Epstein-Barr virus infection in males with X-linked lymphoproliferative disease. Am J Med Genet 40(3):294–297. https://doi.org/10.1002/ajmg.1320400309

Grimbacher B, Holland SM, Gallin JI, Greenberg F, Hill SC, Malech HL, Miller JA, O'Connell AC, Puck JM (1999) Hyper-IgE syndrome with recurrent infections--an autosomal dominant multisystem disorder. N Engl J Med 340(9):692–702. https://doi.org/10.1056/NEJM199903043400904

Group, The International Nijmegen Breakage Syndrome Study (2000) Nijmegen breakage syndrome. Arch Dis Child 82(5):400–406. https://doi.org/10.1136/adc.82.5.400

Guennoun A, Bougarn S, Khan T, Mackeh R, Rahman M, Al-Ali F, Ata M et al (2021) A novel STK4 mutation impairs T cell immunity through dysregulation of cytokine-induced adhesion and chemotaxis genes. J Clin Immunol 41(8):1839–1852. https://doi.org/10.1007/s10875-021-01115-2

Gupta M, Aluri J, Desai M, Lokeshwar M, Taur P, Lenardo M, Bergerson J et al (2018) Clinical, immunological, and molecular findings in four cases of B cell expansion with NF-KB and T cell anergy disease for the first time from India. Front Immunol 9. https://www.frontiersin.org/article/10.3389/fimmu.2018.01049

Haapaniemi EM, Kaustio M, Rajala HLM, van Adrichem AJ, Kainulainen L, Glumoff V, Doffinger R et al (2015) Autoimmunity, hypogammaglobulinemia, lymphoproliferation, and mycobacterial disease in patients with activating mutations in STAT3. Blood 125(4):639–648. https://doi.org/10.1182/blood-2014-04-570101

Hacein-Bey-Abina S, Von Kalle C, Schmidt M, McCormack MP, Wulffraat N, Leboulch P, Lim A et al (2003) LMO2-associated clonal T cell proliferation in two patients after gene therapy for SCID-X1. Science 302(5644):415–419. https://doi.org/10.1126/science.1088547

Hadjadj J, Castro CN, Tusseau M, Stolzenberg M-C, Mazerolles F, Aladjidi N, Armstrong M et al (2020) Early-onset autoimmunity associated with SOCS1 haploinsufficiency. Nat Commun 11(1):5341. https://doi.org/10.1038/s41467-020-18925-4

Hahn CN, Chong C-E, Carmichael CL, Wilkins EJ, Brautigan PJ, Li X-C, Babic M et al (2011) Heritable GATA2 mutations associated with familial myelodysplastic syndrome and acute myeloid leukemia. Nat Genet 43(10):1012–1017. https://doi.org/10.1038/ng.913

Haines BB, Ryu CJ, Chen J (2006) Recombination activating genes (RAG) in lymphoma development. Cell Cycle 5(9):913–916. https://doi.org/10.4161/cc.5.9.2732

Halacli SO, Ayvaz DC, Sun-Tan C, Erman B, Uz E, Yilmaz DY, Ozgul K, Tezcan İ, Sanal O (2015) STK4 (MST1) deficiency in two siblings with autoimmune cytopenias: a novel mutation. Clin Immunol 161(2):316–323. https://doi.org/10.1016/j.clim.2015.06.010

Harris RE, Baehner RL, Gleiser S, Weaver DD, Hodes ME (1981) Cartilage-hair hypoplasia, defective T-cell function, and diamond-Blackfan anemia in an Amish child. Am J Med Genet 8(3):291–297. https://doi.org/10.1002/ajmg.1320080306

Harris TJ, Grosso JF, Yen H-R, Xin H, Kortylewski M, Albesiano E, Hipkiss EL et al (2007) Cutting edge: an in vivo requirement for STAT3 signaling in TH17 development and TH17-dependent autoimmunity. J Immunol (Baltimore, Md.: 1950) 179(7):4313–4317. https://doi.org/10.4049/jimmunol.179.7.4313

Harrison C, Ketchen A-M, Redhead NJ, O'Sullivan MJ, Melton DW (2002) Replication failure, genome instability, and increased cancer susceptibility in mice with a point mutation in the DNA ligase I gene. Cancer Res 62(14):4065–4074

Hassock S, Vetrie D, Giannelli F (1999) Mapping and characterization of the X-linked dyskeratosis congenita (DKC) gene. Genomics 55(1):21–27. https://doi.org/10.1006/geno.1998.5600

Hay BN, Martin JE, Karp B, Davis J, Darnell D, Solomon B, Turner M, Holland SM, Puck JM (2004) Familial immunodeficiency with cutaneous vasculitis, myoclonus, and cognitive impairment. Am J Med Genet A 125A(2):145–151. https://doi.org/10.1002/ajmg.a.20595

Hayden MS, Ghosh S (2011) NF-KB in immunobiology. Cell Res 21(2):223–244. https://doi.org/10.1038/cr.2011.13

Heiss NS, Knight SW, Vulliamy TJ, Klauck SM, Wiemann S, Mason PJ, Poustka A, Dokal I (1998) X-linked dyskeratosis congenita is caused by mutations in a highly conserved gene with putative nucleolar functions. Nat Genet 19(1):32–38. https://doi.org/10.1038/ng0598-32

Hellmig S, Bartscht T, Fischbach W, Ott SJ, Rosenstiel P, Klapper W, Fölsch UR, Schreiber S (2009) Germline variations of the MALT1 gene as risk factors in the development of primary gastric B-cell lymphoma. Eur J Cancer 45(10):1865–1870. https://doi.org/10.1016/j.ejca.2009.03.010

Hendriks J, Gravestein LA, Tesselaar K, van Lier RA, Schumacher TN, Borst J (2000) CD27 is required for generation and long-term maintenance of T cell immunity. Nat Immunol 1(5):433–440. https://doi.org/10.1038/80877

Henter JI, Elinder G, Lübeck PO, Ost A (1993) Myelodysplastic syndrome following epipodophyllotoxin therapy in familial hemophagocytic lymphohistiocytosis. Pediatr Hematol Oncol 10(2):163–168. https://doi.org/10.3109/08880019309016551

Hernandez PA, Gorlin RJ, Lukens JN, Taniuchi S, Bohinjec J, Francois F, Klotman ME, Diaz GA (2003) Mutations in the chemokine receptor gene CXCR4 are associated with WHIM syndrome, a combined immunodeficiency disease. Nat Genet 34(1):70–74. https://doi.org/10.1038/ng1149

Heusinkveld LE, Majumdar S, Gao J-L, McDermott DH, Murphy PM (2019) WHIM syndrome: from pathogenesis towards personalized medicine and cure. J Clin Immunol 39(6):532–556. https://doi.org/10.1007/s10875-019-00665-w
Hiel JA, Weemaes CM, van Engelen BG, Smeets D, Ligtenberg M, van Der Burgt I, van Den Heuvel LP, Cerosaletti KM, Gabreëls FJ, Concannon P (2001) Nijmegen breakage syndrome in a Dutch patient not resulting from a defect in NBS1. J Med Genet 38(6):19. https://doi.org/10.1136/jmg.38.6.e19
Hislop AD, Kuo M, Drake-Lee AB, Akbar AN, Bergler W, Hammerschmitt N, Khan N et al (2005) Tonsillar homing of Epstein-Barr virus–specific CD8+ T cells and the virus-host balance. J Clin Invest 115(9):2546–2555. https://doi.org/10.1172/JCI24810
Horne AC, Ramme KG, Rudd E, Zheng C, Wali Y, Al-Lamki Z, Gürgey A, Yalman N, Nordenskjöld M, Henter J-I (2008) Characterization of PRF1, STX11 and UNC13D genotype-phenotype correlations in familial hemophagocytic lymphohistiocytosis. Br J Haematol 143(1):75–83. https://doi.org/10.1111/j.1365-2141.2008.07315.x
Horwitz MS, Corey SJ, Leighton Grimes H, Tidwell T (2013) ELANE mutations in cyclic and severe congenital neutropenia—genetics and pathophysiology. Hematol Oncol Clin North Am 27(1):19–41. https://doi.org/10.1016/j.hoc.2012.10.004
Horwitz MS, Laurino MY, Keel SB (2019) Normal peripheral blood neutrophil numbers accompanying ELANE whole gene deletion mutation. Blood Adv 3(16):2470–2473. https://doi.org/10.1182/bloodadvances.2019000498
Hsu A, Monica Gramatges M, Williams C, Merritt BY, Tarek Elghetany M, Holland SM, Bertuch A (2013) GATA2 mutations in nonsyndromic pediatric myelodysplastic syndrome. Blood 122(21):2778. https://doi.org/10.1182/blood.V122.21.2778.2778
Hsu AP, McReynolds LJ, Holland SM (2015) GATA2 deficiency. Curr Opin Allergy Clin Immunol 15(1):104–109. https://doi.org/10.1097/ACI.0000000000000126
Hsu AP, Sampaio EP, Khan J, Calvo KR, Lemieux JE, Patel SY, Frucht DM et al (2011) Mutations in GATA2 are associated with the autosomal dominant and sporadic monocytopenia and mycobacterial infection (MonoMAC) syndrome. Blood 118(10):2653–2655. https://doi.org/10.1182/blood-2011-05-356352
Hu Z, Ma D (2018) The precision prevention and therapy of HPV-related cervical cancer: new concepts and clinical implications. Cancer Med 7(10):5217–5236. https://doi.org/10.1002/cam4.1501
Huck K, Feyen O, Niehues T, Rüschendorf F, Hübner N, Laws H-J, Telieps T et al (2009) Girls homozygous for an IL-2–inducible T cell kinase mutation that leads to protein deficiency develop fatal EBV-associated lymphoproliferation. J Clin Invest 119(5):1350–1358. https://doi.org/10.1172/JCI37901
Iijima K, Komatsu K, Matsuura S, Tauchi H (2004) The Nijmegen breakage syndrome gene and its role in genome stability. Chromosoma 113(2):53–61. https://doi.org/10.1007/s00412-004-0298-0
Ijspeert H, Warris A, van der Flier M, Reisli I, Keles S, Chishimba S, van Dongen JJM, van Gent DC, van der Burg M (2013) Clinical spectrum of LIG4 deficiency is broadened with severe dysmaturity, primordial dwarfism, and neurological abnormalities. Hum Mutat 34(12): 1611–1614. https://doi.org/10.1002/humu.22436
Imai K, Morio T, Zhu Y, Jin Y, Itoh S, Kajiwara M, Yata J-I, Mizutani S, Ochs HD, Nonoyama S (2004) Clinical course of patients with WASP gene mutations. Blood 103(2):456–464. https://doi.org/10.1182/blood-2003-05-1480
Imashuku S, Miyagawa A, Chiyonobu T, Ishida H, Yoshihara T, Teramura T, Kuriyama K et al (2002) Epstein-Barr virus-associated T-lymphoproliferative disease with hemophagocytic syndrome, followed by fatal intestinal B lymphoma in a young adult female with WHIM syndrome. Ann Hematol 81(8):470–473. https://doi.org/10.1007/s00277-002-0489-9
Inazawa J, Itoh N, Abe T, Nagata S (1992) Assignment of the human Fas antigen gene (Fas) to 10q24.1. Genomics 14(3):821–822. https://doi.org/10.1016/s0888-7543(05)80200-9

Introne W, Boissy RE, Gahl WA (1999) Clinical, molecular, and cell biological aspects of Chediak-Higashi syndrome. Mol Genet Metab 68(2):283–303. https://doi.org/10.1006/mgme.1999.2927
Isobe M, Emanuel BS, Givol D, Oren M, Croce CM (1986) Localization of gene for human P53 tumour antigen to band 17p13. Nature 320(6057):84–85. https://doi.org/10.1038/320084a0
Jabara HH, Ohsumi T, Chou J, Massaad MJ, Benson H, Megarbane A, Chouery E et al (2013) A homozygous mucosa-associated lymphoid tissue 1 (MALT1) mutation in a family with combined immunodeficiency. J Allergy Clin Immunol 132(1):151–158. https://doi.org/10.1016/j.jaci.2013.04.047
Jacobs C, Huang Y, Masud T, Lu W, Westfield G, Giblin W, Sekiguchi JAM (2011) A hypomorphic Artemis human disease allele causes aberrant chromosomal rearrangements and tumorigenesis. Hum Mol Genet 20(4):806–819. https://doi.org/10.1093/hmg/ddq524
Jacquemont S, Bocéno M, Rival JM, Méchinaud F, David A (2002) High risk of malignancy in mosaic variegated aneuploidy syndrome. Am J Med Genet 109(1):17–21. https://doi.org/10.1002/ajmg.10281
Jiao H, Tóth B, Erdos M, Fransson I, Rákóczi E, Balogh I, Magyarics Z et al (2008) Novel and recurrent STAT3 mutations in hyper-IgE syndrome patients from different ethnic groups. Mol Immunol 46(1):202–206. https://doi.org/10.1016/j.molimm.2008.07.001
Jin Y-Y, Wu J, Chen T-X, Chen J (2019) When WAS gene diagnosis is needed: seeking clues through comparison between patients with Wiskott-Aldrich syndrome and idiopathic thrombocytopenic purpura. Front Immunol 10. https://www.frontiersin.org/article/10.3389/fimmu.2019.01549
Jiricny J (2006) The multifaceted mismatch-repair system. Mol Cell Biol 7(5):335–346. https://doi.org/10.1038/nrm1907
Jørgensen SE, Al-Mousawi A, Assing K, Hartling U, Grosen D, Fisker N, Nielsen C, Jakobsen MA, Mogensen TH (2021) STK4 deficiency impairs innate immunity and interferon production through negative regulation of TBK1-IRF3 signaling. J Clin Immunol 41(1):109–124. https://doi.org/10.1007/s10875-020-00891-7
Jun JE, Wilson LE, Vinuesa CG, Lesage S, Blery M, Miosge LA, Cook MC et al (2003) Identifying the MAGUK protein Carma-1 as a central regulator of humoral immune responses and atopy by genome-wide mouse mutagenesis. Immunity 18(6):751–762. https://doi.org/10.1016/S1074-7613(03)00141-9
Kalra R, Dale D, Freedman M, Bonilla MA, Weinblatt M, Ganser A, Bowman P et al (1995) Monosomy 7 and activating RAS mutations accompany malignant transformation in patients with congenital neutropenia. Blood 86(12):4579–4586
Kaplan J, De Domenico I, Ward DMV (2008) Chediak-Higashi syndrome. Curr Opin Hematol 15(1):22–29. https://doi.org/10.1097/MOH.0b013e3282f2bcce
Karow JK, Chakraverty RK, Hickson ID (1997) The Bloom's syndrome gene product is a 3′-5′ DNA helicase*. J Biol Chem 272(49):30611–30614. https://doi.org/10.1074/jbc.272.49.30611
Kelley CM, Ikeda T, Koipally J, Avitahl N, Wu L, Georgopoulos K, Morgan BA (1998) Helios, a novel dimerization partner of Ikaros expressed in the earliest hematopoietic progenitors. Curr Biol 8(9):508–515. https://doi.org/10.1016/s0960-9822(98)70202-7
Klein M, Roder J, Haliotis T, Korec S, Jett JR, Herberman RB, Katz P, Fauci AS (1980) Chédiak-Higashi gene in humans. II. The selectivity of the defect in natural-killer and antibody-dependent cell-mediated cytotoxicity function. J Exp Med 151(5):1049–1058. https://doi.org/10.1084/jem.151.5.1049
Klirlat AK, Sela PB, Barak Y, Jammalieh J, Katzurti E (1989) Congenital dysgranulopoietic neutropenia in two siblings: clinical, ultrastructural, and in vitro bone marrow culture studies. Pediatr Hematol Oncol 6(4):293–305. https://doi.org/10.3109/08880018909034301
Klug CA, Morrison SJ, Masek M, Hahm K, Smale ST, Weissman IL (1998) Hematopoietic stem cells and lymphoid progenitors express different Ikaros isoforms, and Ikaros is localized to heterochromatin in immature lymphocytes. Proc Natl Acad Sci U S A 95(2):657–662. https://doi.org/10.1073/pnas.95.2.657

Knight ZA, Gonzalez B, Feldman ME, Zunder ER, Goldenberg DD, Williams O, Loewith R et al (2006) A pharmacological map of the PI3-K family defines a role for P110α in insulin signaling. Cell 125(4):733–747. https://doi.org/10.1016/j.cell.2006.03.035

Korkmaz B, Horwitz MS, Jenne DE, Gauthier F (2010) Neutrophil elastase, proteinase 3, and cathepsin G as therapeutic targets in human diseases. Pharmacol Rev 62(4):726–759. https://doi.org/10.1124/pr.110.002733

Kratz CP, Niemeyer CM, Jüttner E, Kartal M, Weninger A, Schmitt-Graeff A, Kontny U et al (2008) Childhood T-cell non-Hodgkin's lymphoma, colorectal carcinoma and brain tumor in association with Café-au-Lait spots caused by a novel homozygous PMS2 mutation. Leukemia 22(5):1078–1080. https://doi.org/10.1038/sj.leu.2405008

Kucuk ZY, Zhang K, Filipovich L, Bleesing JJH (2016) CTP synthase 1 deficiency in successfully transplanted siblings with combined immune deficiency and chronic active EBV infection. J Clin Immunol 36(8):750–753. https://doi.org/10.1007/s10875-016-0332-z

Kuehn HS, Boisson B, Cunningham-Rundles C, Reichenbach J, Stray-Pedersen A, Gelfand EW, Maffucci P et al (2016) Loss of B cells in patients with heterozygous mutations in IKAROS. N Engl J Med 374(11):1032–1043. https://doi.org/10.1056/NEJMoa1512234

Kuehn HS, Nunes-Santos CJ, Rosenzweig SD (2021a) IKAROS-associated diseases in 2020: genotypes, phenotypes, and outcomes in primary immune deficiency/inborn errors of immunity. J Clin Immunol 41(1):1–10. https://doi.org/10.1007/s10875-020-00936-x

Kuehn HS, Nunes-Santos CJ, Rosenzweig SD (2021b) Germline IKZF1 mutations and their impact on immunity: IKAROS-associated diseases and pathophysiology. Expert Rev Clin Immunol 17(4):407–416. https://doi.org/10.1080/1744666X.2021.1901582

Kulinski JM, Leonardo SM, Mounce BC, Malherbe L, Gauld SB, Tarakanova VL (2012) Ataxia telangiectasia mutated kinase controls chronic gammaherpesvirus infection. J Virol 86(23): 12826–12837. https://doi.org/10.1128/JVI.00917-12

Kumánovics A, Perkins SL, Gilbert H, Cessna MH, Augustine NH, Hill HR (2010) Diffuse large B cell lymphoma in hyper-IgE syndrome due to STAT3 mutation. J Clin Immunol 30(6): 886–893. https://doi.org/10.1007/s10875-010-9452-z

Kung C, Pingel JT, Heikinheimo M, Klemola T, Varkila K, Yoo LI, Vuopala K et al (2000) Mutations in the tyrosine phosphatase CD45 gene in a child with severe combined immunodeficiency disease. Nat Med 6(3):343–345. https://doi.org/10.1038/73208

Kurz ARM, Pruenster M, Rohwedder I, Ramadass M, Schäfer K, Harrison U, Gouveia G et al (2016) MST1-dependent vesicle trafficking regulates neutrophil transmigration through the vascular basement membrane. Am Soc Clin Invest 1. https://doi.org/10.1172/JCI87043

Kwan S-P, Lehner T, Hagemann T, Lu B, Blaese M, Ochs H, Wedgwood R, Ott J, Craig IW, Rosen FS (1991) Localization of the gene for the Wiskott-Aldrich syndrome between two flanking markers, TIMP and DXS255, on Xp11.22–Xp11.3. Genomics 10(1):29–33. https://doi.org/10.1016/0888-7543(91)90480-3

Rosée L, Paul ACH, Hines M, von Bahr Greenwood T, Machowicz R, Berliner N, Birndt S et al (2019) Recommendations for the management of hemophagocytic lymphohistiocytosis in adults. Blood 133(23):2465–2477. https://doi.org/10.1182/blood.2018894618

Lau YL, Wong SN, Lawton WM (1992) Takayasu's arteritis associated with Wiskott-Aldrich syndrome. J Paediatr Child Health 28(5):407–409. https://doi.org/10.1111/j.1440-1754.1992.tb02703.x

Lavin MF (2008) Ataxia-Telangiectasia: from a rare disorder to a paradigm for cell signalling and cancer. Nat Rev. Mol Cell Biol 9(10):759–769. https://doi.org/10.1038/nrm2514

Le Deist F, Emile JF, Rieux Laucat F, Benkerrou M, Roberts I, Brousse N, Fischer A (1996) Clinical, immunological, and pathological consequences of Fas-deficient conditions. Lancet (London, England) 348(9029):719–723. https://doi.org/10.1016/S0140-6736(96)02293-3

Lee C-L, Chuang C-K, Chiu H-C, Ru-Yi T, Lo Y-T, Chang Y-H, Lin H-Y, Lin S-P (2021) The first SHORT syndrome in a Taiwanese boy: a case report and review of the literature. Mol Genet Metabolism Rep 27(June):100768. https://doi.org/10.1016/j.ymgmr.2021.100768

Lee PPW, Chen T-X, Jiang L-P, Chen J, Chan K-W, Lee T-L, Ho MHK et al (2009) Clinical and molecular characteristics of 35 Chinese children with Wiskott-Aldrich syndrome. J Clin Immunol 29(4):490–500. https://doi.org/10.1007/s10875-009-9285-9

Lee PP, Woodbine L, Gilmour KC, Bibi S, Cale CM, Amrolia PJ, Veys PA, Davies EG, Jeggo PA, Jones A (2013) The many faces of artemis-deficient combined immunodeficiency — two patients with DCLRE1C mutations and a systematic literature review of genotype–phenotype correlation. Clin Immunol 149(3, Part B):464–474. https://doi.org/10.1016/j.clim.2013.08.006

Lee PY, Platt CD, Weeks S, Grace RF, Maher G, Gauthier K, Devana S et al (2020) Immune dysregulation and multisystem inflammatory syndrome in children (MIS-C) in individuals with haploinsufficiency of SOCS1. J Allergy Clin Immunol 146(5):1194–1200. https://doi.org/10.1016/j.jaci.2020.07.033

Leeksma OC, de Miranda NF, Veelken H (2017) Germline mutations predisposing to diffuse large B-cell lymphoma. Blood Cancer J 7(2):e532–e532. https://doi.org/10.1038/bcj.2017.15

Li F-Y, Chaigne-Delalande B, Kanellopoulou C, Davis JC, Matthews HF, Douek DC, Cohen JI, Uzel G, Su HC, Lenardo MJ (2011) Second messenger role for Mg2+ revealed by human T-cell immunodeficiency. Nature 475(7357):471–476. https://doi.org/10.1038/nature10246

Li F-Y, Chaigne-Delalande B, Helen S, Uzel G, Matthews H, Lenardo MJ (2014) XMEN disease: a new primary immunodeficiency affecting Mg2+ regulation of immunity against Epstein-Barr virus. Blood 123(14):2148–2152. https://doi.org/10.1182/blood-2013-11-538686

Li R, Zheng Y, Li Y, Zhang R, Wang F, Yang D, Ma Y, Xinlin M, Cao Z, Gao Z (2018) Common variable immunodeficiency with genetic defects identified by whole exome sequencing. Biomed Res Int 2018(September):3724630. https://doi.org/10.1155/2018/3724630

Lichter P, Walczak H, Weitz S, Behrmann I, Krammer PH (1992) The human APO-1 (APT) antigen maps to 10q23, a region that is syntenic with mouse chromosome 19. Genomics 14(1):179–180. https://doi.org/10.1016/s0888-7543(05)80302-7

Linder S, Higgs H, Hüfner K, Schwarz K, Pannicke U, Aepfelbacher M (2000) The polarization defect of Wiskott-Aldrich syndrome macrophages is linked to dislocalization of the Arp2/3 complex. J Immunol (Baltimore, Md.: 1950) 165(1):221–225. https://doi.org/10.4049/jimmunol.165.1.221

Link DC (2019) Mechanisms of leukemic transformation in congenital neutropenia. Curr Opin Hematol 26(1):34–40. https://doi.org/10.1097/MOH.0000000000000479

Linka RM, Risse SL, Bienemann K, Werner M, Linka Y, Krux F, Synaeve C et al (2012) Loss-of-function mutations within the IL-2 inducible kinase ITK in patients with EBV-associated lymphoproliferative diseases. Leukemia 26(5):963–971. https://doi.org/10.1038/leu.2011.371

Lino, Rosa CN, Ghosh S (2021) Epstein–Barr virus in inborn immunodeficiency—more than infection. Cancers 13(19):4752. https://doi.org/10.3390/cancers13194752

Liu T, Zhang L, Joo D, Sun S-C (2017) NF-KB signaling in inflammation. Signal Transduct Target Ther 2(1):1–9. https://doi.org/10.1038/sigtrans.2017.23

Lo B, Zhang K, Lu W, Zheng L, Zhang Q, Kanellopoulou C, Yu Z et al (2015) Patients with LRBA deficiency show CTLA4 loss and immune dysregulation responsive to abatacept therapy. Science 349(6246):436–440. https://doi.org/10.1126/science.aaa1663

Lopez-Herrera G, Tampella G, Pan-Hammarström Q, Herholz P, Trujillo-Vargas CM, Phadwal K, Simon AK et al (2012) Deleterious mutations in LRBA are associated with a syndrome of immune deficiency and autoimmunity. Am J Hum Genet 90(6):986–1001. https://doi.org/10.1016/j.ajhg.2012.04.015

Lorenzi R, Brickell PM, Katz DR, Kinnon C, Thrasher AJ (2000) Wiskott-Aldrich syndrome protein is necessary for efficient IgG-mediated phagocytosis. Blood 95(9):2943–2946. https://doi.org/10.1182/blood.V95.9.2943.009k17_2943_2946

Lougaris V, Chou J, Beano A, Wallace JG, Baronio M, Gazzurelli L, Lorenzini T et al (2019) A monoallelic activating mutation in RAC2 resulting in a combined immunodeficiency. J Allergy Clin Immunol 143(4):1649–1653. https://doi.org/10.1016/j.jaci.2019.01.001

Lu HY, Bauman BM, Arjunaraja S, Dorjbal B, Milner JD, Snow AL, Turvey SE (2018) The CBM-opathies—a rapidly expanding spectrum of human inborn errors of immunity caused by mutations in the CARD11-BCL10-MALT1 complex. Front Immunol 0. https://doi.org/10.3389/fimmu.2018.02078

Lu J, Wei Q, Bondy ML, Yu T-K, Li D, Brewster A, Shete S, Sahin A, Meric-Bernstam F, Wang LE (2006) Promoter polymorphism (−786t>C) in the endothelial nitric oxide synthase gene is associated with risk of sporadic breast cancer in non-Hispanic white women age younger than 55 years. Cancer 107(9):2245–2253. https://doi.org/10.1002/cncr.22269

Lucas CL, Chandra A, Nejentsev S, Condliffe AM, Okkenhaug K (2016) PI3Kδ and primary immunodeficiencies. Nat Rev Immunol 16(11):702–714. https://doi.org/10.1038/nri.2016.93

Lucas KG, Ungar D, Comito M, Groh B (2008) Epstein Barr virus induced lymphoma in a child with IPEX syndrome. Pediatr Blood Cancer 50(5):1056–1057. https://doi.org/10.1002/pbc.21341

Lum SH, Bonney D, Cheesman E, Wrignt NB, Hughes S, Wynn R (2016) Successful curative therapy with rituximab and allogeneic haematopoietic stem cell transplantation for MALT lymphoma associated with STK4-mutated CD4+ lymphocytopenia. Pediatr Blood Cancer 63(9):1657–1659. https://doi.org/10.1002/pbc.26048

Lundin KE, Hamasy A, Backe PH, Moens LN, Falk-Sörqvist E, Elgstøen KB, Mørkrid L et al (2015) Susceptibility to infections, without concomitant hyper-IgE, reported in 1976, is caused by hypomorphic mutation in the phosphoglucomutase 3 (PGM3) gene. Clin Immunol 161(2): 366–372. https://doi.org/10.1016/j.clim.2015.10.002

Łyszkiewicz M, Ziętara N, Frey L, Pannicke U, Stern M, Liu Y, Fan Y et al (2020) Human FCHO1 deficiency reveals role for Clathrin-mediated endocytosis in development and function of T cells. Nat Commun 11(1):1031. https://doi.org/10.1038/s41467-020-14809-9

Ma CA, Stinson JR, Zhang Y, Abbott JK, Weinreich MA, Hauk PJ, Reynolds PR et al (2017) Germline hypomorphic CARD11 mutations in severe atopic disease. Nat Genet 49(8): 1192–1201. https://doi.org/10.1038/ng.3898

Ma CS, Hare NJ, Nichols KE, Dupré L, Andolfi G, Roncarolo M-G, Adelstein S, Hodgkin PD, Tangye SG (2005) Impaired humoral immunity in X-linked lymphoproliferative disease is associated with defective IL-10 production by CD4+ T cells. J Clin Invest 115(4):1049–1059. https://doi.org/10.1172/JCI23139

Ma Q, Jones D, Borghesani PR, Segal RA, Nagasawa T, Kishimoto T, Bronson RT, Springer TA (1998) Impaired B lymphopoiesis, myelopoiesis, and derailed cerebellar neuron migration in CXCR4- and SDF-1-deficient mice. Proc Natl Acad Sci U S A 95(16):9448–9453. https://doi.org/10.1073/pnas.95.16.9448

Maas C, Lüftinger R, Krois W, Matthes-Martin S, Bayer G, Boztug K, Metzelder M (2018) EBV-positive B-cell lymphoma manifestation of the liver in an infant with RAG1 severe combined immunodeficiency disease. Pediatr Blood Cancer 65(9):e27258. https://doi.org/10.1002/pbc.27258

Mache CJ, Slavc I, Schmid C, Hoefler G, Urban CE, Schwinger W, Winter E, Hulla W, Zenz W, Holter W (1994) Familial hemophagocytic lymphohistiocytosis associated with disseminated T-cell lymphoma: a report of two siblings. Ann Hematol 69(2):85–91. https://doi.org/10.1007/BF01698488

Maffucci P, Chavez J, Jurkiw TJ, O'Brien PJ, Abbott JK, Reynolds PR, Worth A et al (n.d.) Biallelic mutations in DNA ligase 1 underlie a spectrum of immune deficiencies. J Clin Invest 128(12):5489–5504. https://doi.org/10.1172/JCI99629

Makaryan V, Zeidler C, Bolyard AA, Skokowa J, Rodger E, Kelley ML, Boxer LA et al (2015) The diversity of mutations and clinical outcomes for ELANE-associated neutropenia. Curr Opin Hematol 22(1):3–11. https://doi.org/10.1097/MOH.0000000000000105

Mäkitie O, Pukkala E, Teppo L, Kaitila I (1999) Increased incidence of cancer in patients with cartilage-hair hypoplasia. J Pediatr 134(3):315–318. https://doi.org/10.1016/S0022-3476(99)70456-7

Mäkitie O, Vakkilainen S (2020) Cartilage-hair hypoplasia – anauxetic dysplasia Spectrum disorders. GeneReviews® [Internet]. University of Washington, Seattle. https://www.ncbi.nlm.nih.gov/books/NBK84550/

Malu S, De Ioannes P, Kozlov M, Greene M, Francis D, Hanna M, Pena J et al (2012) Artemis C-terminal region facilitates V(D)J recombination through its interactions with DNA ligase IV and DNA-PKcs. J Exp Med 209(5):955–963. https://doi.org/10.1084/jem.20111437

Mansour S, Connell F, Steward C, Ostergaard P, Brice G, Smithson S, Lunt P et al (2010) Emberger syndrome—primary lymphedema with myelodysplasia: report of seven new cases. Am J Med Genet A 152A(9):2287–2296. https://doi.org/10.1002/ajmg.a.33445

Marrone A, Walne A, Tamary H, Masunari Y, Kirwan M, Beswick R, Vulliamy T, Dokal I (2007) Telomerase reverse-transcriptase homozygous mutations in autosomal recessive Dyskeratosis Congenita and Hoyeraal-Hreidarsson syndrome. Blood 110(13):4198–4205. https://doi.org/10.1182/blood-2006-12-062851

Marsh JC, Will AJ, Hows JM, Sartori P, Darbyshire PJ, Williamson PJ, Oscier DG, Dexter TM, Testa NG (1992) 'Stem cell' origin of the hematopoietic defect in Dyskeratosis Congenita. Blood 79(12):3138–3144

Marsh RA, Bleesing JJ, Chandrakasan S, Jordan MB, Davies SM, Filipovich AH (2014) Reduced-intensity conditioning hematopoietic cell transplantation is an effective treatment for patients with SLAM-associated protein deficiency/X-linked lymphoproliferative disease type 1. Biol Blood Marrow Transplant 20(10):1641–1645. https://doi.org/10.1016/j.bbmt.2014.06.003

Martin E, Minet N, Boschat A-C, Sanquer S, Sobrino S, Lenoir C, de Villartay JP et al (n.d.) Impaired lymphocyte function and differentiation in CTPS1-deficient patients result from a hypomorphic homozygous mutation. JCI Insight 5(5):e133880. https://doi.org/10.1172/jci.insight.133880

Martin E, Palmic N, Sanquer S, Lenoir C, Hauck F, Mongellaz C, Fabrega S et al (2014) CTP synthase 1 deficiency in humans reveals its central role in lymphocyte proliferation. Nature 510(7504):288–292. https://doi.org/10.1038/nature13386

Martinez, Real CA, Pinheiro LV, Rossi DH, Camargo MG, de Lourdes Setsuko Ayrizono M, Leal RF, Coy CSR (2016) Adenocarcinoma of the right colon in a patient with bloom syndrome. Case Rep Surg 2016:3176842. https://doi.org/10.1155/2016/3176842

Martinod K, Witsch T, Farley K, Gallant M, Remold-O'Donnell E, Wagner DD (2016) Neutrophil elastase-deficient mice form neutrophil extracellular traps in an experimental model of deep vein thrombosis. J Thrombosis Haemostasis: JTH 14(3):551–558. https://doi.org/10.1111/jth.13239

Martins VC, Busch K, Juraeva D, Blum C, Ludwig C, Rasche V, Lasitschka F et al (2014) Cell competition is a tumour suppressor mechanism in the thymus. Nature 509(7501):465–470. https://doi.org/10.1038/nature13317

Marzollo A, Maestrini G, La Starza R, Elia L, Malfona F, Pierini T, Parenzan CT et al (2020) A novel germline variant in PIK3R1 results in SHORT syndrome associated with TAL/LMO T-cell acute lymphoblastic leukemia. Am J Hematol 95(12):E335–E338. https://doi.org/10.1002/ajh.25998

Massaad MJ, Ramesh N, Geha RS (2013) Wiskott-Aldrich syndrome: a comprehensive review. Ann N Y Acad Sci 1285(1):26–43. https://doi.org/10.1111/nyas.12049

Matsuda-Lennikov M, Biancalana M, Zou J, Ravell JC, Zheng L, Kanellopoulou C, Jiang P et al (2019) Magnesium transporter 1 (MAGT1) deficiency causes selective defects in N-linked glycosylation and expression of immune-response genes. J Biol Chem 294(37):13638–13656. https://doi.org/10.1074/jbc.RA119.008903

McBride KA, Ballinger ML, Killick E, Kirk J, Tattersall MHN, Eeles RA, Thomas DM, Mitchell G (2014) Li-Fraumeni syndrome: cancer risk assessment and clinical management. Nat Rev Clin Oncol 11(5):260–271. https://doi.org/10.1038/nrclinonc.2014.41

McBride OW, Merry D, Givol D (1986) The gene for human P53 cellular tumor antigen is located on chromosome 17 short arm (17p13). Proc Natl Acad Sci U S A 83(1):130–134. https://doi.org/10.1073/pnas.83.1.130

McKinnon ML, Rozmus J, Fung S-Y, Hirschfeld AF, Del Bel KL, Thomas L, Marr N et al (2014) Combined immunodeficiency associated with homozygous MALT1 mutations. J Allergy Clin Immunol 133(5):1458–1462. https://doi.org/10.1016/j.jaci.2013.10.045

McKinnon PJ (2012) ATM and the molecular pathogenesis of ataxia telangiectasia. Annu Rev Pathol 7:303–321. https://doi.org/10.1146/annurev-pathol-011811-132509

Meininger I, Krappmann D (2016) Lymphocyte signaling and activation by the CARMA1-BCL10-MALT1 signalosome. Biol Chem 397(12):1315–1333. https://doi.org/10.1515/hsz-2016-0216

Mellon I, Rajpal DK, Koi M, Boland CR, Champe GN (1996) Transcription-coupled repair deficiency and mutations in human mismatch repair genes. April 272:557. https://doi.org/10.1126/science.272.5261.557

Mensa-Vilaró A, García-Morato MB, de la Calle-Martin O, Franco-Jarava C, Martínez-Saavedra MT, González-Granado LI, González-Roca E et al (2019) Unexpected relevant role of gene mosaicism in patients with primary immunodeficiency diseases. J Allergy Clin Immunol 143(1): 359–368. https://doi.org/10.1016/j.jaci.2018.09.009

Michalovich D, Nejentsev S (2018) Activated PI3 kinase delta syndrome: from genetics to therapy. Front Immunol 9. https://www.frontiersin.org/article/10.3389/fimmu.2018.00369

Miki H, Takenawa T (1998) Direct binding of the Verprolin-homology domain in N-WASP to actin is essential for cytoskeletal reorganization. Biochem Biophys Res Commun 243(1):73–78. https://doi.org/10.1006/bbrc.1997.8064

Miller K, Müller W, Winkler L, Hadam MR, Ehrich JHH, Flatz SD (1990) Mitotic disturbance associated with mosaic aneuploidies. Hum Genet 84(4):361–364. https://doi.org/10.1007/BF00196235

Milner JD, Vogel TP, Forbes L, Ma CA, Stray-Pedersen A, Niemela JE, Lyons JJ et al (2015) Early-onset Lymphoproliferation and autoimmunity caused by germline STAT3 gain-of-function mutations. Blood 125(4):591–599. https://doi.org/10.1182/blood-2014-09-602763

Mitchell JR, Wood E, Collins K (1999) A telomerase component is defective in the human disease Dyskeratosis Congenita. Nature 402(6761):551–555. https://doi.org/10.1038/990141

Molnár A, Wu P, Largespada DA, Vortkamp A, Scherer S, Copeland NG, Jenkins NA, Bruns G, Georgopoulos K (1996) The Ikaros gene encodes a family of Lymphocyte-restricted zinc finger DNA binding proteins, highly conserved in human and mouse. J Immunol (Baltimore, Md.: 1950) 156(2):585–592

Moreno E (2014) Darwinian tumour suppression. Nature 509(7501):435–436. https://doi.org/10.1038/nature13337

Morgan B, Sun L, Avitahl N, Andrikopoulos K, Ikeda T, Gonzales E, Wu P, Neben S, Georgopoulos K (1997) Aiolos, a lymphoid restricted transcription factor that interacts with Ikaros to regulate lymphocyte differentiation. EMBO J 16(8):2004–2013. https://doi.org/10.1093/emboj/16.8.2004

Morgan NV, Goddard S, Cardno TS, McDonald D, Rahman F, Barge D, Ciupek A et al (2011) Mutation in the TCRα subunit constant gene (*TRAC*) leads to a human immunodeficiency disorder characterized by a lack of TCRαβ^{+} T cells. J Clin Invest 121(2):695–702. https://doi.org/10.1172/JCI41931

Moshous D, Callebaut I, de Chasseval R, Corneo B, Cavazzana-Calvo M, Le Deist F, Tezcan I et al (2001) Artemis, a novel DNA double-Strand break repair/V(D)J recombination protein, is mutated in human severe combined immune deficiency. Cell 105(2):177–186. https://doi.org/10.1016/s0092-8674(01)00309-9

Moshous D, Li L, Chasseval R, Philippe N, Jabado N, Cowan MJ, Fischer A, de Villartay JP (2000) A new gene involved in DNA double Strand break repair and V(D)J recombination is located on human chromosome 10p. Hum Mol Genet 9(4):583–588. https://doi.org/10.1093/hmg/9.4.583

Moshous D, Martin E, Carpentier W, Lim A, Callebaut I, Canioni D, Hauck F et al (2013) Whole-exome sequencing identifies Coronin-1A deficiency in 3 siblings with immunodeficiency and EBV-associated B-cell lymphoproliferation. J Allergy Clin Immunol 131(6):1594–1603. https://doi.org/10.1016/j.jaci.2013.01.042

Moshous D, Pannetier C, de Chasseval R, le Deist F, Cavazzana-Calvo M, Romana S, Macintyre E et al (2003) Partial T and B lymphocyte immunodeficiency and predisposition to lymphoma in patients with Hypomorphic mutations in *Artemis*. J Clin Invest 111(3):381–387. https://doi.org/10.1172/JCI16774

Müller A, Schackert HK, Lange B, Rüschoff J, Füzesi L, Willert J, Burfeind P et al (2006) A novel MSH2 germline mutation in homozygous state in two brothers with colorectal cancers diagnosed at the age of 11 and 12 years. Am J Med Genet A 140A(3):195–199. https://doi.org/10.1002/ajmg.a.31070

Muller M-L, Meeths M, Chiang S, Tesi B, Entesarian M, Nilsson D, Wood S et al (2014) An N-terminal missense mutation in STX11 causative of FHL4 abrogates syntaxin-11 binding to munc18-2. Front Immunol 4. https://www.frontiersin.org/article/10.3389/fimmu.2013.00515

Munitic I, Kuka M, Allam A, Scoville JP, Ashwell JD (2013) CD70 deficiency impairs effector CD8 T cell generation and viral clearance but Is dispensable for the recall response to Lymphocytic choriomeningitis virus. J Immunol (Baltimore, Md.: 1950) 190(3):1169–1179. https://doi.org/10.4049/jimmunol.1202353

Murray JE, Bicknell LS, Yigit G, Duker AL, van Kogelenberg M, Haghayegh S, Wieczorek D et al (2014) Extreme growth failure is a common presentation of ligase IV deficiency. Hum Mutat 35(1):76–85. https://doi.org/10.1002/humu.22461

Nademi Z, Wynn RF, Slatter M, Hughes SM, Bonney D, Qasim W, Latour S et al (2019) Hematopoietic stem cell transplantation for cytidine triphosphate synthase 1 (CTPS1) deficiency. Bone Marrow Transplant 54(1):130–133. https://doi.org/10.1038/s41409-018-0246-x

Nagasawa T, Hirota S, Tachibana K, Takakura N, Nishikawa S, Kitamura Y, Yoshida N, Kikutani H, Kishimoto T (1996) Defects of B-cell lymphopoiesis and bone-marrow Myelopoiesis in mice lacking the CXC chemokine PBSF/SDF-1. Nature 382(6592):635–638. https://doi.org/10.1038/382635a0

Narra MB, Shearer WT (2009) Cartilage-hair hypoplasia and severe allergy. J Allergy Clin Immunol 123(6):1418–1419; author reply 1419-1420. https://doi.org/10.1016/j.jaci.2009.02.036

Narumi S, Amano N, Ishii T, Katsumata N, Muroya K, Adachi M, Toyoshima K et al (2016) SAMD9 mutations cause a novel multisystem disorder, MIRAGE syndrome, and are associated with loss of chromosome 7. Nat Genet 48(7):792–797. https://doi.org/10.1038/ng.3569

Nayak RC, Trump LR, Aronow BJ, Myers K, Mehta P, Kalfa T, Wellendorf AM et al (2015) Pathogenesis of *ELANE*-mutant severe neutropenia revealed by induced pluripotent stem cells. J Clin Invest 125(8):3103–3116. https://doi.org/10.1172/JCI80924

Nehme NT, Schmid JP, Debeurme F, André-Schmutz I, Lim A, Nitschke P, Rieux-Laucat F et al (2012) MST1 mutations in autosomal recessive primary immunodeficiency characterized by defective naive T-cell survival. Blood 119(15):3458–3468. https://doi.org/10.1182/blood-2011-09-378364

Neven B, Magerus-Chatinet A, Florkin B, Gobert D, Lambotte O, De Somer L, Lanzarotti N et al (2011) A survey of 90 patients with autoimmune lymphoproliferative syndrome related to TNFRSF6 mutation. Blood 118(18):4798–4807. https://doi.org/10.1182/blood-2011-04-347641

Nichols KE, Hom J, Gong S-Y, Ganguly A, Ma CS, Cannons JL, Tangye SG, Schwartzberg PL, Koretzky GA, Stein PL (2005a) Regulation of NKT cell development by SAP, the protein defective in XLP. Nat Med 11(3):340–345. https://doi.org/10.1038/nm1189

Nichols KE, Ma CS, Cannons JL, Schwartzberg PL, Tangye SG (2005b) Molecular and cellular pathogenesis of X-linked lymphoproliferative disease. Immunol Rev 203(February):180–199. https://doi.org/10.1111/j.0105-2896.2005.00230.x

Nicolaides NC, Carter KC, Shell BK, Papadopoulos N, Vogelstein B, Kinzler KW (1995) Genomic organization of the human PMS2 gene family. Genomics 30(2):195–206. https://doi.org/10.1006/geno.1995.9885

Niehues T, Perez-Becker R, Schuetz C (2010) More than just SCID--the phenotypic range of combined immunodeficiencies associated with mutations in the recombinase activating genes (RAG) 1 and 2. Clin Immunol (Orlando, Fla.) 135(2):183–192. https://doi.org/10.1016/j.clim.2010.01.013

Nikolov NP, Shimizu M, Cleland S, Bailey D, Aoki J, Strom T, Schwartzberg PL, Candotti F, Siegel RM (2010) Systemic autoimmunity and defective Fas ligand secretion in the absence of the Wiskott-Aldrich syndrome protein. Blood 116(5):740–747. https://doi.org/10.1182/blood-2009-08-237560

Nolte MA, van Olffen RW, van Gisbergen Klaas PJM, van Lier RAW (2009) Timing and tuning of CD27-CD70 interactions: the impact of signal strength in setting the balance between adaptive responses and immunopathology. Immunol Rev 229(1):216–231. https://doi.org/10.1111/j.1600-065X.2009.00774.x

Nunes-Santos CJ, Kuehn HS, Rosenzweig SD (2020) IKAROS family zinc finger 1-associated diseases in primary immunodeficiency patients. Immunol Allergy Clin N Am 40(3):461–470. https://doi.org/10.1016/j.iac.2020.04.004

O'Driscoll M, Cerosaletti KM, Girard PM, Dai Y, Stumm M, Kysela B, Hirsch B et al (2001) DNA ligase IV mutations identified in patients exhibiting developmental delay and immunodeficiency. Mol Cell 8(6):1175–1185. https://doi.org/10.1016/s1097-2765(01)00408-7

Oh H, Ghosh S (2013) NF-KB: roles and regulation in different CD4(+) T-cell subsets. Immunol Rev 252(1):41–51. https://doi.org/10.1111/imr.12033

Okano M, Gross TG (1996) Epstein-Barr virus-associated hemophagocytic syndrome and fatal infectious mononucleosis. Am J Hematol 53(2):111–115. https://doi.org/10.1002/(SICI)1096-8652(199610)53:2<111::AID-AJH8>3.0.CO;2-2

Okkenhaug K (2013) Signaling by the phosphoinositide 3-kinase family in immune cells. Annu Rev Immunol 31:675–704. https://doi.org/10.1146/annurev-immunol-032712-095946

Oliveira JB, Bleesing JJ, Dianzani U, Fleisher TA, Jaffe ES, Lenardo MJ, Rieux-Laucat F et al (2010) Revised diagnostic criteria and classification for the autoimmune lymphoproliferative syndrome (ALPS): report from the 2009 NIH international workshop. Blood 116(14):e35–e40. https://doi.org/10.1182/blood-2010-04-280347

Orange JS, Ramesh N, Remold-O'Donnell E, Sasahara Y, Koopman L, Byrne M, Bonilla FA, Rosen FS, Geha RS, Strominger JL (2002) Wiskott-Aldrich syndrome protein is required for NK cell cytotoxicity and colocalizes with actin to NK cell-activating immunologic synapses. Proc Natl Acad Sci U S A 99(17):11351–11356. https://doi.org/10.1073/pnas.162376099

Orkin SH (2000) Diversification of haematopoietic stem cells to specific lineages. Nat Rev Genet 1(1):57–64. https://doi.org/10.1038/35049577

O'Shea JJ, Holland SM, Staudt LM (2013) JAKs and STATs in immunity, immunodeficiency, and cancer. N Engl J Med 368(2):161–170. https://doi.org/10.1056/NEJMra1202117

Ostergaard P, Simpson MA, Connell FC, Steward CG, Brice G, Woollard WJ, Dafou D et al (2011) Mutations in GATA2 cause primary lymphedema associated with a predisposition to acute myeloid leukemia (Emberger syndrome). Nat Genet 43(10):929–931. https://doi.org/10.1038/ng.923

Parolini O, Berardelli S, Riedl E, Bello-Fernandez C, Strobl H, Majdic O, Knapp W (1997) Expression of Wiskott-Aldrich syndrome protein (WASP) gene during hematopoietic differentiation. Blood 90(1):70–75. https://doi.org/10.1182/blood.V90.1.70

Parry EM, Alder JK, Lee SS, Phillips JA, Loyd JE, Duggal P, Armanios M (2011) Decreased dyskerin levels as a mechanism of telomere shortening in X-linked dyskeratosis congenita. J Med Genet 48(5):327–333. https://doi.org/10.1136/jmg.2010.085100

Pearson TS (2016) More than ataxia: hyperkinetic movement disorders in childhood autosomal recessive ataxia syndromes. Tremor Other Hyperkinetic Mov (New York, N.Y.) 6:368. https://doi.org/10.7916/D8H70FSS

Peng Z, Liao Z, Dziegielewska B, Matsumoto Y, Thomas S, Wan Y, Yang A, Tomkinson AE (2012) Phosphorylation of serine 51 regulates the interaction of human DNA ligase I with replication factor C and its participation in DNA replication and repair. J Biol Chem 287(44):36711–36719. https://doi.org/10.1074/jbc.M112.383570

Péron S, Metin A, Gardès P, Alyanakian M-A, Sheridan E, Kratz CP, Fischer A, Durandy A (2008) Human PMS2 deficiency is associated with impaired immunoglobulin class switch recombination. J Exp Med 205(11):2465–2472. https://doi.org/10.1084/jem.20080789

Perou CM, Moore KJ, Nagle DL, Misumi DJ, Woolf EA, McGrail SH, Holmgren L et al (1996) Identification of the murine beige gene by YAC complementation and positional cloning. Nat Genet 13(3):303–308. https://doi.org/10.1038/ng0796-303

Perry GS, Spector BD, Schuman LM, Mandel JS, Anderson VE, McHugh RB, Hanson MR, Fahlstrom SM, Krivit W, Kersey JH (1980) The Wiskott-Aldrich syndrome in the United States and Canada (1892-1979). J Pediatr 97(1):72–78. https://doi.org/10.1016/s0022-3476(80)80133-8

Pilarski R (2009) Cowden syndrome: a critical review of the clinical literature. J Genet Couns 18(1):13–27. https://doi.org/10.1007/s10897-008-9187-7

Pitsikas P, Lee D, Rainbow AJ (2007) Reduced host cell reactivation of oxidative DNA damage in human cells deficient in the mismatch repair gene HMSH2. Mutagenesis 22(3):235–243. https://doi.org/10.1093/mutage/gem008

Platt CD, Fried AJ, Hoyos-Bachiloglu R, Usmani GN, Schmidt B, Whangbo J, Chiarle R, Chou J, Geha RS (2017) Combined immunodeficiency with EBV positive B cell lymphoma and epidermodysplasia verruciformis due to a novel homozygous mutation in RASGRP1. Clin Immunol (Orlando, Fla.) 183(October):142–144. https://doi.org/10.1016/j.clim.2017.08.007

PMS2 PMS1 Homolog 2, Mismatch Repair System Component [Homo Sapiens (Human)] - Gene - NCBI (n.d.) Accessed 27 June 2022. https://www.ncbi.nlm.nih.gov/gene?Db=gene&Cmd=ShowDetailView&TermToSearch=5395

Polprasert C, Takeuchi Y, Kakiuchi N, Yoshida K, Assanasen T, Sitthi W, Bunworasate U et al (2019) Frequent germline mutations of HAVCR2 in sporadic subcutaneous panniculitis-like T-cell lymphoma. Blood Adv 3(4):588–595. https://doi.org/10.1182/bloodadvances.2018028340

Porter CC, Baturin D, Choudhary R, DeGregori J (2011) Relative fitness of hematopoietic progenitors influences leukemia progression. Leukemia 25(5):891–895. https://doi.org/10.1038/leu.2011.22

Price S, Shaw PA, Seitz A, Joshi G, Davis J, Niemela JE, Perkins K et al (2014) Natural history of autoimmune lymphoproliferative syndrome associated with FAS gene mutations. Blood 123(13):1989–1999. https://doi.org/10.1182/blood-2013-10-535393

Punwani D, Kawahara M, Yu J, Sanford U, Roy S, Patel K, Carbonaro DA et al (2017) Lentivirus mediated correction of artemis-deficient severe combined immunodeficiency. Hum Gene Ther 28(1):112–124. https://doi.org/10.1089/hum.2016.064

Punwani D, Pelz B, Yu J, Arva NC, Schafernak K, Kondratowicz K, Makhija M, Puck JM (2015a) Coronin-1A: immune deficiency in humans and mice. J Clin Immunol 35(2):100–107. https://doi.org/10.1007/s10875-015-0130-z

Punwani D, Wang H, Chan AY, Cowan MJ, Mallott J, Sunderam U, Mollenauer M et al (2015b) Combined immunodeficiency due to MALT1 mutations, treated by hematopoietic cell transplantation. J Clin Immunol 35(2):135–146. https://doi.org/10.1007/s10875-014-0125-1

Purtilo DT, Grierson HL (1991) Methods of detection of new families with X-linked lymphoproliferative disease. Cancer Genet Cytogenet 51(2):143–153. https://doi.org/10.1016/0165-4608(91)90127-G

Radwan N, El-Owaidy R, El-Sayed ZA, Abdel-Baky A, El-Haddad A, Rashad H, Khorshed EN et al (2020) A case of STK4 deficiency with complications evoking mycobacterial infection. J Clin Immunol 40(4):665–669. https://doi.org/10.1007/s10875-020-00783-w

Ravell JC, Chauvin SD, He T, Lenardo M (2020a) An update on XMEN disease. J Clin Immunol 40(5):671–681. https://doi.org/10.1007/s10875-020-00790-x

Ravell JC, Matsuda-Lennikov M, Chauvin SD, Zou J, Biancalana M, Deeb SJ, Price S et al (2020b) Defective glycosylation and multisystem abnormalities characterize the primary immunodeficiency XMEN disease. J Clin Invest 130(1):507–522. https://doi.org/10.1172/JCI131116

Ravell J, Chaigne-Delalande B, Lenardo M (2014) XMEN disease: a combined immune deficiency with magnesium defect. Curr Opin Pediatr 26(6):713–719. https://doi.org/10.1097/MOP.0000000000000156

Rawat A, Singh A, Dobbs K, Pala F, Delmonte OM, Vignesh P, Jindal AK et al (2021) Skewed TCR alpha, but not beta, gene rearrangements and lymphoma associated with a pathogenic TRAC variant. J Clin Immunol 41(6):1395–1399. https://doi.org/10.1007/s10875-021-01047-x

Rezaei N, Hedayat M, Aghamohammadi A, Nichols KE (2011a) Primary immunodeficiency diseases associated with increased susceptibility to viral infections and malignancies. J Allergy Clin Immunol 127(6):1329–1341. https://doi.org/10.1016/j.jaci.2011.02.047

Rezaei N, Mahmoudi E, Aghamohammadi A, Das R, Nichols KE (2011b) X-linked lymphoproliferative syndrome: a genetic condition typified by the triad of infection, immunodeficiency and lymphoma. Br J Haematol 152(1):13–30. https://doi.org/10.1111/j.1365-2141.2010.08442.x

Ridanpää M, van Eenennaam H, Pelin K, Chadwick R, Johnson C, Yuan B, vanVenrooij W et al (2001) Mutations in the RNA component of RNase MRP cause a pleiotropic human disease, cartilage-hair hypoplasia. Cell 104(2):195–203. https://doi.org/10.1016/s0092-8674(01)00205-7

Risma K (2012) A loss of naiveté. Blood 119(15):3371–3372. https://doi.org/10.1182/blood-2012-02-410399

Rivero-Lezcano OM, Marcilla A, Sameshima JH, Robbins KC (1995) Wiskott-Aldrich syndrome protein physically associates with Nck through Src homology 3 domains. Mol Cell Biol 15(10):5725–5731. https://doi.org/10.1128/MCB.15.10.5725

Rodrigues NP, Janzen V, Forkert R, Dombkowski DM, Boyd AS, Orkin SH, Enver T, Vyas P, Scadden DT (2005) Haploinsufficiency of GATA-2 perturbs adult hematopoietic stem-cell homeostasis. Blood 106(2):477–484. https://doi.org/10.1182/blood-2004-08-2989

Rodrigues NP, Tipping AJ, Wang Z, Enver T (2012) GATA-2 mediated regulation of normal hematopoietic stem/progenitor cell function, myelodysplasia and myeloid leukemia. Int J Biochem Cell Biol 44(3):457–460. https://doi.org/10.1016/j.biocel.2011.12.004

Rodriguez R, Fournier B, Cordeiro DJ, Winter S, Izawa K, Martin E, Boutboul D et al (2019) Concomitant PIK3CD and TNFRSF9 deficiencies cause chronic active Epstein-Barr virus infection of T cells. J Exp Med 216(12):2800–2818. https://doi.org/10.1084/jem.20190678

Roifman CM, Yiping G, Cohen A (2006) Mutations in the RNA component of RNase mitochondrial RNA processing might cause Omenn syndrome. J Allergy Clin Immunol 117(4):897–903. https://doi.org/10.1016/j.jaci.2006.01.003

Rommel PC, Oliveira TY, Nussenzweig MC, Robbiani DF (2017) RAG1/2 induces genomic insertions by mobilizing DNA into RAG1/2-independent breaks. J Exp Med 214(3):815–831. https://doi.org/10.1084/jem.20161638

Rosenberg PS, Alter BP, Link DC, Stein S, Rodger E, Bolyard AA, Aprikyan AA et al (2008) Neutrophil elastase mutations and risk of leukaemia in severe congenital neutropenia. Br J Haematol 140(2):210–213. https://doi.org/10.1111/j.1365-2141.2007.06897.x

Rotulo GA, Beaupain B, Rialland F, Paillard C, Nachit O, Galambrun C, Gandemer V et al (2020) HSCT may lower leukemia risk in ELANE neutropenia: a before–after study from the French severe congenital neutropenia registry. Bone Marrow Transplant 55(8):1614–1622. https://doi.org/10.1038/s41409-020-0800-1

Rudd E, Göransdotter Ericson K, Zheng C, Uysal Z, Özkan A, Gürgey A, Fadeel B, Nordenskjöld M, Henter J-I (2006) Spectrum and clinical implications of syntaxin 11 gene mutations in familial haemophagocytic lymphohistiocytosis: association with disease-free remissions and haematopoietic malignancies. J Med Genet 43(4):e14. https://doi.org/10.1136/jmg.2005.035253

Ruschak PJ, Kauh YC, Luscombe HA (1981) Cowden's disease associated with immunodeficiency. Arch Dermatol 117(9):573–575. https://doi.org/10.1001/archderm.1981.01650090055027

Sahoo SS, Pastor VB, Goodings C, Voss RK, Kozyra EJ, Szvetnik A, Noellke P et al (2021) Clinical evolution, genetic landscape and trajectories of clonal hematopoiesis in SAMD9/SAMD9L syndromes. Nat Med 27(10):1806–1817. https://doi.org/10.1038/s41591-021-01511-6

Salzer E, Cagdas D, Hons M, Mace EM, Garncarz W, Petronczki ÖY, Platzer R et al (2016) RASGRP1 deficiency causes immunodeficiency with impaired cytoskeletal dynamics. Nat Immunol 17(12):1352–1360. https://doi.org/10.1038/ni.3575

Sasahara Y, Rachid R, Byrne MJ, de la Fuente MA, Abraham RT, Ramesh N, Geha RS (2002) Mechanism of recruitment of WASP to the immunological synapse and of its activation following TCR ligation. Mol Cell 10(6):1269–1281. https://doi.org/10.1016/s1097-2765(02)00728-1

Sato T, Tatsuzawa O, Koike Y, Wada Y, Nagata M, Kobayashi S, Ishizawa A, Miyauchi J, Shimizu K (1999) B-cell lymphoma associated with DiGeorge syndrome. Eur J Pediatr 158(7):609–609. https://doi.org/10.1007/s004310051160

Sayos J, Wu C, Morra M, Wang N, Zhang X, Allen D, van Schaik S et al (1998) The X-linked lymphoproliferative-disease gene product SAP regulates signals induced through the co-receptor SLAM. Nature 395(6701):462–469. https://doi.org/10.1038/26683

Schejter YD, Even-Or E, Shadur B, NaserEddin A, Stepensky P, Zaidman I (2020) The broad clinical spectrum and transplant results of PNP deficiency. J Clin Immunol 40(1):123–130. https://doi.org/10.1007/s10875-019-00698-1

Schimke LF, Sawalle-Belohradsky J, Roesler J, Wollenberg A, Rack A, Borte M, Rieber N et al (2010) Diagnostic approach to the hyper-IgE syndromes: immunologic and clinical key findings to differentiate hyper-IgE syndromes from atopic dermatitis. J Allergy Clin Immunol 126(3): 611–617. https://doi.org/10.1016/j.jaci.2010.06.029

Schipp C, Schlütermann D, Hönscheid A, Nabhani S, Höll J, Oommen PT, Ginzel S et al (2018) EBV negative lymphoma and autoimmune lymphoproliferative syndrome like phenotype extend the clinical spectrum of primary immunodeficiency caused by STK4 deficiency. Front Immunol 9:2400. https://doi.org/10.3389/fimmu.2018.02400

Schlauderer F, Seeholzer T, Desfosses A, Gehring T, Strauss M, Hopfner K-P, Gutsche I, Krappmann D, Lammens K (2018) Molecular architecture and regulation of BCL10-MALT1 filaments. Nat Commun 9(1):4041. https://doi.org/10.1038/s41467-018-06573-8

Schmitz R, Wright GW, Huang DW, Johnson CA, Phelan JD, Wang JQ, Roulland S et al (2018) Genetics and pathogenesis of diffuse large B-cell lymphoma. N Engl J Med 378(15): 1396–1407. https://doi.org/10.1056/NEJMoa1801445

Schneider K, Zelley K, Nichols KE, Garber J (2019) Li-Fraumeni syndrome. GeneReviews® [Internet]. University of Washington, Seattle. https://www.ncbi.nlm.nih.gov/books/NBK1311/

Schuetz C, Barbi G, Barth TTE, Hoenig M, Schulz A, Möeller P, Smeets D et al (2007) ICF syndrome: high variability of the chromosomal phenotype and association with classical Hodgkin lymphoma. Am J Med Genet A 143A(17):2052–2057. https://doi.org/10.1002/ajmg.a.31885

Schuetz C, Huck K, Gudowius S, Megahed M, Feyen O, Hubner B, Schneider DT et al (2008) An immunodeficiency disease with RAG mutations and granulomas. N Engl J Med 358(19): 2030–2038. https://doi.org/10.1056/NEJMoa073966

Schurman SH, Candotti F (2003) Autoimmunity in Wiskott-Aldrich syndrome. Curr Opin Rheumatol 15(4):446–453

Schuster V, Kreth HW (2000) X-linked lymphoproliferative disease is caused by deficiency of a novel SH2 domain-containing signal transduction adaptor protein. Immunol Rev 178 (December):21–28. https://doi.org/10.1034/j.1600-065x.2000.17819.x

Segundo GRS, Takano OA, Moraes LSL, Nadaf MISV, Fernandes SJ, Ochs HD, Torgerson TR (2017) Paternal gonadal mosaicism as cause of a puzzling inheritance pattern of activated PI3-kinase delta syndrome. Ann Allergy Asthma Immunol Off Pub Am Coll Allergy Asthma Immunol 119(6):564–566. https://doi.org/10.1016/j.anai.2017.09.054

Sellars M, Kastner P, Chan S (2011) Ikaros in B cell development and function. World J Biol Chem 2(6):132–139. https://doi.org/10.4331/wjbc.v2.i6.132

Sepulveda FE, Debeurme F, Ménasché G, Kurowska M, Côte M, Schmid JP, Fischer A, de Saint Basile G (2013) Distinct severity of HLH in both human and murine mutants with complete loss of cytotoxic effector PRF1, RAB27A, and STX11. Blood 121(4):595–603. https://doi.org/10.1182/blood-2012-07-440339

Seton-Rogers S (2014) Competition can be a good thing. Nat Rev Cancer 14(6):381–381. https://doi.org/10.1038/nrc3752

Shah S, Wu E, Koneti Rao V (2014) Autoimmune lymphoproliferative syndrome: an update and review of the literature. Curr Allergy Asthma Rep 14(9):462. https://doi.org/10.1007/s11882-014-0462-4

Shaikh AG, Zee DS, Mandir AS, Lederman HM, Crawford TO (2013) Disorders of upper limb movements in ataxia-telangiectasia. PLoS One 8(6):e67042. https://doi.org/10.1371/journal.pone.0067042

Shamriz O, Vilk SR, Wolf DG, Ta-Shma A, Averbuch D, Weintraub M, Stepensky P (2014) Hematopoietic stem cell transplantation conditioning with use of rituximab in EBV related lymphoproliferative disorders. Clin Immunol 151(2):79–83. https://doi.org/10.1016/j.clim.2014.01.007

Shannon-Lowe C, Rickinson AB, Bell AI (2017) Epstein–Barr virus-associated lymphomas. Philos Trans R Soc B Biol Sci 372(1732):20160271. https://doi.org/10.1098/rstb.2016.0271

Sharafian S, Ziaee V, Shahrooei M, Ahadi M, Parvaneh N (2019) A novel STK4 mutation presenting with juvenile idiopathic arthritis and epidermodysplasia Verruciformis. J Clin Immunol 39(1):11–14. https://doi.org/10.1007/s10875-018-0586-8

Sharfe N, Karanxha A, Dadi H, Merico D, Chitayat D, Herbrick J-A, Freeman S, Grinstein S, Roifman CM (2018) Dual loss of P110δ PI3-kinase and SKAP (KNSTRN) expression leads to combined immunodeficiency and multisystem syndromic features. J Allergy Clin Immunol 142(2):618–629. https://doi.org/10.1016/j.jaci.2017.10.033

Sharifi R, Sinclair JC, Gilmour KC, Arkwright PD, Kinnon C, Thrasher AJ, Bobby Gaspar H (2004) SAP mediates specific cytotoxic T-cell functions in X-linked lymphoproliferative disease. Blood 103(10):3821–3827. https://doi.org/10.1182/blood-2003-09-3359

Sharma P, Nicoli E-R, Serra-Vinardell J, Morimoto M, Toro C, Malicdan MCV, Introne WJ (2020) Chediak-Higashi syndrome: a review of the past, present, and future. Drug Dis Today Dis Mod 31:31–36. https://doi.org/10.1016/j.ddmod.2019.10.008

Sherkat R, Sabri MR, Dehghan B, Bigdelian H, Reisi N, Afsharmoghadam N, Rahimi H, Rahmanian N, Klein C (2017) EBV lymphoproliferative-associated disease and primary cardiac T-cell lymphoma in a STK4 deficient patient: a case report. Medicine 96(48):e8852. https://doi.org/10.1097/MD.0000000000008852

Singer GG, Abbas AK (1994) The fas antigen is involved in peripheral but not thymic deletion of T lymphocytes in T cell receptor transgenic mice. Immunity 1(5):365–371. https://doi.org/10.1016/1074-7613(94)90067-1

Singh A, Joshi V, Jindal AK, Mathew B, Rawat A (2019) An updated review on activated PI3 kinase delta syndrome (APDS). Genes Diseases 7(1):67–74. https://doi.org/10.1016/j.gendis.2019.09.015

Skokowa J, Steinemann D, Katsman-Kuipers JE, Zeidler C, Klimenkova O, Klimiankou M, Unalan M et al (2014) Cooperativity of RUNX1 and CSF3R mutations in severe congenital neutropenia: a unique pathway in myeloid leukemogenesis. Blood 123(14):2229–2237. https://doi.org/10.1182/blood-2013-11-538025

Slatter MA, Gennery AR (2010) Primary immunodeficiencies associated with DNA-repair disorders. Expert Rev Mol Med 12(March). https://doi.org/10.1017/S1462399410001419

Sneller MC, Straus SE, Jaffe ES, Jaffe JS, Fleisher TA, Stetler-Stevenson M, Strober W (1992) A novel lymphoproliferative/autoimmune syndrome resembling murine Lpr/Gld disease. J Clin Invest 90(2):334–341. https://doi.org/10.1172/JCI115867

Sneller MC, Wang J, Dale JK, Strober W, Middelton LA, Choi Y, Fleisher TA et al (1997) Clincal, immunologic, and genetic features of an autoimmune lymphoproliferative syndrome associated with abnormal lymphocyte apoptosis. Blood 89(4):1341–1348

Sobacchi C, Marrella V, Rucci F, Vezzoni P, Villa A (2006) RAG-dependent primary immunodeficiencies. Hum Mutat 27(12):1174–1184. https://doi.org/10.1002/humu.20408

Sogkas G, Fedchenko M, Dhingra A, Jablonka A, Schmidt RE, Atschekzei F (2018) Primary immunodeficiency disorder caused by phosphoinositide 3-kinase δ deficiency. J Allergy Clin Immunol 142(5):1650–1653. https://doi.org/10.1016/j.jaci.2018.06.039

Somekh I, Marquardt B, Liu Y, Rohlfs M, Hollizeck S, Karakukcu M, Unal E et al (2018) Novel mutations in RASGRP1 are associated with immunodeficiency, immune dysregulation, and EBV-induced lymphoma. J Clin Immunol 38(6):699–710. https://doi.org/10.1007/s10875-018-0533-8

Somekh I, Thian M, Medgyesi D, Gülez N, Magg T, Duque AG, Stauber T et al (2019) CD137 deficiency causes immune dysregulation with predisposition to lymphomagenesis. Blood 134(18):1510–1516. https://doi.org/10.1182/blood.2019000644

Spinner MA, Sanchez LA, Hsu AP, Shaw PA, Zerbe CS, Calvo KR, Arthur DC et al (2014) GATA2 deficiency: a protean disorder of hematopoiesis, lymphatics, and immunity. Blood 123(6):809–821. https://doi.org/10.1182/blood-2013-07-515528

Stepensky P, Keller B, Buchta M, Kienzler A-K, Elpeleg O, Somech R, Cohen S et al (2013) Deficiency of caspase recruitment domain family, member 11 (CARD11), causes profound combined immunodeficiency in human subjects. J Allergy Clin Immunol 131(2):477–485. https://doi.org/10.1016/j.jaci.2012.11.050

Stepensky P, Weintraub M, Yanir A, Revel-Vilk S, Krux F, Huck K, Linka RM et al (2011) IL-2-inducible T-cell kinase deficiency: clinical presentation and therapeutic approach. Haematologica 96(3):472–476. https://doi.org/10.3324/haematol.2010.033910

Stray-Pedersen A, Jouanguy E, Crequer A, Bertuch AA, Brown BS, Jhangiani SN, Muzny DM et al (2014) Compound heterozygous CORO1A mutations in siblings with a mucocutaneous-immunodeficiency syndrome of epidermodysplasia verruciformis-HPV, molluscum contagiosum and granulomatous tuberculoid leprosy. J Clin Immunol 34(7):871–890. https://doi.org/10.1007/s10875-014-0074-8

Spegarova S, Jarmila DL, Mohamad SMB, Engelhardt KR, Doody G, Shrimpton J, Rensing-Ehl A et al (2020) Germline TET2 loss of function causes childhood immunodeficiency and lymphoma. Blood 136(9):1055–1066. https://doi.org/10.1182/blood.2020005844

Suarez F, Mahlaoui N, Canioni D, Andriamanga C, d'Enghien CD, Brousse N, Jais J-P, Fischer A, Hermine O, Stoppa-Lyonnet D (2015) Incidence, presentation, and prognosis of malignancies in ataxia-telangiectasia: a report from the French National Registry of primary immune deficiencies. J Clin Oncol Off J Am Soc Clin Oncol 33(2):202–208. https://doi.org/10.1200/JCO.2014.56.5101

Sullivan KE, Mullen CA, Blaese RM, Winkelstein JA (1994) A multiinstitutional survey of the Wiskott-Aldrich syndrome. J Pediatr 125(6 Pt 1):876–885. https://doi.org/10.1016/s0022-3476(05)82002-5

Sun L, Liu A, Georgopoulos K (1996) Zinc finger-mediated protein interactions modulate Ikaros activity, a molecular control of lymphocyte development. EMBO J 15(19):5358–5369

Sun S-C (2012) The noncanonical NF-KB pathway. Immunol Rev 246(1):125–140. https://doi.org/10.1111/j.1600-065X.2011.01088.x

Suri D, Rikhi R, Jindal AK, Rawat A, Sudhakar M, Vignesh P, Gupta A et al (2021) Wiskott Aldrich syndrome: a multi-institutional experience from India. Front Immunol 12. https://www.frontiersin.org/article/10.3389/fimmu.2021.627651

Symons M, Derry JMJ, Karlak B, Jiang S, Lemahieu V, McCormick F, Francke U, Abo A (1996) Wiskott–Aldrich syndrome protein, a novel effector for the GTPase CDC42Hs, is implicated in actin polymerization. Cell 84(5):723–734. https://doi.org/10.1016/S0092-8674(00)81050-8

Tang BL, Low DY, Hong W (1998) Syntaxin 11: a member of the syntaxin family without a carboxyl terminal transmembrane domain. Biochem Biophys Res Commun 245(2):627–632. https://doi.org/10.1006/bbrc.1998.8490

Tassone L, Notarangelo LD, Bonomi V, Savoldi G, Sensi A, Soresina A, Smith CIE et al (2009) Clinical and genetic diagnosis of warts, hypogammaglobulinemia, infections, and myelokathexis syndrome in 10 patients. J Allergy Clin Immunol 123(5):1170–1173. https://doi.org/10.1016/j.jaci.2008.12.1133

Taylor AM, Metcalfe JA, Thick J, Mak YF (1996) Leukemia and lymphoma in ataxia telangiectasia. Blood 87(2):423–438

Thaventhiran JED, Allen HL, Burren OS, Rae W, Greene D, Staples E, Zhang Z et al (2020) Whole-genome sequencing of a sporadic primary immunodeficiency cohort. Nature 583(7814):90–95. https://doi.org/10.1038/s41586-020-2265-1

Thouenon R, Moreno-Corona N, Poggi L, Durandy A, Kracker S (2021) Activated PI3Kinase delta syndrome—a multifaceted disease. Front Pediatr 9. https://www.frontiersin.org/article/10.3389/fped.2021.652405

Tien F-M, Hou H-A, Tsai C-H, Tang J-L, Chiu Y-C, Chen C-Y, Kuo Y-Y et al (2018) GATA2 zinc finger 1 mutations are associated with distinct clinico-biological features and outcomes different from GATA2 zinc finger 2 mutations in adult acute myeloid leukemia. Blood Cancer J 8(9):87. https://doi.org/10.1038/s41408-018-0123-2

Toita N, Hatano N, Ono S, Yamada M, Kobayashi R, Kobayashi I, Kawamura N et al (2007) Epstein-Barr virus-associated B-cell lymphoma in a patient with DNA ligase IV (LIG4) syndrome. Am J Med Genet Part A 143A(7):742–745. https://doi.org/10.1002/ajmg.a.31644

Toro C, Nicoli E-R, Malicdan MC, Adams DR, Introne WJ (2018) Chediak-Higashi syndrome. GeneReviews® [Internet]. University of Washington, Seattle. https://www.ncbi.nlm.nih.gov/books/NBK5188/

Torres JM, Martinez-Barricarte R, García-Gómez S, Mazariegos MS, Itan Y, Boisson B, Álvarez R et al (2014) Inherited BCL10 deficiency impairs hematopoietic and nonhematopoietic immunity. J Clin Invest 124(12):5239–5248. https://doi.org/10.1172/JCI77493

Toyoda H, Ido M, Hayashi T, Gabazza EC, Suzuki K, Jun B, Tanaka S et al (2004) Impairment of IL-12-dependent STAT4 nuclear translocation in a patient with recurrent mycobacterium avium infection. J Immunol 172(6):3905–3912. https://doi.org/10.4049/jimmunol.172.6.3905

Trimbath JD, Petersen GM, Erdman SH, Ferre M, Luce MC, Giardiello FM (2001) Café-Au-Lait spots and early onset colorectal neoplasia: a variant of HNPCC? Familial Cancer 1(2):103–108. https://doi.org/10.1023/A:1013881832014

Trück J, Kelly DF, Taylor JM, Kienzler AK, Lester T, Seller A, Pollard AJ, Patel SY (2016) Variable phenotype and discrete alterations of immune phenotypes in CTP synthase 1 deficiency: report of 2 siblings. J Allergy Clin Immunol 138(6):1722–1725. https://doi.org/10.1016/j.jaci.2016.04.059

Tsai F-Y, Keller G, Kuo FC, Weiss M, Chen J, Rosenblatt M, Alt FW, Orkin SH (1994) An early haematopoietic defect in mice lacking the transcription factor GATA-2. Nature 371(6494):221–226. https://doi.org/10.1038/371221a0

Tsilifis C, Freeman AF, Gennery AR (2021) STAT3 hyper-IgE syndrome-an update and unanswered questions. J Clin Immunol 41(5):864–880. https://doi.org/10.1007/s10875-021-01051-1

Tuijnenburg P, Allen HL, Burns SO, Greene D, Jansen MH, Staples E, Stephens J et al (2018) Loss-of-function nuclear factor KB subunit 1 (NFKB1) variants are the most common monogenic cause of common variable immunodeficiency in Europeans. J Allergy Clin Immunol 142(4):1285–1296. https://doi.org/10.1016/j.jaci.2018.01.039

Turvey SE, Durandy A, Fischer A, Fung S-Y, Geha RS, Gewies A, Giese T et al (2014) The CARD11-BCL10-MALT1 (CBM) signalosome complex: stepping into the limelight of human primary immunodeficiency. J Allergy Clin Immunol 134(2):276–284. https://doi.org/10.1016/j.jaci.2014.06.015

Unal S, Cerosaletti K, Uckan-Cetinkaya D, Cetin M, Gumruk F (2009) A novel mutation in a family with DNA ligase IV deficiency syndrome. Pediatr Blood Cancer 53(3):482–484. https://doi.org/10.1002/pbc.22031

Vakkilainen S, Taskinen M, Klemetti P, Pukkala E, Mäkitie O (2019) A 30-year prospective follow-up study reveals risk factors for early death in cartilage-hair hypoplasia. Front Immunol 10. https://www.frontiersin.org/article/10.3389/fimmu.2019.01581

Vallabhapurapu S, Karin M (2009) Regulation and function of NF-KappaB transcription factors in the immune system. Annu Rev Immunol 27:693–733. https://doi.org/10.1146/annurev.immunol.021908.132641

Varon, Raymonda, Ilja Demuth, and Krystyna H. Chrzanowska. 2017. Nijmegen Breakage Syndrome. GeneReviews® [Internet]. University of Washington, Seattle. https://www.ncbi.nlm.nih.gov/books/NBK1176/

Venkataraman G, McClain KL, Stefania P, Koneti Rao V, Jaffe ES (2010) Development of disseminated histiocytic sarcoma in a patient with autoimmune lymphoproliferative syndrome and associated Rosai-Dorfman disease. Am J Surg Pathol 34(4):589–594. https://doi.org/10.1097/PAS.0b013e3181d5ddf8

Vignesh P, Rawat A, Kumrah R, Singh A, Gummadi A, Sharma M, Kaur A et al (2020) Clinical, immunological, and molecular features of severe combined immune deficiency: a multi-institutional experience from India. Front Immunol 11:619146. https://doi.org/10.3389/fimmu.2020.619146

Villarino AV, Kanno Y, Ferdinand JR, O'Shea JJ (2015) Mechanisms of Jak/STAT signaling in immunity and disease. J Immunol (Baltimore, Md.: 1950) 194(1):21–27. https://doi.org/10.4049/jimmunol.1401867

Vinh DC, Patel SY, Uzel G, Anderson VL, Freeman AF, Olivier KN, Spalding C et al (2010) Autosomal dominant and sporadic monocytopenia with susceptibility to mycobacteria, fungi, papillomaviruses, and myelodysplasia. Blood 115(8):1519–1529. https://doi.org/10.1182/blood-2009-03-208629

Voce DJ, Schmitt AM, Uppal A, McNerney ME, Bernal GM, Cahill KE, Wahlstrom JS et al (2015) Nfkb1 is a haploinsufficient DNA damage-specific tumor suppressor. Oncogene 34(21):2807–2813. https://doi.org/10.1038/onc.2014.211

Volk T, Pannicke U, Reisli I, Bulashevska A, Ritter J, Björkman A, Alejandro A. Schäffer, et al (2015) DCLRE1C (ARTEMIS) mutations causing phenotypes ranging from atypical severe combined immunodeficiency to mere antibody deficiency. Hum Mol Genet 24(25):7361–7372. https://doi.org/10.1093/hmg/ddv437

Wang Z, Cui D, Weiquan L (2010) NBS1 8360G > C polymorphism is associated with breast cancer risk: a meta-analysis. Breast Cancer Res Treat 123(2):557–561. https://doi.org/10.1007/s10549-010-0772-8

Wang L, Aschenbrenner D, Zeng Z, Cao X, Mayr D, Mehta M, Capitani M et al (2021) Gain-of-function variants in SYK cause immune dysregulation and systemic inflammation in humans and mice. Nat Genet 53(4):500–510. https://doi.org/10.1038/s41588-021-00803-4

Watson AR, Evans DI, Marsden HB, Miller V, Rogers PA (1981) Purine nucleoside phosphorylase deficiency associated with a fatal lymphoproliferative disorder. Arch Dis Child 56(7):563–565. https://doi.org/10.1136/adc.56.7.563

Webster ADB, Barnes DE, Lindahl T, Arlett CF, Lehmann AR (1992) Growth retardation and immunodeficiency in a patient with mutations in the DNA ligase I gene. Lancet 339(8808):1508–1509. https://doi.org/10.1016/0140-6736(92)91266-B

West RR, Hsu AP, Holland SM, Cuellar-Rodriguez J, Hickstein DD (2014) Acquired ASXL1 mutations are common in patients with inherited GATA2 mutations and correlate with myeloid transformation. Haematologica 99(2):276–281. https://doi.org/10.3324/haematol.2013.090217

Whiteside D, McLeod R, Graham G, Steckley JL, Booth K, Somerville MJ, Andrew SE (2002) A homozygous germ-line mutation in the human MSH2 gene predisposes to hematological malignancy and multiple Café-Au-Lait spots. Cancer Res 62(2):359–362

Wiedemann HP, McGuire J, Dwyer JM, Sabetta J, Gee JB, Smith GJ, Loke J (1984) Progressive immune failure in dyskeratosis congenita. Report of an adult in whom Pneumocystis carinii and fatal disseminated candidiasis developed. Arch Intern Med 144(2):397–399

Wimmer K, Etzler J (2008) Constitutional mismatch repair-deficiency syndrome: have we so far seen only the tip of an iceberg? Hum Genet 124(2):105–122. https://doi.org/10.1007/s00439-008-0542-4

Winter S, Martin E, Boutboul D, Lenoir C, Boudjemaa S, Petit A, Picard C, Fischer A, Leverger G, Latour S (2018) Loss of RASGRP1 in humans impairs T-cell expansion leading to Epstein-Barr virus susceptibility. EMBO Mol Med 10(2):188–199. https://doi.org/10.15252/emmm.201708292

Wlodarski MW, Collin M, Horwitz MS (2017) GATA2 deficiency and related myeloid neoplasms. Semin Hematol 54(2):81–86. https://doi.org/10.1053/j.seminhematol.2017.05.002

Woellner C, Gertz EM, Schäffer AA, Lagos M, Perro M, Glocker E-O, Pietrogrande MC et al (2010) Mutations in STAT3 and diagnostic guidelines for hyper-IgE syndrome. J Allergy Clin Immunol 125(2):424–432. https://doi.org/10.1016/j.jaci.2009.10.059

Woods N-B, Bottero V, Schmidt M, von Kalle C, Verma IM (2006) Therapeutic gene causing lymphoma. Nature 440(7088):1123–1123. https://doi.org/10.1038/4401123a

Xia M, David L, Teater M, Onder O, Elenitoba-Johnson KSJ, Fontan L, Wu H, Melnick A (2020) BCL10 gain-of-function mutations aberrantly induce canonical and non-canonical NF-kb activation and resistance to Ibrutinib in ABC-DLBCL. Blood 136(November):2–3. https://doi.org/10.1182/blood-2020-141363

Yamashita M, Kuehn HS, Okuyama K, Okada S, Inoue Y, Mitsuiki N, Imai K et al (2021) A variant in human AIOLOS impairs adaptive immunity by interfering with IKAROS. Nat Immunol 22(7):893–903. https://doi.org/10.1038/s41590-021-00951-z

Yoshida N, Sakaguchi H, Muramatsu H, Okuno Y, Song C, Dovat S, Shimada A et al (2017) Germline IKAROS mutation associated with primary immunodeficiency that progressed to T-cell acute lymphoblastic leukemia. Leukemia 31(5):1221–1223. https://doi.org/10.1038/leu.2017.25

Yoshimi A, Kamachi Y, Imai K, Watanabe N, Nakadate H, Kanazawa T, Ozono S et al (2013) Wiskott-Aldrich syndrome presenting with a clinical picture mimicking juvenile myelomonocytic leukaemia. Pediatr Blood Cancer 60(5):836–841. https://doi.org/10.1002/pbc.24359

Zhang J, Shehabeldin A, da Cruz LA, Butler J, Somani AK, McGavin M, Kozieradzki I et al (1999) Antigen receptor-induced activation and cytoskeletal rearrangement are impaired in Wiskott-Aldrich syndrome protein-deficient lymphocytes. J Exp Med 190(9):1329–1342. https://doi.org/10.1084/jem.190.9.1329

Zhang Q, Lenardo MJ, Baltimore D (2017) 30 years of NF-KB: A blossoming of relevance to human pathobiology. Cell 168(1–2):37–57. https://doi.org/10.1016/j.cell.2016.12.012

Zhang S-J, Ma L-Y, Huang Q-H, Guo L, Bai-Wei G, Gao X-D, Shi J-Y et al (2008) Gain-of-function mutation of GATA-2 in acute myeloid transformation of chronic myeloid leukemia. Proc Natl Acad Sci U S A 105(6):2076–2081. https://doi.org/10.1073/pnas.0711824105

Zhang Y, Ji B, Li J, Li Y, Zhang M, Ban B (2020) SHORT syndrome in two Chinese girls: a case report and review of the literature. Molecular Genetics & Genomic Medicine 8(9):e1385. https://doi.org/10.1002/mgg3.1385

Zhang Y, Yu X, Ichikawa M, Lyons JJ, Datta S, Lamborn IT, Jing H et al (2014) Autosomal recessive phosphoglucomutase 3 (PGM3) mutations link glycosylation defects to atopy, immune deficiency, autoimmunity, and neurocognitive impairment. J Allergy Clin Immunol 133(5):1400–1409. https://doi.org/10.1016/j.jaci.2014.02.013

Zhao B, Li L, Lei Q, Guan K-L (2010) The hippo-YAP pathway in organ size control and tumorigenesis: an updated version. Genes Dev 24(9):862–874. https://doi.org/10.1101/gad.1909210

Zhou H, Han Z, Chen L, Liu L, Xu X, Zou P (2010) The interferon gamma Receptor-1 (IFNGR1) – 56C/T gene polymorphism is associated with increased risk of chronic lymphocytic leukemia. Blood 116(21):4617. https://doi.org/10.1182/blood.V116.21.4617.4617

Pathogenesis and Consequences of a Disruption to the Intestinal Barrier Functions in Patients with Hematologic Malignancies

Elio Haroun, Seah H. Lim, and Dibyendu Dutta

Abstract

The intestinal barrier is a complex structure that plays a pivotal role in the regulation of the influx of luminal contents into the systemic circulation and maintains homeostasis of immune cells, microbes, and metabolites. Disruption in the intestinal barrier functions has been implicated in the development of many systemic diseases such as nonalcoholic steatohepatitis, inflammatory bowel disease, multiple sclerosis, atherosclerosis, type 2 diabetes mellitus, and cancer. Therefore, the intestinal barrier presents a potential target for therapeutic intervention to alter the course and mitigate complications associated with these diseases. Patients with hematologic malignancies are particularly at high risk for developing a disruption to the intestinal barrier functions due to the disease processes and to therapeutic interventions, either to treat the underlying disease or to support the patient following treatment. In this chapter, we not only discuss the pathogenesis and nature of the intestinal barrier function disruption in patients with hematologic malignancies but also examine the associated complications. We explore methods available to modulate the complications and their associated challenges.

Keywords

Antibiotics · Antineoplatic agents · Dysbiosis · Paracellular pathway · Tight junctions

E. Haroun · S. H. Lim · D. Dutta (✉)
Division of Hematology and Oncology, Department of Medicine, SUNY Upstate Medical University, Syracuse, NY, USA
e-mail: duttad@upstate.edu

© The Author(s), under exclusive license to Springer Nature Switzerland AG 2024
Interdisciplinary Cancer Research, https://doi.org/10.1007/16833_2024_253
Published online: 16 May 2024

1 Introduction

In addition to regulating metabolism and maintaining microbial homeostasis and immune responses, the gastrointestinal (GI) mucosal epithelium forms an important physical and functional barrier separating the host internal milieu from potentially harmful and toxic entities. Although epithelial cells are linked together by the apical junction complex to form a barrier, this seal is not completely impermeable, and it allows for selective transport of luminal content into the systemic circulation. The dynamics of the barrier function is multifactorial where tight junction components, mucosal immune system, and microbial products and metabolites communicate to create an effective intestinal barrier. This physiologic state may be perturbed by intestinal dysbiosis, disease, or medications. A disruption in the barrier functions may occur due to mucosal injury, changes in cellular molecules, or dysregulated immune responses. Such disruptive changes in the mucosal barrier functions are associated with different systemic diseases.

The relationship between the disruption of the intestinal barrier functions in various disease states and disease development is bidirectional. While the origin of a disease may trigger increased intestinal permeability, changes in the GI tract due to either altered physiology or treatment of a disease may also induce a breach in the intestinal barrier functions. There is a growing body of literature demonstrating this bidirectionality in multiple diseases such as inflammatory bowel disease (Michielan and D'Incà 2015), cardiovascular disease (Masenga et al. 2022), celiac disease (Pizzuti et al. 2004), nonalcoholic steatohepatitis (Plaza-Díaz et al. 2020), diabetes mellitus (Yuan et al. 2021), and autism (Fiorentino et al. 2016).

Patients with hematologic malignancies are susceptible to disruption in their intestinal barrier functions. The complexity of treating patients with hematologic malignancies not only involves side effects of the antineoplastic therapy and supportive care given to the patients following treatment for the underlying diseases but also the associated bone marrow failure syndrome in patients with acute leukemia. Furthermore, treatment-induced impairment to the intestinal barrier functions exposes patients to a range of complications such as infections, nausea, fatigue, and culture-negative fever. Since the proper maintenance of the intestinal barrier functions is important for health, it is not surprising that its disruption may be responsible for some of the morbidities and mortalities in these patients. In the chapter, we discuss the pathogenesis of a disruption in the intestinal barrier functions in relation to the structure of the intestinal barrier, clinical consequences of this complication in relation to the functions of the barrier, and methods to prevent/alleviate the complications based on current understanding of the intestinal barrier functions.

2 Antineoplastic Therapy-Induced Disruption to the Intestinal Barrier

The intestinal epithelial cells form the most superficial layer of the multilayer intestinal barrier (Fig. 1). They prevent the traverse of intraluminal microbes and regulate the transport of ingested substances into the systemic circulation. These

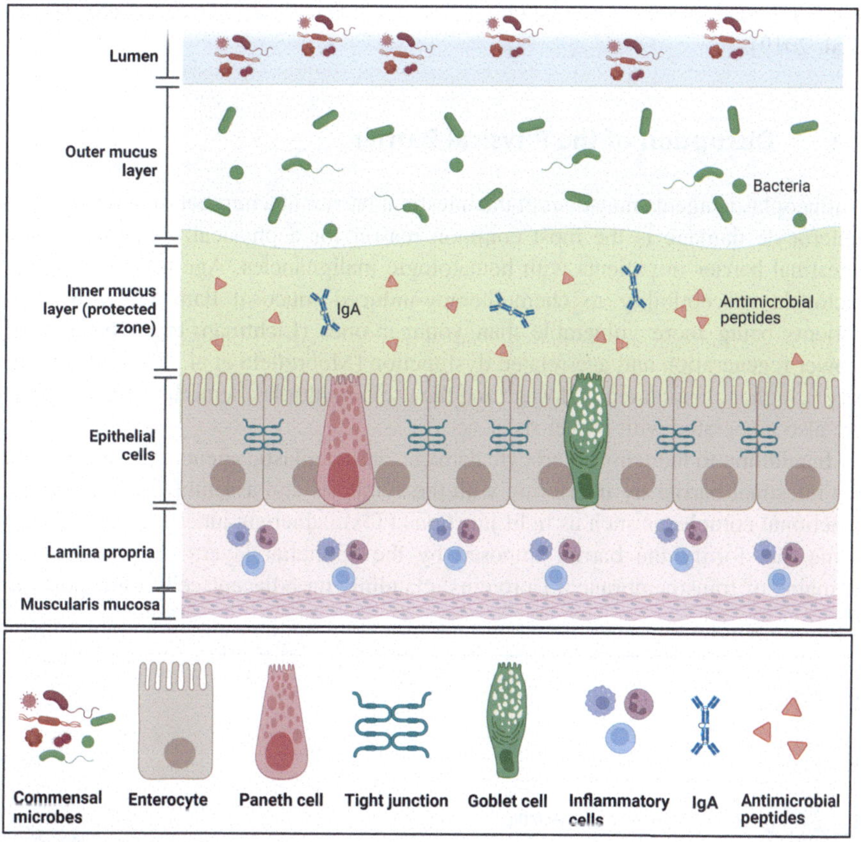

Fig. 1 The interacting layers of the intestinal barrier. The luminal layer is made up of commensal organisms that help prevent the colonization of pathogenic microbes. The mucus layer acts as a barrier to keep intestinal microorganisms from contacting the mucosal epithelium. This mucus layer serves not only as a physical barrier but also contains antimicrobial peptides and IgA, enhancing protection of the mucosal epithelium. Finally, the mucosal epithelium itself forms a selectively permeable barrier to the contents of the lumen

cells have a high cell turnover, with a lifespan of 3–5 days. It is, therefore, not surprising that enterocyte damage occurs frequently after antineoplastic therapy since most antineoplastic agents target cells with high turnover by interfering with DNA replication and cell growth. These antineoplastic agents often induce intestinal barrier injury and increased permeability (Melichar and Zezulová 2011), dysbiosis (Montassier et al. 2015), and neutropenia (Crawford et al. 2004). The clinical consequence of mucosal damage is manifested in the form of mucositis, which is often associated with diarrhea, inflammation, ulceration, nausea, malnutrition, and sepsis. Since intestinal fatty acid binding protein (iFABP) is an intracellular protein in the enterocytes, mucosal injury can be determined by measuring serum iFABP that is released into the systemic circulation when enterocyte damage occurs (Dutta et al. 2019).

2.1 Disruption of the Physical Barrier

Antineoplastic agents may disrupt the intestinal barrier in a number of ways (Fig. 2). Enterocyte damage is the most common reason for a physical disruption to the intestinal barrier in patients with hematologic malignancies. Age is a determining factor in susceptibility to chemotherapy-induced mucosal damage, with older patients being more vulnerable than younger ones (Lichtman 1995) because of slower regeneration and age-related dysfunction (Moorefield et al. 2017). Additionally, a higher chemotherapy dosage and the use of multiple chemotherapeutic agents are also associated with higher risks.

In addition to inducing enterocyte damage, antineoplastic agents may also disrupt the intestinal barrier by interfering with the synthesis and assembly of intercellular junctional complexes such as tight junctions (TJs), adherens junctions, and desmosome that fortify the barrier imposed by the epithelial layer. The extracellular domains of transmembrane TJ proteins, claudins, on adjacent cells form pores to

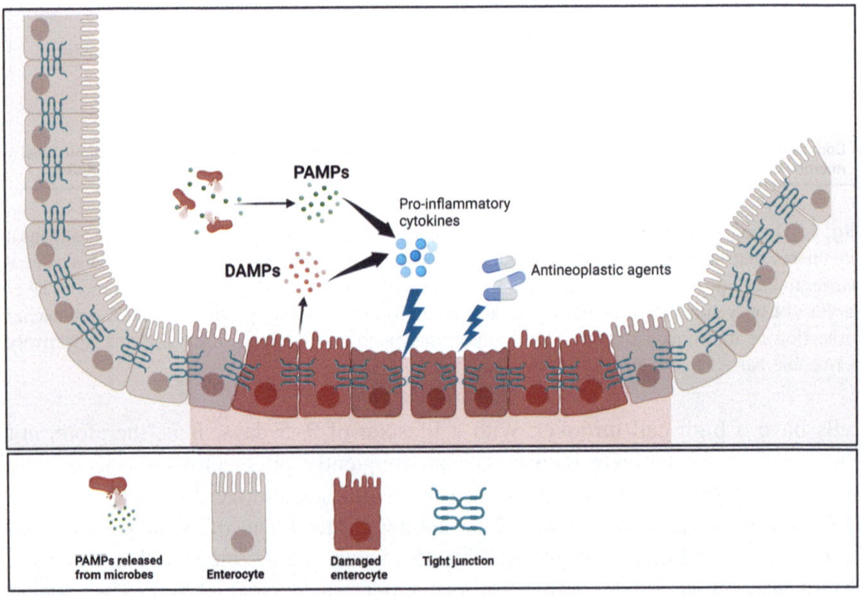

Fig. 2 Antineoplastic agents disrupt the intestinal barrier by causing enterocyte damage and interfering with the synthesis and assembly of intercellular junctional complexes. This disruption is further exacerbated by pro-inflammatory cytokines, which are released in response to endogenous damage-associated molecular patterns (DAMPs) from chemotherapy-induced enterocyte injury, as well as pathogen-associated molecular patterns (PAMPs) from intestinal microbes. Besides inducing direct cellular injury, these pro-inflammatory cytokines also alter tight junction (TJ) function

regulate the pore pathway during paracellular transport. Other integral TJ proteins including the tight junction–associated MARVEL protein (TAMP) family members, such as occludin and tricellulin (Raleigh et al. 2010), and periplasmic scaffolding proteins, such as the zonula occludens (ZO) protein family, regulate and maintain the leak pathway (France and Turner 2017). Oxaliplatin, a platinum-containing alkylating agent used either in combination with gemcitabine (GemOx) or with rituximab and gemcitabine (R-GemOx), for treating lymphoma (Ozdemir et al. 2015; Shen et al. 2018), disrupts TJs by causing dysbiosis, inducing morphological changes in $TLR4^{+}$ cells and increasing expression of immunogenic damage-associated molecular patterns (DAMPs) (Stojanovska et al. 2018). Methotrexate (MTX), a folic acid antagonist, commonly used in the treatment of acute leukemia and lymphoma compromises the integrity of TJs, by altering the localization and cellular expression of TJ protein. MTX treatment results in the reduction in occludin and claudin-1 proteins without changes in gene expression levels. The effects of MTX on intestinal barrier functions are driven by the mitogen-activated protein kinase (MAPK) and nuclear factor kappa B (NF-κB) pathways (Beutheu Youmba et al. 2012).

The intestinal epithelial cells are sensitive to injury, and hence disruption of the intestinal barrier, by pro-inflammatory cytokines such as TNFα, interferon-γ (IFN γ), interleukin 1 (IL-1), and IL-6 (Ma et al. 2004; Logan et al. 2007). Pro-inflammatory cytokines are released in response to Toll-like receptor (TLR) 4 activation by endogenous damage-associated molecular patterns (DAMPs) released from chemotherapy-induced enterocyte injury and pathogen-associated molecular patterns (PAMPs) from intestinal microbes. In addition to inducing direct cellular injury, TNFα also alters TJ function (Logan et al. 2007), by reducing the expression and disrupting the distribution of ZO-1, occludin, claudin-4, and claudin-5, intimately involved in maintaining the integrity of the barrier (M Logan et al. 2008). IL-1 and IL-6 also cause changes in TJs (Williams 2001; Poritz et al. 2004); adding IL-1 to cell cultures led to a decrease in occludin expression (Ma et al. 2005), while IL-6 treatment caused changes in the cytoskeleton, increasing permeability through ZO-1 redistribution, actin structure remodeling, and increased actin contractility (Marcus et al. 1996).

Other treatment-related complications associated with cytokine-induced disruption to the intestinal barrier include the cytokine release syndrome (CRS) following the administration of chimeric antigen receptor (CAR T) cell therapy and cytokine storms that occur in patients receiving allogeneic hematopoietic stem cell transplant (allo-HSCT) utilizing posttransplant high-dose cyclophosphamide as a strategy for prophylaxis of graft-versus-host disease (GVHD). CAR T therapy is currently approved by the US Food and Drug Administration (FDA) for treating multiple hematologic malignancies including B cell acute lymphoblastic leukemia (Sheykhhasan et al. 2022), B cell lymphomas (Chavez and Locke 2018), and multiple myeloma (Rendo et al. 2022). CRS occurs commonly after the infusions of the CAR T. In patients who are treated with posttransplant high dose cyclophosphamide following allo-HSCT (Luznik et al. 2010), high levels of cytokines are released during the first 72 h, especially in peripheral blood stem cell recipients and haploidentical transplants.

The high levels of circulating pro-inflammatory cytokines may induce enterocyte damage and disruption to the intestinal barrier.

2.2 Increased Epithelial Permeability

Injury to the DNA in crypt stem cells due to chemoradiation causes the cells to go into apoptosis. While the primary assault is targeted toward stem cells, chemotherapy does not spare enterocytes, immune cells, and mesenchymal cells. The degree of apoptosis and the local cryptal variations differ depending on doses, administration routes, and type of chemotherapeutics (Ijiri and Potten 1983). The resulting effect involves mucosal inflammation, ulceration, disruption of TJs, and an increased permeability that leads to the translocation of luminal microbial components into the lamina propria.

Increased permeability is detected in the clinics by analyzing the ratios of inert probes of various molecular masses such as lactulose:mannitol (L:M) or lactulose: rhamnose (L:R) and recovery of ^{51}Cr-EDTA following oral dosing. For example, polyethylene glycol (PEG) 400 g/mol excretion in urine increased 88% in MTX (20 mg/m^2)-treated acute lymphoblastic leukemia (ALL) patients (Lifschitz and Mahoney 1989). Similarly, ^{51}Cr-EDTA excretion increased fourfold in lymphoma patients after melphalan (200 mg/m^2) treatment (Selby et al. 1987). Other serological surrogate markers of intestinal permeability are zonulin, an integral part of intercellular TJs (Fasano et al. 2000), citrulline, an amino acid produced by small intestine enterocytes (Crenn et al. 2008) that reflects the loss of small intestine mass and lipopolysaccharide (LPS) that denotes systemic inflammation and endotoxemia due to compromised TJs (Munford 2016). However, none of these methods can effectively differentiate between the paracellular pathways involved in increased permeability.

The paracellular pathway for transporting luminal content across the epithelium can be categorized into the pore pathway, the leak pathway, and the unrestricted pathway (Zuo et al. 2020) (Fig. 3). The pore pathway is a high-conductance route that is charge- and size-selective (upper limit of 6 Å to 8 Å diameter) and is influenced by physiological factors like Na^+-glucose cotransport and claudin 2 expression. The leak pathway allows the movement of larger molecules and is regulated by tricellular TJs and proteins like occludin and ZO-1, with modulations by myosin light-chain kinase (MLCK) expression and immune mediators like TNF and IL-1β (Buschmann et al. 2013). The unrestricted pathway is independent of TJs where epithelial cell damage associated with pathological states result in flux of microbes and microbial content across the epithelium. The differentiation of the type of paracellular pathway can be performed by quantifying the ratios of creatinine (6 Å diameter), fluorescein isothiocyanate (FITC)-labeled dextran 4 kDa (28 Å diameter), and rhodamine-dextran 70 kDa (120 Å diameter) in serum or plasma (Chanez-Paredes et al. 2021). Creatinine can travel through all three paracellular pathways, 4 kDa dextran can traverse through both the leak and unrestricted pathways, and the passage of 70 kDa dextran is restricted only to the unrestricted

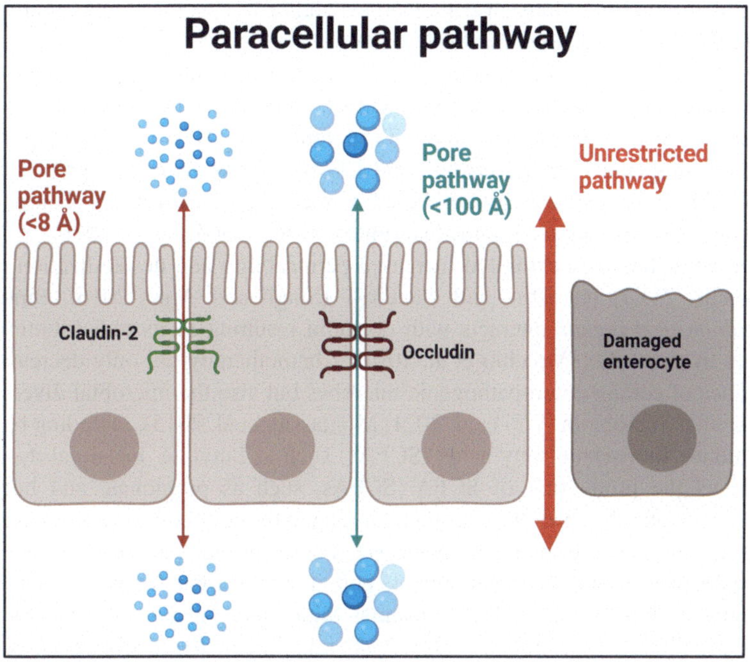

Fig. 3 Mechanisms of intestinal barrier functions. Movement of intraluminal content across tight junctions via two distinct routes: the pore pathway and the leak pathway. The pore pathway is characterized by its high capacity and charge selectivity, allowing the passage of small molecules with a diameter of less than 8 Å. In contrast, the leak pathway, which is low in capacity and not charge selective, permits the transit of molecules up to 100 Å in diameter. In cases of epithelial damage, an alternate route comes into play. This tight junction-independent pathway, which is neither size selective nor charge selective, becomes the primary means of increased intestinal permeability

pathway. Thus, a concomitant increase in 4 kDa dextran and 70 kDa dextran flux can be due to enhanced unrestricted pathway permeability, and an increase in 4 kDa dextran flux without accompanying changes in 70 kDa dextran flux may represent leak pathway. While this method is reproducible in animal models, it has not been adapted for clinical evaluation.

2.3 Disruption in Intestinal Microbial Homeostasis

Commensal intestinal microbiota that resides in the most luminal layer of the intestinal barrier, along with mucin produced by Goblet cells and antimicrobial peptides (AMP) by Paneth cells that occupy the deepest layer of the barrier, provide a functional barrier preventing the entry of pathogenic microbes into the systemic circulation. It, therefore, follows that any disruption to the balanced intestinal microbial community, either in compositions and/or in densities, may compromise the integrity of the intestinal barrier functions.

In addition to inducing direct enterocyte damage, some antineoplastic agents used to treat hematologic malignancies may induce changes in the intestinal microbial composition that are pathogenic to the TJ proteins (Fig. 4). An increased presence of enteropathogenic *Escherichia coli* and *Clostridium perfringens*, both of which disrupt TJ proteins and increase epithelial permeability, has been noted in patients with chemotherapy-induced intestinal injury (Singh and Aijaz 2016). *E. coli*-infected epithelial cells exhibited increased permeability and destabilization of the TJ proteins ZO-1, occludin, and claudin-1 (Singh and Aijaz 2016). Similarly, *Clostridium perfringens* interacts with occludin resulting in its redistribution and removal from the TJs (Vecchio et al. 2021). Chemotherapy not only decreases the abundance of certain enteropathogenic microbes but also the microbial diversity of the intestinal microbiota (Wei et al. 2021; Montassier et al. 2015), including bacteria that produce short-chain fatty acids (SCFAs) such as butyrate. Intestinal dysbiosis may affect the production of SCFA. SCFAs, such as propionate and butyrate, produced by microbes prevent microbial binding to the epithelium and help maintain barrier function and immune homeostasis. For example, butyrate is crucial for enterocyte health and TJ development, and it was shown to have potent anti-inflammatory effects that reduce intestinal injury and preserve the mucus layer (Louis and Flint 2017; Sun et al. 2021; Yan and Ajuwon 2017; Silva et al. 2019).

Antineoplastic agents may also exhibit direct effects on both the Goblet cells and the Paneth cells. Consequently, there is a thinning of the mucus layer to increase the opportunity for the luminal microbes to come in close proximity to the intestinal epithelium and a decrease in the production of antimicrobial peptides, with the results of a decreased bacterial diversity and domination of pathogenic species such as Proteobacteria, Enterobacteriales, *Escherichia* spp., and *Bacteroides* spp. (Eriguchi et al. 2012; Hayase et al. 2017) (Fig. 4). Dysbiosis causes a reduction in SCFAs including butyrate that is essential for the formation of TJs (Peng et al. 2009; Wang et al. 2012) and activation of peroxisome proliferator-activated receptor gamma (PPAR-γ). PPAR-γ enhances mucosal epithelial oxygen consumption that reduces the emanation of oxygen from the epithelial surface and helps maintain an anaerobic condition in the lumen required by obligate anaerobes to colonize (Litvak et al. 2018). A shift in epithelial oxygenation due to dysbiosis promotes a switch from obligate to facultative anaerobe proliferation that uses oxygen (Litvak et al. 2018) and increases susceptibility to infections. Histone deacetylases (HDACs) are inhibited by SCFAs that suppress NF$\kappa\beta$ activity and pro-inflammatory innate immune responses by downregulating tumor necrosis factor (TNF) (Vinolo et al. 2011; Usami et al. 2008). Many antimicrobials such as paromomycin (Heinsen et al. 2015) and clindamycin (Zaura et al. 2015) diminish butyrate producers and thus generate an environment that is conducive to the translocation of luminal bacteria into the central circulation. This promotes inflammation and increase the risk of severe infection.

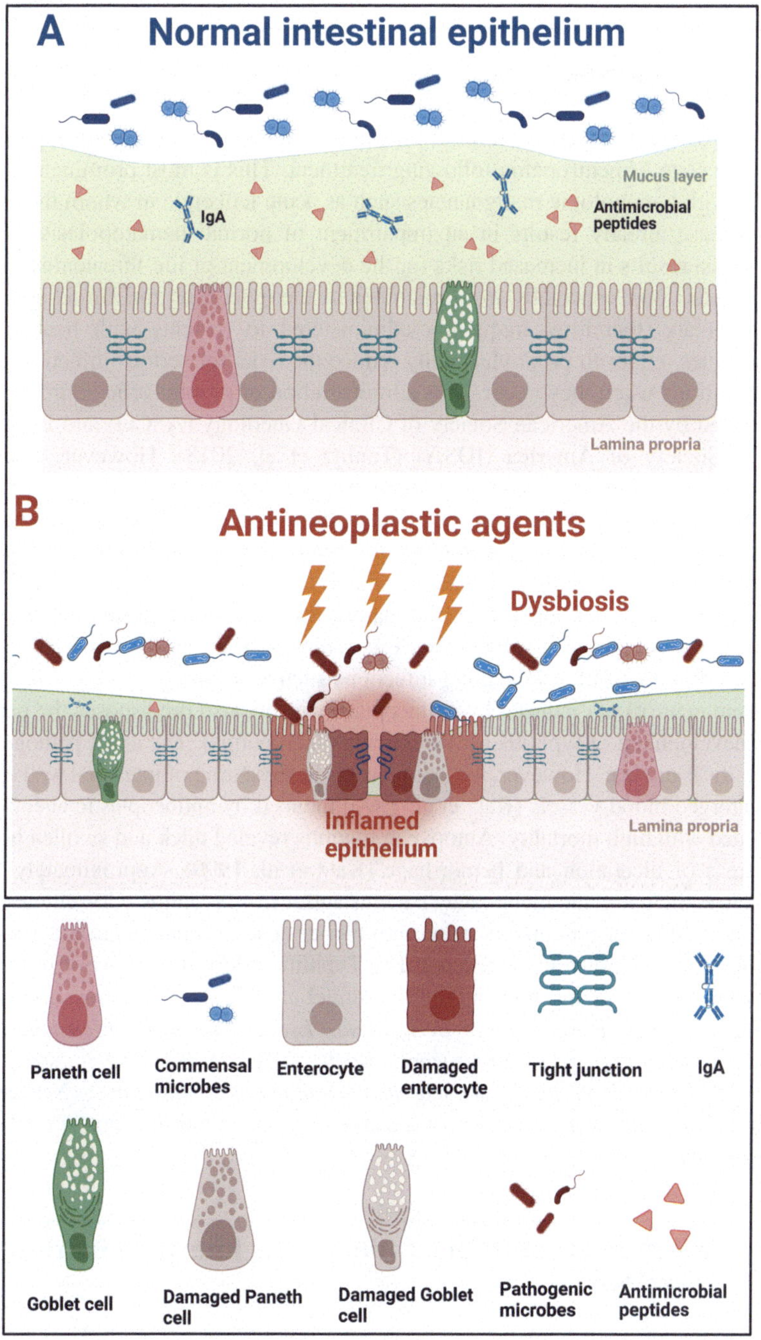

Fig. 4 (**a**) Normally functioning intestinal epithelium; (**b**) impact of antineoplastic agents on intestinal flora: Antineoplastic agents induce dysbiosis, altering the composition of intestinal microbes in a way that negatively affects tight junction (TJ) proteins. Additionally, these agents may directly impact goblet and Paneth cells. This leads to the thinning of the mucus layer, increasing the likelihood of luminal microbes approaching the intestinal epithelium closely. Furthermore, there is a reduction in the production of antimicrobial peptides and IgA. As a result, there is a decrease in bacterial diversity and an increase in the prevalence of pathogenic species

2.4 Neutropenia

By virtue of their properties, most antineoplastic agents induce myelosuppression, creating a state of neutropenia following treatment. This is most prominent among patients with hematologic malignancies such as acute leukemia in whom the underlying disease already results in an impairment of normal hematopoiesis. Severe neutropenia results in increased risks for the development of life-threatening opportunistic infections, especially from microbes of the enteric origin. Broad-spectrum antibiotics are, therefore, frequently administered to patients with hematologic malignancies, either to prophylactically reduce the risks for serious infections or to treat infections when they occur. This administration of prophylactic antimicrobials is endorsed by the American Society of Clinical Oncology (ASCO) and Infectious Disease Society of America (IDSA) (Taplitz et al. 2018). However, although antimicrobials reduce the risks for infections, they adversely affect the intestinal microbial communities. They alter the intestinal flora and diminish the diversity of the microbiota, paving the way for opportunistic microbes to proliferate (Fig. 4), increase barrier permeability, and alter immune function. Traversing luminal bacteria due to increased permeability further damages the intestinal barrier and amplifies a vicious cycle that exacerbates bacterial translocation and inflammation that increases the risk of life-threatening infections such as sepsis.

Severe neutropenia induced by antineoplastic agents also predisposes the patients to the development of typhlitis (or necrotizing enterocolitis). The exact pathogenesis of typhlitis is unclear. However, it usually occurs in the setting of intestinal wall edema and engorged blood vessels (Katz et al. 1990) induced by antineoplastic agents. It is associated with high mortality. Autopsy in typhlitis reveled thick and swollen bowels with areas of ulceration and hemorrhage (Katz et al. 1990). Approximately 5.3% hospitalized hematologic malignancy patients suffer from typhlitis (Gorschlüter et al. 2005). The risks for typhlitis increase with higher dose of antineoplastic agents that cause mucositis (Nesher and Rolston 2013). Typhlitis is linked to high risk of enteric-originated septicemia and high mortality (Altınel et al. 2012). The most commonly isolated microbes *Pseudomonas aeruginosa*, *Escherichia coli*, *Klebsiella* spp., *viridans* group *streptococci*, *Enterococci*, *Bacteroides* spp., *Clostridium* spp., *Candida* spp., *Clostridium septicum*, and *Stenotrophomonas maltophilia* (Nesher and Rolston 2013) confirm the role of enteric microbes' translocation in typhlitis pathogenesis.

3 Infection-Induced Disruption in the Intestinal Barrier Functions

Opportunistic infections occur commonly among patients with hematologic malignancies. Multiple factors contribute to this risk, and they include the development of severe neutropenia, extended use of prophylactic antibiotics, prolonged hospital stays, chemotherapy and immunosuppressive treatment for HSCT, and transplant-associated complications like GVHD. *Clostridioides difficile* infection

(CDI) and *Cytomegalovirus* (CMV) colitis, particularly after allo-HSCT, are not uncommon. Neutropenia, along with dysbiosis due to chemotherapy and the use of antimicrobials increase the risk of CDI. Dysbiosis favors the overgrowth of *C. difficile.* Moreover, antibiotic-induced dysbiosis promoted the colonization of less prominent mucin-degrading microbes such as *Ruminoccocus* and *Bacteroides* (Engevik et al. 2021). These bacteria create a gradient of mucin monosaccharides that chemoattracts *C. difficile* to the mucus layer. Once at the intestinal mucus layer, *C. difficile* degrades mucin using cell surface protein Cwp84 and facilitates adhesion to the epithelium by surface layer protein A (SlpA) (Smits et al. 2016). Upon colonization, *C. difficile* produce pro-inflammatory and cytoskeleton disorganizing enterotoxin A and cytotoxin B that damage enterocytes by adding single sugar molecules to proteins (Voth and Ballard 2005). This leads to the formation of actin filaments that disrupt F-actin at the luminal and basal ends of the cell and trigger the detachment of occludin, ZO-1, and ZO-2 from the lateral TJ membranes. These events trigger a disruption to the intestinal barrier functions. Between 7 and 14% of patients with hematologic malignancies experience CDI (McGlone et al. 2012), and 11–31% suffer from recurrent CDI (Willems et al. 2012; Kamboj et al. 2014). This includes patients with acute leukemia, multiple myeloma, Hodgkin's disease, and non-Hodgkin's lymphoma.

In addition to CDI, CMV colitis is common in immunosuppressed patients. CMV is generally latent in immunocompetent adults (Rafailidis et al. 2008). However, dysbiosis, inflammation, and injury of the gut due to hematologic malignancy treatment can reactivate latent CMV. Reactivation of CMV and colitis can occur due to the use of high-dose steroids, fludarabine, alemtuzumab (anti-CD52 monoclonal antibody), bortezomib (proteasome inhibitor), and rituximab (anti-CD20 monoclonal antibody) (Marchesi et al. 2018). All these treatments can cause dysbiosis. Acute primary CMV infection is associated with an altered ratio of *Firmicutes* to *Bacteroidetes* (Le-Trilling et al. 2023). With increased dysbiosis, there was crypt hyperplasia and an increase in colonic pro-inflammatory cytokine levels. Moreover, intestinal epithelial cells and crypt cells are direct targets of CMV causing increased cell death. CMV leads to a decrease in the transepithelial electrical resistance and an increase in intestinal epithelial cells' permeability by apoptosis of enterocytes and damage to the TJs via an IL-6-mediated pathway (Maidji et al. 2017). With increased apoptosis of intestinal epithelial cells and disruption of the TJs, the barrier integrity is compromised. This leads to increased translocation of bacteria across the intestinal barrier and a higher likelihood of secondary bacteremia.

4 Disruption in the Intestinal Barrier Functions due to GVHD Following allo-HSCT

One of the most formidable complications arising from allo-HSCT is GVHD. In acute GVHD, allo-reactive donor T cells induce damage in host tissues, especially the skin, intestines, and liver, and elicit acute inflammatory reactions. Intestinal barrier damage associated with GVHD has two distinct phases, initiation and propagation. The initiation of the damage is attributed to high-dose chemoradiation conditioning treatment for allo-HSCT. The luminal microbiota contributes to the pathogenesis of GVHD (Heidt and Vossen 1992) that has been demonstrated in animal models where LPS-induced TLR4 signaling contributed to GVHD following major histocompatibility complex (MHC) mismatch HSCT (Cooke et al. 1998, 2001) and the severity of GVHD correlated with the level of TNF production triggered by LPS induction (Cooke et al. 1998). An altered pattern recognition receptor (PRR) signaling and toll-like receptors (TLRs) pathways have been implicated in GVHD (van der Velden et al. 2009). Increased damage to the intestinal epithelium allows microbiota-derived pathogen-associated-molecular patterns (PAMPs) such as LPS to interact with epithelium-derived PRRs and activate TLR 2/4-Myeloid differentiation primary response protein 88 (MYD88) (Kinashi and Hase 2021). PRRs also interact with damage-associated molecular patterns (DAMPs), such as extracellular adenosine triphosphate (ATP), cytoplasmic calreticulin, high mobility group box 1 (HMGB1) proteins, and endogenous nucleic acids that are released from injured intestinal epithelium (Dutta and Lim 2020) to activate macrophages and promote the maturation of dendritic cells, leading to inflammation and disruption of epithelial TJ barriers. Disruption of the barrier creates a pro-inflammatory environment in the host that allows systemic exposure of luminal microbial components. It not only increases the risk of infection but also activates donor cell TLR4 and MyD88 to further exacerbate and induce systemic GVHD.

The propagation phase of acute GVHD is dependent on the magnitude of the damage to the mucosal barrier that causes effector T cell activation and infiltration. Acute GVHD severity is associated with sustained intestinal barrier loss associated with mucus layer thinning and the presence of bacterial RNA in the colon lamina propria (Ara et al. 2020). It has been observed that the magnitude of intestinal barrier loss is correlated with disease severity (Nalle et al. 2019). In acute GVHD, there is epithelial apoptosis within intestinal crypts and loss of the number of intestinal stem cells (ISCs) per crypt (Jansen et al. 2022). Severe colonic crypt loss is associated with delayed recovery, persistence of symptoms, and development of more serious steroid-refractory GVHD. The extent of intestinal permeability and progression of the disease is correlated with a reduction in antimicrobial peptides in the colon and an increase in plasma lipopolysaccharide-binding protein (LBP) (Ye et al. 2012). By using the ^{51}Cr-EDTA absorption test, it was found that patients with mild acute GVHD had relatively improved intestinal barrier function and reduced gastrointestinal toxicity compared to patients with severe acute GVHD and the magnitude of intestinal barrier disruption is a significant predictor of acute GVHD severity (Tyszka et al. 2021). In mice, altered myosin light chain kinase 210 (MLCK210) expression significantly induced intestinal damage and GVHD progression (Nalle et al. 2019). MLCK210 regulates the skeleton of intestinal TJs, endothelial retraction, and apoptosis. MLCK210-dependent changes in intestinal TJ permeability triggered cytolytic effector CD8$^+$ T cell expansion and infiltration in the target organs. MLCK activation phosphorylates perijunctional MLC and occludin endocytosis (Horowitz et al. 2023). Occludin endocytosis also occurs during TNF-induced barrier loss. Mice lacking MLCK210, while still susceptible to GVHD in MHC mismatched transplant scenarios, exhibit protection against prolonged loss of intestinal barrier function and severe GVHD as evident from lessened tissue damage, reduction of the number of $CD8^+$ T cells in the gut, and improved survival rates (Nalle et al. 2019). It is hypothesized that the decrease in $CD8^+$ T cells infiltration to the target organs in $MLCK^{-/-}$ mice is due to the suppression of cytokine storm that is propagated by intestinal barrier dysfunction.

5 Consequences of Intestinal Barrier Breach in Hematologic Malignancies

Compromise of the intestinal barrier function could contribute to the development and progression of hematologic malignancies. Impaired ability to restrict the exposure of intestinal luminal contents to the circulation leads to a multitude of complications.

5.1 Dysbiosis-Induced Permeability as a Driver of Malignancy

Dysbiosis leads to increased permeability that affects the mucosa-associated lymphoid tissue (MALT). *Helicobacter pylori* is linked to MALT lymphoma (Yamamoto and Schiestl 2014), and *Chlamydophila psittaci* is associated with ocular MALT lymphoma (Aigelsreiter et al. 2011).

The make-up of the gut microbiota is key in preserving the number of myeloid cells in the bone marrow (Manzo an Bhatt 2015). Patients with acute myeloid leukemia (AML) and acute lymphoblastic leukemia (ALL) exhibit dysbiosis and weakened intestinal barriers (Blijlevens et al. 2004; Sundström et al. 1998; Song and Gyarmati 2019). ALL patients' fecal microbiota demonstrates reduced diversity (Rajagopala et al. 2016) and is characterized by higher levels of *Enterococcaceae*, *Porphyromonadaceae*, and *Bacteroidetes* (especially *Bacteroides fragilis*), while showing lower levels of *Blautia*, *Erysipelotrichiales*, *Lachnospiraceae*, and *Clostridiales* (Bai et al. 2017; Chua et al. 2020). An abundance of *Proteobacteria* predicts neutropenic fever, and high levels of *Enterococcaceae* and *Streptococcaceae* are strong indicators of infectious complications in ALL patients (Hakim et al. 2018). In multiple myeloma, greater gut microbiota diversity correlates with a lower risk of disease relapse (Peled et al. 2017). ALL patients with infectious complications often show an increase in *Brevundimonas diminuta* and *Agrobacterium tumefaciens*, while completely lacking *Faecalibacterium prausnitzii* (a major butyrate producer) (Nearing et al. 2019). Similar trends are observed in non-Hodgkin lymphoma with infectious complications (Montassier et al. 2016). Since changes in microbial ecosystem in the intestine result in a decreased production of SCFAs that are essential for the development and health of enterocytes and TJs, it is not surprising that reduction in SCFAs aid in the disruption of intestinal barrier function and progression of the disease. For example, the depletion of *Faecalibacterium* damages the intestinal barrier in AML. There was a decrease in the expression of TJ proteins such as claudin-1 and ZO-1. This accelerates LPS leakage into the blood, promotes inflammatory response, and exacerbates leukemia progression (Wang et al. 2022). Similarly, endotoxin and genotoxins such as colibactin and cytolethal distending toxin (CDT) leak into the circulation due to the damage to the intestinal barrier integrity, induce inflammation, damage host DNA, and trigger tumor development (Guerra et al. 2011).

5.2 Life-Threatening Infections

Patients with hematological malignancies are at a higher risk of sepsis and bacteremia. During such infection, a severe state of immunosuppression affects both cellular effectors of the innate and adaptative immune systems persists (Delano and Ward 2016a, b). In addition, there is a state of immune dysfunction, including hyperinflammation (Rittirsch et al. 2008), complement activation, fibrinolytic and clotting system stimulation (Huber-Lang et al. 2001), severe oxidative stress (Santos et al. 2012), and molecular alterations, resulting in multiple organ damage and death (Delano and Ward 2016b). The increased risk of septicemia and bacteremia in hematologic malignancy patients is due to several factors: immunosuppression caused by the underlying disease, the antineoplastic treatments that cause immunosuppression, and the use of prophylactic antimicrobials. Patients with acute leukemia often present prolonged and profound neutropenia. Multiple myeloma (Lemiale et al. 2020; Williams et al. 2004; Torres et al. 2015) and non-Hodgkin lymphoma (Torres et al. 2015; Guarana et al. 2019; Lemiale et al. 2020) patients also experience high rates of sepsis. Allo-HSCT recipients who are severely immunocompromised due to the underlying disease and those who receive the conditioning regimen and suffer from the treatment-related complications such as GVHD are particularly vulnerable to septicemia and bacteremia.

In addition to the physical and functional barriers, normal intestinal microbiota also serves a critical defense against pathogens. Their disruption can increase the risk of serious life-threatening infections. Approximately 31% of hematologic malignancy patients suffer from sepsis and have an increased risk of bacteremia (Chebl et al. 2021). Mucositis-induced epithelium, TJ damage, and pro-inflammatory environment promote the translocation of pathogenic microbes into the circulation. Moreover, dysbiosis generates an environment conducive to the excess growth of normally sparse taxa such as gram-positive cocci (*Staphylococci* and *Streptococci*) and gram-negative bacilli (*Klebsiella*, *Pseudomonas aeruginosa*, and *E. coli*), fungi (*Candida*), parasites, and viruses (Pop-Began et al. 2014) while simultaneously decreasing the abundance of obligate anaerobes (Bacteroidetes and Firmicutes) (Ubeda et al. 2010). The excessive growth and invasion of these pathogenic microbes via the bloodstream lead to the loss of control of immune response and systemic activation of the coagulation cascade that can result in death.

Bacteremia in hematologic malignancy patients primarily involves three antibiotic-resistant strains vancomycin-resistant enterococci (VRE), extended-spectrum beta-lactamase (ESBL) organisms, and carbapenem-resistant enterococci (CRE). The increase in these bacteria in the circulation is attributed to dysbiosis and damage to the intestinal barrier functions due to the underlying disease and both antineoplastic treatment and the use of prophylactic broad-spectrum antibiotics. For example, intravenous vancomycin and cephalosporin are associated with increased risks of VRE colonization (Heisel et al. 2017). The excessive use of such broad-spectrum antibiotics increases the risk of VRE, ESBL, and CRE bacteremia.

VRE bacteremia increases the risk of bloodstream infections and increased mortality (Chiang et al. 2017; Worth et al. 2007). The primary strains of VRE are *Enterococcus faecium* and *Enterococcus faecalis*. Other VRE clinical isolates involve *Enterococcus gallinarum*, *Enterococcus casseliflavus*, *Enterococcus avium*, and *Enterococcus raffinosus* (Faron et al. 2016). VRE colonization is detected in up to 60% of AML patients and predicted for a much higher risk of VRE bacteremia (Heisel et al. 2017). Between 10% and 15% allo-HSCT receivers suffer from VRE bacteremia (Benamu and Deresinski 2018).

Chemotherapy-induced neutropenia, disruption of the mucosal barrier, excessive use of antibiotics, and allo-HSCT increase the risk of ESBL bacteremia. Changes in the intestinal microenvironment create a selection pressure that promotes the growth of ESBL-producing bacteria. ESBLs are enzymes produced by certain gram-negative bacteria such as Enterobacteriaceae family (*Escherichia coli* and *Klebsiella pneumoniae*), *Pseudomonas* spp., *Burkholderia* spp., and *Acinetobacter* spp. (Paterson and Bonomo 2005) that hydrolyze extended-spectrum cephalosporin. This renders beta-lactam antibiotics such as ceftazidime, ceftriaxone, cefotaxime, and oxyimino monobactam ineffective (Bradford 2001; Paterson and Bonomo 2005). ESBL-producing bacteria are also resistant to aminoglycosides and fluoroquinolones.

CRE infections are present in approximately 73% of leukemia patients, mostly AML (63%) (Lalaoui et al. 2020). The widespread use of antibiotics such as carbapenem antibiotics is one of the primary reasons for the emergence of CRE infections. The vast majority of carbapenem-resistant isolates (49%) are *Klebsiella* spp. (Pouch and Satlin 2017).

5.3 Disorders of Gut-Brain Interaction (DGBI)

Patients with hematologic malignancies commonly experience a spectrum of functional gastrointestinal disorders, currently known as disorders of gut-brain interaction (DGBI), including fatigue, nausea, and vomiting (Drossman et al. 2018). Chemotherapy-induced mucositis is also associated with symptoms such as nausea, vomiting, diarrhea, and pain. These symptoms increase over the course of cancer treatment and significantly affect the quality of life. The gut-brain axis involved in DGBI has a bidirectional relationship where endocrine, humoral, metabolic, and immune responses are all interconnected (Agirman et al. 2021). Dysbiosis of the intestine, either as a driver of malignancy or consequence of cancer treatment, is associated with chronic fatigue and depression (Carabotti et al. 2015) that trigger the expression of these pro-inflammatory cytokines and cause symptoms of neuroinflammation. Intestinal microbiota-derived increased the expression of pro-inflammatory cytokines such as IL-1β, IL-6, TNF-α, and IFN-γ are linked with depression (Appleton 2018). Circulating pro-inflammatory factors can disrupt epithelial TJs and compromise the blood-brain barrier (BBB) integrity (Parker et al. 2019). For example, low-grade inflammation due to increased intestinal permeability and translocation of endotoxins are partly responsible for DGBI (Safadi et al. 2022). Lower levels of *Bifidobacteria* and higher levels of aerobic bacteria are found in patients with chronic fatigue (Logan et al. 2003).

In addition to altered microbial composition, changes in the microbe-derived metabolites also influence DGBI. For example, SCFAs not only help in the maintenance of mucosal TJs, but they also regulate BBB permeability and microglia function that eventually modulate neuroinflammatory responses (Erny et al. 2015). Tryptophan metabolites derived from the intestinal microbes target the aryl hydrocarbon receptor to strengthen the intestinal barrier (Scott et al. 2020) and also mitigate central nervous system (CNS) inflammation by inducing IFN-I signaling in astrocytes (Rothhammer et al. 2016). Changes in tryptophan metabolism due to dysbiosis cause an increase in tryptophan entry into the brain that boosts the 5-hydroxytryptamine (5-HT) neurotransmission that is responsible for fatigue, nausea, and vomiting (Drossman et al. 2018).

6 Therapeutic Strategies to Reduce Disruption to Intestinal Barrier Function

Numerous studies have examined strategies to fortify the intestinal barrier. The selection of a specific strategy should be based on the individual clinical context, even though these methods can theoretically be applicable to all patients.

6.1 Antibiotic Stewardship

Prioritizing antibiotic stewardship is a critical strategy to address the disruption of the intestinal barrier functions by intestinal dysbiosis in patients with hematologic malignancies, as antibiotics are the most frequent cause of intestinal dysbiosis in this population. Broad-spectrum antibiotics are commonly used in practice. For example, a study of 251 adult patients (55% with hematologic malignancies and/or had undergone an HSCT) who visited an outpatient cancer center with symptoms of upper respiratory tract infections found that one third of patients were given broad-spectrum antibiotics, despite having an absolute neutrophil count of more than 500/μL and 75% were positive for respiratory viruses (Krantz et al. 2020).

A survey investigating the antimicrobial practice in febrile neutropenia in 194 centers in Europe and Asia further exemplifies the importance of antibiotic stewardship in this group of patients (Verlinden et al. 2020). In this study, two third of the centers adopted piperacillin/tazobactam as first-line therapy for febrile patients and a third used other empiric combination therapy. However, 10% of the centers adopted carbapenems as their first-line therapy. Furthermore, despite recommendations from the European Conference on Infections in Leukemia to use an escalation approach of non-carbapenem beta-lactams in stable febrile patients without a history of colonization/infection with resistant bacteria and only add a glycopeptide if fever persists after 2–3 days of the first-line empiric therapy, empiric glycopeptide was used in more than 60% of the centers.

These practices should be reevaluated, especially since multiple studies have shown that shorter treatment courses for febrile neutropenia appear safe. Three published studies have indicated that shorter courses of broad-spectrum antibiotic treatments for febrile neutropenia episodes, even without a microbiological diagnosis, do not increase mortality (Snyder et al. 2017; Aguilar-Guisado et al. 2017; Rearigh et al. 2020), supporting the safety of such a practice. There are limited studies assessing the impact of antibiotic stewardship in patients with hematologic malignancies and HSCT recipients, but available data show its effectiveness in reducing antibiotic use and improving patient outcomes (Aitken et al. 2019). Furthermore, antibiotic stewardship has been found to decrease the incidence of infections and colonization with antibiotic-resistant bacteria, as well as *C. difficile* infections (Baur et al. 2017).

6.2 Short-Chain Fatty Acids

If SCFAs are needed for the development of TJ proteins, it follows that the use of SCFAs in these patients may alleviate the complication of a breach in the intestinal barrier functions. Fermentation of dietary microbiota–accessible carbohydrates (MACs) by gut microbes, which serve as an energy source and are indigestible in the host's colon, produces organic acids such as acetate, propionate, butyrate, and valerate. When dietary MACs are lacking, bacteria that degrade mucus proliferate by consuming mucin. This disrupts the microbial balance in a way that leads to the thinning of the mucus layer and impacts the generation of the SCFAs crucial for the health and metabolism of enterocytes, as well as for the development of TJs. Studies in inflammatory bowel diseases (IBD) showed that increased intestinal permeability was associated with a shortfall of butyrate (Li et al. 2021). Butyrate also plays an anti-inflammatory role in the regulation of inflammatory reactions triggered by locally invasive bacteria. The oral supplementation of SCFAs is currently under investigation for its potential benefits in hematological and oncological diseases.

6.2.1 L-Glutamine

L-Glutamine, a "conditionally essential" amino acid, has a crucial role in maintaining the intestinal barrier function. L-Glutamine supports the structural integrity of the gut epithelial cells and TJ development and serves as an energy source for the intestinal epithelium (Wang et al. 2015). Improvement in gut barrier function in both animal studies and some human research has been noted with L-glutamine supplementation, especially in cases of gut-related disorders such as IBD and short bowel syndrome (Kim and Kim 2017). L-Glutamine has also been explored in clinical trials as a potential treatment for chemoradiation-induced mucositis (Thomsen and Vitetta

2018). However, the outcomes of these trials have been inconsistent, possibly due to variations in the dosages and timing of L-glutamine administration.

6.2.2 Pre- and Probiotics

Supplementation with pre- and probiotics is a promising method to enhance intestinal barrier functions (Rosenfeldt et al. 2004; Madsen et al. 2001). These supplements have been shown to work by boosting the expression of TJ proteins, diminishing intestinal inflammation, and altering the local immune responses (Butel 2014). Studies have shown that they influence intestinal tissue dendritic cells to produce IL-10, reduce IFN-γ release from intestinal T cells and decrease LPS-induced production of the pro-inflammatory IL-12 (de Moreno de LeBlanc et al. 2011). Although not regulated by the US FDA, the most extensively researched products for safety include *Lactobacillus*, *Bifidobacterium*, and *Saccharomyces boulardii*. Intake of the prebiotic inulin notably enhances the levels of *Faecalibacterium prausnitzii*, *Anaerostipes*, *Bifidobacterium adolescentis*, and *Bifidobacterium bifidum* significantly in the gut (Hryckowian et al. 2018). Butyrate is produced by *Faecalibacterium prausnitzii*, and the two *Bifidobacterium* spp. aid in the production of this compound. The fermentation pathway used by *Bifidobacterium* also generates acetate and lactate that are transformed into butyrate by *Eubacterium hallii*, *Anaerostipes caccae*, and *Anaerostipes hadrus* (Baxter et al. 2019). However, some concerns arise regarding these products; there exists wide variability in the available preparations, which range from single- or multi-strain formulations; they also present a risk of fungemia and bacteremia when used in immunocompromised patients. Studies regarding this issue have shown controversial results: Ladas et al. in their retrospective review identified 59 cases of fungemia linked to the use of probiotics (Rannikko et al. 2021). Conversely, a study found that administration of *Lactobacillus* from day −7 of the conditioning regimen through day +114 of 31 pediatric and adolescent patients undergoing allogeneic hematopoietic stem cell transplant did not lead to *Lactobacillus* bacteremia (Ladas et al. 2016). Dietary modifications are also being explored as potential treatment in various hematologic and oncologic diseases.

6.2.3 Fecal Microbiota Transplant

Fecal microbiota transplant (FMT) has been effectively utilized in treating recurrent or refractory CDI (Rohlke and Stollman 2012) and has the potential to improve treatment outcomes in hematologic malignancy. While its applicability has been restrictive due to the invasive nature of the treatment, it has recently been overcome with the introduction of FMT capsules. Beyond reestablishing the intestinal microbial balance, FMT strengthens the intestinal barrier by providing a means to replenish the microbial community. The breakdown of the intestinal barrier and dysbiosis in hematologic malignancy patients due to the effects of treatment can be minimized by FMT. A balanced intestinal microbial composition would, therefore, help in maintaining the intestinal barrier function and development of TJs through the production of the SCFAs. In addition to fortifying enterocyte health and improving intestinal barrier functions, since commensal intestinal microbiota also play a vital role in immune homeostasis, FMT can improve hematologic malignancy treatment efficacy. The primary concern with FMT is the risk of introducing infections, especially in cancer patients who are often immunocompromised due to their treatment. For instance, a case of fatal extended-spectrum beta-lactamases (ESBL) *E. coli* septicemia occurred in a patient with MDS following preemptive FMT (DeFilipp et al. 2019).

6.2.4 Akkermansia Muciniphila

An alternative approach could involve supplementing or enhancing the intestinal microbiome with beneficial microbes. The gram-negative, mucin-degrading anaerobe *Akkermansia muciniphila* could be suitable for this. *A. muciniphila* strengthens the mucosal layer and is crucial for maintaining the integrity of the gut barrier (Dao et al. 2016). It influences immune balance by stimulating TLR2-expressing cells and has a minor impact on TLR4, with negligible effects on TLR5, TLR9, or NOD2 receptors. Numerous studies have found an inverse correlation between the presence of *A. muciniphila* in the gut and various disease states (Dao et al. 2016; Routy et al. 2018; Depommier et al. 2019; Grander et al. 2018; Li et al. 2017; Ponziani et al. 2019). Supplementation with *A. muciniphila* can restore gut barrier functionality, reduce inflammation, and offer protection against cancer, obesity, type 2 diabetes, hypertension, and liver diseases. In a human trial focusing on metabolic syndrome, participants who received live or pasteurized *A. muciniphila* for 3 months exhibited improvements in several metabolic factors, including insulin sensitivity (Depommier et al. 2019). They also had better gut barrier functions, evidenced by lower plasma LPS levels, reduced white blood cell counts, and decreased inflammatory markers. The role of *A. muciniphila* in enhancing ICI therapy was demonstrated in cases of non–small cell lung cancer (NSCLC) and renal cell carcinoma (RCC), where patients with a higher intestinal concentration of *A. muciniphila* had a better response to anti-PD-1 treatment (Routy et al. 2018). Mice with RET melanoma unresponsive to PD-1 blockade due to antibiotic treatment showed a positive response following natural gut recolonization and *A. muciniphila* supplementation.

Unfortunately, an overgrowth or overabundance of *A muciniphila* may be detrimental. In a hormone receptor–positive breast cancer mouse model, high levels of *A. muciniphila* has been associated with promoting metastasis in adenocarcinoma (Buchta Rosean et al. 2019). The excessive breakdown of intestinal mucin by the bacterium has been shown to trigger colitis caused by increased epithelial exposure to intestinal bacteria, as observed in *Citrobacter rodentium*–induced colitis in germ-free mice (Desai et al. 2016).

Therefore, before attempting any isolated supplementation or enrichment with *A. muciniphila* to improve cancer therapy outcomes, it is essential to determine the optimal dosage that enhances treatment response without causing adverse effects.

6.2.5 Keratinocyte Growth Factors

Keratinocyte growth factors (KGFs) are cytokines known to foster the growth and differentiation of epithelial cells, including the intestinal epithelium. KGFs foster the proliferation and differentiation of gut epithelial cells and uphold the structural integrity of TJs that constitute the gut barrier. This role is key to their function in preserving intestinal barrier function (Athar and Gentile 2009). Their potential therapeutic role in various diseases and conditions linked to epithelial cell dysfunction has been investigated in multiple studies. In fact, KGFs have been found to enhance gut barrier function and aid in the healing of gut epithelium in individuals with IBD (Ido et al. 2005). They have also shown promise in boosting the survival and growth of gut epithelial cells in cases of short bowel syndrome by restoring intestinal barrier functions (Johnson et al. 2000; Yang et al. 2003).

In patients receiving high dose chemotherapy and autologous hematopoietic stem cell transplant, KGFs have been studied as a potential approach for the prevention and treatment of oral mucositis (Tsirigotis et al. 2008). A lower occurrence of severe oral mucositis and reduced rate of septicemia was noted in patients treated with recombinant KGF. KGF is currently approved for preventing chemotherapy-induced mucositis. However, its widespread use has been constrained by the high benefit-to-cost ratio.

7 Conclusion

The disruption of the intestinal barrier is a crucial factor in hematologic malignancies, often playing a bidirectional role. While the hematologic malignancies may be triggered by increased intestinal permeability, changes in the GI tract either due to altered physiology or due to treatment of a hematologic malignancues may also induce a breach in the intestinal barrier functions. Treatments such as antineoplastic agents and antimicrobials, along with consequences like increased risk of opportunistic infections and graft-versus-host disease (GVHD), all contribute to the altered intestinal barrier function by causing enterocyte damage, disrupting tight junction complexes, and unbalancing the intestinal microbial homeostasis and the epithelium's normal defenses. This increased permeability may not only contribute to the development and progression of hematologic malignancies but also lead to life-threatening infections and disorders of gut-brain interaction (DGBI).

In response to these challenges, various therapeutic strategies are being employed. These include antibiotic stewardship, fecal microbial transplantation, growth factors, and supplementation with beneficial products. These approaches aim to maintain the intestinal flora, preserve the normal functioning of the multilayered intestinal barrier, and promote intestinal epithelial growth and differentiation.

However, the complexity and interrelated nature of these issues underscore the need for interdisciplinary research in this field. Collaborative efforts spanning oncology, gastroenterology, microbiology, immunology, and pharmacology are essential to enhance our understanding and improve patient care in hematologic malignancies. Through such integrative research, more effective and targeted interventions can be developed, potentially leading to better outcomes for patients suffering from these challenging conditions.

Acknowledgment Figures created with Bioender.com.

Conflict of Interest None.

Funding None.

References

Agirman G, Yu KB, Hsiao EY (2021) Signaling inflammation across the gut-brain axis. Science 374:1087–1092

Aguilar-Guisado M, Espigado I, Martín-Peña A, Gudiol C, Royo-Cebrecos C, Falantes J, Vázquez-López L, Montero MI, Rosso-Fernández C, de la Luz Martino M, Parody R, González-Campos J, Garzón-López S, Calderón-Cabrera C, Barba P, Rodríguez N, Rovira M, Montero-Mateos E, Carratalá J, Pérez-Simón JA et al (2017) Optimisation of empirical antimicrobial therapy in patients with haematological malignancies and febrile neutropenia (how long study): an open-label, randomised, controlled phase 4 trial. Lancet Haematol 4:e573–e583

Aigelsreiter A, Gerlza T, Deutsch AJA, Leitner E, Beham-Schmid C, Beham A, Popper H, Borel N, Pospischil A, Raderer M, Kessler HH, Neumeister P (2011) Chlamydia psittaci infection in nongastrointestinal extranodal MALT lymphomas and their precursor lesions. Am J Clin Pathol 135:70–75

Aitken SL, Nagel JL, Abbo L, Alegria W, Barreto JN, Dadwal S, Freifeld AG, Jain R, Pergam SA, Tverdek FP, Seo SK, Antimicrobial Stewardship in Cancer Consortium ASCC (2019) Antimicrobial stewardship in cancer patients: the time is now. J Natl Compr Cancer Netw 17:772–775

Altınel E, Yarali N, Isık P, Bay A, Kara A, Tunc B (2012) Typhlitis in acute childhood Leukemia. Med Princ Pract 21:36–39

Appleton J (2018) The gut-brain Axis: influence of microbiota on mood and mental health. Integr Med Clin J 17:28–32

Ara T, Hashimoto D, Hayase E, Noizat C, Kikuchi R, Hasegawa Y, Matsuda K, Ono S, Matsuno Y, Ebata K, Ogasawara R, Takahashi S, Ohigashi H, Yokoyama E, Matsuo K, Sugita J, Onozawa M, Okumura R, Takeda K, Teshima T (2020) Intestinal goblet cells protect against GVHD after allogeneic stem cell transplantation via Lypd8. Sci Transl Med 12:eaaw0720

Athar U, Gentile TC (2009) Keratinocyte growth factor. Expert Opin Biol Ther 9:779–787

Bai L, Zhou P, Li D, Ju X (2017) Changes in the gastrointestinal microbiota of children with acute lymphoblastic leukaemia and its association with antibiotics in the short term. J Med Microbiol 66:1297–1307

Baur D, Gladstone BP, Burkert F, Carrara E, Foschi F, Döbele S, Tacconelli E (2017) Effect of antibiotic stewardship on the incidence of infection and colonisation with antibiotic-resistant bacteria and Clostridium difficile infection: a systematic review and meta-analysis. Lancet Infect Dis 17:990–1001

Baxter NT, Schmidt AW, Venkataraman A, Kim KS, Waldron C, Schmidt TM (2019) Dynamics of human gut microbiota and short-chain fatty acids in response to dietary interventions with three fermentable fibers. MBio 10:e02566-18

Benamu E, Deresinski S (2018) Vancomycin-resistant enterococcus infection in the hematopoietic stem cell transplant recipient: an overview of epidemiology, management, and prevention. F1000Research 7:3

Beutheu Youmba S, Belmonte L, Galas L, Boukhettala N, Bôle-Feysot C, Déchelotte P, Coëffier M (2012) Methotrexate modulates tight junctions through NF-κB, MEK, and JNK pathways. J Pediatr Gastroenter Nutr 54:463–470

Blijlevens NMA, van't Land B, Donnelly JP, M'Rabet L, de Pauw BE (2004) Measuring mucosal damage induced by cytotoxic therapy. Support Care Cancer 12:227–233

Bradford PA (2001) Extended-spectrum beta-lactamases in the 21st century: characterization, epidemiology, and detection of this important resistance threat. Clin Microbiol Rev 14:933–951. table of contents

Buchta Rosean C, Bostic RR, Ferey JCM, Feng T-Y, Azar FN, Tung KS, Dozmorov MG, Smirnova E, Bos PD, Rutkowski MR (2019) Preexisting commensal dysbiosis is a host-intrinsic regulator of tissue inflammation and tumor cell dissemination in hormone receptor–positive breast cancer. Cancer Res 79:3662–3675

Buschmann MM, Shen L, Rajapakse H, Raleigh DR, Wang Y, Wang Y, Lingaraju A, Zha J, Abbott E, McAuley EM, Breskin LA, Wu L, Anderson K, Turner JR, Weber CR (2013) Occludin OCEL-domain interactions are required for maintenance and regulation of the tight junction barrier to macromolecular flux. Mol Biol Cell 24:3056–3068

Butel M-J (2014) Probiotics, gut microbiota and health. Med Mal Infect 44:1–8

Carabotti M, Scirocco A, Maselli MA, Severi C (2015) The gut-brain axis: interactions between enteric microbiota, central and enteric nervous systems. Ann Gastroenterol 28:203–209

Chanez-Paredes SD, Abtahi S, Kuo W-T, Turner JR (2021) Differentiating between tight junction-dependent and tight junction-independent intestinal barrier loss in vivo. Methods Mol Biol 2367:249–271

Chavez JC, Locke FL (2018) CAR T cell therapy for B-cell lymphomas. Best Pract Res Clin Haematol 31:135–146

Chebl RB, Safa R, Sabra M, Chami A, Berbari I, Jamali S, Makki M, Tamim H, Dagher GA (2021) Sepsis in patients with haematological versus solid cancer: a retrospective cohort study. BMJ Open 11:e038349

Chiang H-Y, Perencevich EN, Nair R, Nelson RE, Samore M, Khader K, Chorazy ML, Herwaldt LA, Blevins A, Ward MA, Schweizer ML (2017) Incidence and outcomes associated with infections caused by vancomycin-resistant enterococci in the United States: systematic literature review and meta-analysis. Infect Control Hosp Epidemiol 38:203–215

Chua LL, Rajasuriar R, Lim YAL, Woo YL, Loke P, Ariffin H (2020) Temporal changes in gut microbiota profile in children with acute lymphoblastic leukemia prior to commencement-, during-, and post-cessation of chemotherapy. BMC Cancer 20:151

Cooke KR, Hill GR, Crawford JM, Bungard D, Brinson YS, Delmonte J, Ferrara JL (1998) Tumor necrosis factor- alpha production to lipopolysaccharide stimulation by donor cells predicts the severity of experimental acute graft-versus-host disease. J Clin Invest 102:1882–1891

Cooke KR, Gerbitz A, Crawford JM, Teshima T, Hill GR, Tesolin A, Rossignol DP, Ferrara JLM (2001) LPS antagonism reduces graft-versus-host disease and preserves graft-versus-leukemia activity after experimental bone marrow transplantation. J Clin Invest 107:1581–1589

Crawford J, Dale DC, Lyman GH (2004) Chemotherapy-induced neutropenia. Cancer 100:228–237

Crenn P, Messing B, Cynober L (2008) Citrulline as a biomarker of intestinal failure due to enterocyte mass reduction. Clin Nutr 27:328–339

Dao MC, Everard A, Aron-Wisnewsky J, Sokolovska N, Prifti E, Verger EO, Kayser BD, Levenez F, Chilloux J, Hoyles L, MICRO-Obes Consortium, Dumas M-E, Rizkalla SW, Doré J, Cani PD, Clément K (2016) Akkermansia muciniphila and improved metabolic health during a dietary intervention in obesity: relationship with gut microbiome richness and ecology. Gut 65:426–436

de Moreno de LeBlanc A, del Carmen S, Zurita-Turk M, Santos Rocha C, van de Guchte M, Azevedo V, Miyoshi A, LeBlanc JG (2011) Importance of IL-10 modulation by probiotic microorganisms in gastrointestinal inflammatory diseases. ISRN Gastroenterol 2011:1–11

DeFilipp Z, Bloom PP, Torres Soto M, Mansour MK, Sater MRA, Huntley MH, Turbett S, Chung RT, Chen Y-B, Hohmann EL (2019) Drug-Resistant *E. coli* Bacteremia transmitted by fecal microbiota transplant. N Engl J Med 381:2043–2050

Delano MJ, Ward PA (2016a) Sepsis-induced immune dysfunction: can immune therapies reduce mortality? J Clin Invest 126:23–31

Delano MJ, Ward PA (2016b) The immune system's role in sepsis progression, resolution, and long-term outcome. Immunol Rev 274:330–353

Depommier C, Everard A, Druart C, Plovier H, Van Hul M, Vieira-Silva S, Falony G, Raes J, Maiter D, Delzenne NM, de Barsy M, Loumaye A, Hermans MP, Thissen J-P, de Vos WM, Cani PD (2019) Supplementation with Akkermansia muciniphila in overweight and obese human volunteers: a proof-of-concept exploratory study. Nat Med 25:1096–1103

Desai MS, Seekatz AM, Koropatkin NM, Kamada N, Hickey CA, Wolter M, Pudlo NA, Kitamoto S, Terrapon N, Muller A, Young VB, Henrissat B, Wilmes P, Stappenbeck TS, Núñez G, Martens EC (2016) A dietary fiber-deprived gut microbiota degrades the colonic mucus barrier and enhances pathogen susceptibility. Cell 167:1339–1353.e21

Drossman DA, Tack J, Ford AC, Szigethy E, Törnblom H, Van Oudenhove L (2018) Neuromodulators for functional gastrointestinal disorders (disorders of gut-brain interaction): a Rome foundation working team report. Gastroenterology 154:1140–1171.e1

Dutta D, Lim SH (2020) Bidirectional interaction between intestinal microbiome and cancer: opportunities for therapeutic interventions. Biomark Res 8:31

Dutta D, Methe B, Amar S, Morris A, Lim SH (2019) Intestinal injury and gut permeability in sickle cell disease. J Transl Med 17:183

Engevik MA, Engevik AC, Engevik KA, Auchtung JM, Chang-Graham AL, Ruan W, Luna RA, Hyser JM, Spinler JK, Versalovic J (2021) Mucin-degrading microbes release monosaccharides that chemoattract clostridioides difficile and facilitate colonization of the human intestinal mucus layer. ACS Infect Dis 7:1126–1142

Eriguchi Y, Takashima S, Oka H, Shimoji S, Nakamura K, Uryu H, Shimoda S, Iwasaki H, Shimono N, Ayabe T, Akashi K, Teshima T (2012) Graft-versus-host disease disrupts intestinal microbial ecology by inhibiting Paneth cell production of α-defensins. Blood 120:223–231

Erny D, Hrabě de Angelis AL, Jaitin D, Wieghofer P, Staszewski O, David E, Keren-Shaul H, Mahlakoiv T, Jakobshagen K, Buch T, Schwierzeck V, Utermöhlen O, Chun E, Garrett WS, McCoy KD, Diefenbach A, Staeheli P, Stecher B, Amit I, Prinz M (2015) Host microbiota constantly control maturation and function of microglia in the CNS. Nat Neurosci 18:965–977

Faron ML, Ledeboer NA, Buchan BW (2016) Resistance mechanisms, epidemiology, and approaches to screening for vancomycin-resistant enterococcus in the health care setting. J Clin Microbiol 54:2436–2447

Fasano A, Not T, Wang W, Uzzau S, Berti I, Tommasini A, Goldblum SE (2000) Zonulin, a newly discovered modulator of intestinal permeability, and its expression in coeliac disease. Lancet 355:1518–1519

Fiorentino M, Sapone A, Senger S, Camhi SS, Kadzielski SM, Buie TM, Kelly DL, Cascella N, Fasano A (2016) Blood–brain barrier and intestinal epithelial barrier alterations in autism spectrum disorders. Mol Autism 7(1). https://doi.org/10.1186/s13229-016-0110-z

France MM, Turner JR (2017) The mucosal barrier at a glance. J Cell Sci 130:307–314

Gorschlüter M, Mey U, Strehl J, Ziske C, Schepke M, Schmidt-Wolf IGH, Sauerbruch T, Glasmacher A (2005) Neutropenic enterocolitis in adults: systematic analysis of evidence quality. Eur J Haematol 75:1–13

Grander C, Adolph TE, Wieser V, Lowe P, Wrzosek L, Gyongyosi B, Ward DV, Grabherr F, Gerner RR, Pfister A, Enrich B, Ciocan D, Macheiner S, Mayr L, Drach M, Moser P, Moschen AR, Perlemuter G, Szabo G, Cassard AM et al (2018) Recovery of ethanol-induced Akkermansia muciniphila depletion ameliorates alcoholic liver disease. Gut 67:891–901

Guarana M, Nucci M, Nouér SA (2019) Shock and early death in hematologic patients with febrile neutropenia. Antimicrob Agents Chemother 63:e01250-19

Guerra L, Guidi R, Frisan T (2011) Do bacterial genotoxins contribute to chronic inflammation, genomic instability and tumor progression? FEBS J 278:4577–4588

Hakim H, Dallas R, Wolf J, Tang L, Schultz-Cherry S, Darling V, Johnson C, Karlsson EA, Chang T-C, Jeha S, Pui C-H, Sun Y, Pounds S, Hayden RT, Tuomanen E, Rosch JW (2018) Gut microbiome composition predicts infection risk during chemotherapy in children with acute lymphoblastic Leukemia. Clin Infect Dis 67:541–548

Hayase E, Hashimoto D, Nakamura K, Noizat C, Ogasawara R, Takahashi S, Ohigashi H, Yokoi Y, Sugimoto R, Matsuoka S, Ara T, Yokoyama E, Yamakawa T, Ebata K, Kondo T, Hiramine R, Aizawa T, Ogura Y, Hayashi T, Mori H et al (2017) R-Spondin1 expands Paneth cells and prevents dysbiosis induced by graft-versus-host disease. J Exp Med 214:3507–3518

Heidt PJ, Vossen JM (1992) Experimental and clinical gnotobiotics: influence of the microflora on graft-versus-host disease after allogeneic bone marrow transplantation. J Med 23:161–173

Heinsen F-A, Knecht H, Neulinger SC, Schmitz RA, Knecht C, Kühbacher T, Rosenstiel PC, Schreiber S, Friedrichs AK, Ott SJ (2015) Dynamic changes of the luminal and mucosa-associated gut microbiota during and after antibiotic therapy with paromomycin. Gut Microbes 6:243–254

Heisel RW, Sutton RR, Mascara GP, Winger DG, Weber DR, Lim SH, Oleksiuk L-M (2017) Vancomycin-resistant enterococci in acute myeloid leukemia and myelodysplastic syndrome patients undergoing induction chemotherapy with idarubicin and cytarabine. Leuk Lymphoma 58:2565–2572

Horowitz A, Chanez-Paredes SD, Haest X, Turner JR (2023) Paracellular permeability and tight junction regulation in gut health and disease. Nat Rev Gastroenterol Hepatol 20:417–432

Hryckowian AJ, Van Treuren W, Smits SA, Davis NM, Gardner JO, Bouley DM, Sonnenburg JL (2018) Microbiota-accessible carbohydrates suppress Clostridium difficile infection in a murine model. Nat Microbiol 3:662–669

Huber-Lang MS, Sarma JV, McGuire SR, Lu KT, Guo RF, Padgaonkar VA, Younkin EM, Laudes IJ, Riedemann NC, Younger JG, Ward PA (2001) Protective effects of anti-C5a peptide antibodies in experimental sepsis. FASEB J 15:568–570

Ido A, Numata M, Kodama M, Tsubouchi H (2005) Mucosal repair and growth factors: recombinant human hepatocyte growth factor as an innovative therapy for inflammatory bowel disease. J Gastroenterol 40:925–931

Ijiri K, Potten CS (1983) Response of intestinal cells of differing topographical and hierarchical status to ten cytotoxic drugs and five sources of radiation. Br J Cancer 47:175–185

Jansen SA, Nieuwenhuis EES, Hanash AM, Lindemans CA (2022) Challenges and opportunities targeting mechanisms of epithelial injury and recovery in acute intestinal graft-versus-host disease. Mucosal Immunol 15:605–619

Johnson WF, DiPalma CR, Ziegler TR, Scully S, Farrell CL (2000) Keratinocyte growth factor enhances early gut adaptation in a rat model of short bowel syndrome. Vet Surg 29:17–27

Kamboj M, Xiao K, Kaltsas A, Huang Y-T, Sun J, Chung D, Wu S, Sheahan A, Sepkowitz K, Jakubowski AA, Papanicolaou G (2014) Clostridium difficile infection after allogeneic hematopoietic stem cell transplant: strain diversity and outcomes associated with NAP1/027. Biol Blood Marrow Transplant 20:1626–1633

Katz JA, Wagner ML, Gresik MV, Mahoney DH, Fernbach DJ (1990) Typhlitis. An 18-year experience and postmortem review. Cancer 65:1041–1047

Kim M-H, Kim H (2017) The roles of glutamine in the intestine and its implication in intestinal diseases. Int J Mol Sci 18:1051

Kinashi Y, Hase K (2021) Partners in leaky gut syndrome: intestinal dysbiosis and autoimmunity. Front Immunol 12:673708

Krantz EM, Zier J, Stohs E, Ogimi C, Sweet A, Marquis S, Klaassen J, Pergam SA, Liu C (2020) Antibiotic prescribing and respiratory viral testing for acute upper respiratory infections among adult patients at an ambulatory cancer center. Clin Infect Dis 70:1421–1428

Ladas EJ, Bhatia M, Chen L, Sandler E, Petrovic A, Berman DM, Hamblin F, Gates M, Hawks R, Sung L, Nieder M (2016) The safety and feasibility of probiotics in children and adolescents undergoing hematopoietic cell transplantation. Bone Marrow Transplant 51:262–266

Lalaoui R, Javelle E, Bakour S, Ubeda C, Rolain J-M (2020) Infections due to Carbapenem-resistant bacteria in patients with hematologic malignancies. Front Microbiol 11:1422

Lemiale V, Pons S, Mirouse A, Tudesq J-J, Hourmant Y, Mokart D, Pène F, Kouatchet A, Mayaux J, Nyunga M, Bruneel F, Meert A-P, Borcoman E, Bisbal M, Legrand M, Benoit D, Azoulay E, Darmon M, Zafrani L (2020) Sepsis and septic shock in patients with malignancies:

a Groupe de Recherche Respiratoire en Réanimation Onco-Hématologique Study. Crit Care Med 48:822–829

Le-Trilling VTK, Ebel J-F, Baier F, Wohlgemuth K, Pfeifer KR, Mookhoek A, Krebs P, Determann M, Katschinski B, Adamczyk A, Lange E, Klopfleisch R, Lange CM, Sokolova V, Trilling M, Westendorf AM (2023) Acute cytomegalovirus infection modulates the intestinal microbiota and targets intestinal epithelial cells. Eur J Immunol 53:2249940

Li J, Zhao F, Wang Y, Chen J, Tao J, Tian G, Wu S, Liu W, Cui Q, Geng B, Zhang W, Weldon R, Auguste K, Yang L, Liu X, Chen L, Yang X, Zhu B, Cai J (2017) Gut microbiota dysbiosis contributes to the development of hypertension. Microbiome 5:14

Li G, Lin J, Zhang C, Gao H, Lu H, Gao X, Zhu R, Li Z, Li M, Liu Z (2021) Microbiota metabolite butyrate constrains neutrophil functions and ameliorates mucosal inflammation in inflammatory bowel disease. Gut Microbes 13:1968257

Lichtman SM (1995) Physiological aspects of aging: implications for the treatment of cancer. Drugs Aging 7:212–225

Lifschitz CH, Mahoney DH (1989) Low-dose methotrexate-induced changes in intestinal permeability determined by polyethylene glycol polymers. J Pediatr Gastroenterol Nutr 9:301

Litvak Y, Byndloss MX, Bäumler AJ (2018) Colonocyte metabolism shapes the gut microbiota. Science (New York, NY) 362:eaat9076

Logan AC, Venket Rao A, Irani D (2003) Chronic fatigue syndrome: lactic acid bacteria may be of therapeutic value. Med Hypotheses 60:915–923

Logan RM, Stringer AM, Bowen JM, Yeoh AS-J, Gibson RJ, Sonis ST, Keefe DMK (2007) The role of pro-inflammatory cytokines in cancer treatment-induced alimentary tract mucositis: pathobiology, animal models and cytotoxic drugs. Cancer Treat Rev 33:448–460

Logan RM, Stringer AM, Bowen JM, Gibson RJ, Sonis ST, Keefe DMK (2008) Serum levels of NF-κB and pro-inflammatory cytokines following administration of mucotoxic drugs. Cancer Biol Ther 7:1139–1145

Louis P, Flint HJ (2017) Formation of propionate and butyrate by the human colonic microbiota. Environ Microbiol 19:29–41

Luznik L, Jones RJ, Fuchs EJ (2010) High-dose cyclophosphamide for graft-versus-host disease prevention. Curr Opin Hematol 17:493–499

Ma TY, Iwamoto GK, Hoa NT, Akotia V, Pedram A, Boivin MA, Said HM (2004) TNF-alpha-induced increase in intestinal epithelial tight junction permeability requires NF-kappa B activation. Am J Physiol Gastrointest Liver Physiol 286:G367–G376

Ma TY, Boivin MA, Ye D, Pedram A, Said HM (2005) Mechanism of TNF-α modulation of Caco-2 intestinal epithelial tight junction barrier: role of myosin light-chain kinase protein expression. Am J Physiol Gastrointest Liver Physiol 288:G422–G430

Madsen K, Cornish A, Soper P, McKaigney C, Jijon H, Yachimec C, Doyle J, Jewell L, De Simone C (2001) Probiotic bacteria enhance murine and human intestinal epithelial barrier function. Gastroenterology 121:580–591

Maidji E, Somsouk M, Rivera JM, Hunt PW, Stoddart CA (2017) Replication of CMV in the gut of HIV-infected individuals and epithelial barrier dysfunction. PLoS Pathog 13:e1006202

Manzo VE, Bhatt AS (2015) The human microbiome in hematopoiesis and hematologic disorders. Blood 126:311–318

Marchesi F, Pimpinelli F, Ensoli F, Mengarelli A (2018) Cytomegalovirus infection in hematologic malignancy settings other than the allogeneic transplant. Hematol Oncol 36:381–391

Marcus BC, Wyble CW, Hynes KL, Gewertz BL (1996) Cytokine-induced increases in endothelial permeability occur after adhesion molecule expression. Surgery 120:411–417

Masenga SK, Hamooya B, Hangoma J, Hayumbu V, Ertuglu LA, Ishimwe J, Rahman S, Saleem M, Laffer CL, Elijovich F, Kirabo A (2022) Recent advances in modulation of cardiovascular diseases by the gut microbiota. J Hum Hypertens 36:952–959

McGlone SM, Bailey RR, Zimmer SM, Popovich MJ, Tian Y, Ufberg P, Muder RR, Lee BY (2012) The economic burden of Clostridium difficile. Clin Microbiol Infect 18:282–289

Melichar B, Zezulová M (2011) The significance of altered gastrointestinal permeability in cancer patients. Curr Opin Support Palliat Care 5:47

Michielan A, D'Incà R (2015) Intestinal permeability in inflammatory bowel disease: pathogenesis, clinical evaluation, and therapy of leaky gut. Mediat Inflamm 2015:628157

Montassier E, Gastinne T, Vangay P, Al-Ghalith GA, Bruley des Varannes S, Massart S, Moreau P, Potel G, de La Cochetière MF, Batard E, Knights D (2015) Chemotherapy-driven dysbiosis in the intestinal microbiome. Aliment Pharmacol Ther 42:515–528

Montassier E, Al-Ghalith GA, Ward T, Corvec S, Gastinne T, Potel G, Moreau P, de la Cochetiere MF, Batard E, Knights D (2016) Pretreatment gut microbiome predicts chemotherapy-related bloodstream infection. Genome Med 8:49

Moorefield EC, Andres SF, Blue RE, Van Landeghem L, Mah AT, Santoro MA, Ding S (2017) Aging effects on intestinal homeostasis associated with expansion and dysfunction of intestinal epithelial stem cells. Aging (Albany NY) 9:1898–1915

Munford RS (2016) Endotoxemia—menace, marker, or mistake? J Leukoc Biol 100:687–698

Nalle SC, Zuo L, Ong MLDM, Singh G, Worthylake AM, Choi W, Manresa MC, Southworth AP, Edelblum KL, Baker GJ, Joseph NE, Savage PA, Turner JR (2019) Graft-versus-host disease propagation depends on increased intestinal epithelial tight junction permeability. J Clin Invest 129:902–914

Nearing JT, Connors J, Whitehouse S, Van Limbergen J, Macdonald T, Kulkarni K, Langille MGI (2019) Infectious complications are associated with alterations in the gut microbiome in Pediatric patients with acute lymphoblastic Leukemia. Front Cell Infect Microbiol 9:28

Nesher L, Rolston KVI (2013) Neutropenic enterocolitis, a growing concern in the era of widespread use of aggressive chemotherapy. Clin Infect Dis 56:711–717

Ozdemir E, Aslan A, Turker A, Barista I, Kars A (2015) Gemcitabine in combination with Oxaliplatin (GEMOX) as a salvage regimen in patients with relapsed/refractory Hodgkin's lymphoma. Blood 126:1517

Parker A, Fonseca S, Carding SR (2019) Gut microbes and metabolites as modulators of blood-brain barrier integrity and brain health. Gut Microbes 11:135–157

Paterson DL, Bonomo RA (2005) Extended-spectrum beta-lactamases: a clinical update. Clin Microbiol Rev 18:657–686

Peled JU, Devlin SM, Staffas A, Lumish M, Khanin R, Littmann ER, Ling L, Kosuri S, Maloy M, Slingerland JB, Ahr KF, Porosnicu Rodriguez KA, Shono Y, Slingerland AE, Docampo MD, Sung AD, Weber D, Alousi AM, Gyurkocza B, Ponce DM et al (2017) Intestinal microbiota and relapse after hematopoietic-cell transplantation. J Clin Oncol Off J Am Soc Clin Oncol 35: 1650–1659

Peng L, Li Z-R, Green RS, Holzman IR, Lin J (2009) Butyrate enhances the intestinal barrier by facilitating tight junction assembly via activation of AMP-activated protein kinase in Caco-2 cell monolayers. J Nutr 139:1619–1625

Pizzuti D, Bortolami M, Mazzon E, Buda A, Guariso G, D'Odorico A, Chiarelli S, D'Incà R, De Lazzari F, Martines D (2004) Transcriptional downregulation of tight junction protein ZO-1 in active coeliac disease is reversed after a gluten-free diet. Dig Liver Dis 36:337–341

Plaza-Díaz J, Solís-Urra P, Rodríguez-Rodríguez F, Olivares-Arancibia J, Navarro-Oliveros M, Abadía-Molina F, Álvarez-Mercado AI (2020) The gut barrier, intestinal microbiota, and liver disease: molecular mechanisms and strategies to manage. Int J Mol Sci 21:8351

Ponziani FR, Bhoori S, Castelli C, Putignani L, Rivoltini L, Del Chierico F, Sanguinetti M, Morelli D, Paroni Sterbini F, Petito V, Reddel S, Calvani R, Camisaschi C, Picca A, Tuccitto A, Gasbarrini A, Pompili M, Mazzaferro V (2019) Hepatocellular carcinoma is associated with gut microbiota profile and inflammation in nonalcoholic fatty liver disease. Hepatology 69:107–120

Pop-Began V, Păunescu V, Grigorean V, Pop-Began D, Popescu C (2014) Molecular mechanisms in the pathogenesis of sepsis. J Med Life 7:38–41

Poritz LS, Garver KI, Tilberg AF, Koltun WA (2004) Tumor necrosis factor alpha disrupts tight junction assembly. J Surg Res 116:14–18

Pouch SM, Satlin MJ (2017) Carbapenem-resistant Enterobacteriaceae in special populations: solid organ transplant recipients, stem cell transplant recipients, and patients with hematologic malignancies. Virulence 8:391–402

Rafailidis PI, Mourtzoukou EG, Varbobitis IC, Falagas ME (2008) Severe cytomegalovirus infection in apparently immunocompetent patients: a systematic review. Virol J 5:47

Rajagopala SV, Yooseph S, Harkins DM, Moncera KJ, Zabokrtsky KB, Torralba MG, Tovchigrechko A, Highlander SK, Pieper R, Sender L, Nelson KE (2016) Gastrointestinal microbial populations can distinguish pediatric and adolescent Acute Lymphoblastic Leukemia (ALL) at the time of disease diagnosis. BMC Genomics 17:635

Raleigh DR, Marchiando AM, Zhang Y, Shen L, Sasaki H, Wang Y, Long M, Turner JR (2010) Tight junction-associated MARVEL proteins marveld3, tricellulin, and occludin have distinct but overlapping functions. Mol Biol Cell 21:1200–1213

Rannikko J, Holmberg V, Karppelin M, Arvola P, Huttunen R, Mattila E, Kerttula N, Puhto T, Tamm Ü, Koivula I, Vuento R, Syrjänen J, Hohenthal U (2021) Fungemia and other fungal infections associated with use of *Saccharomyces boulardii* Probiotic Supplements. Emerg Infect Dis 27. Available at: https://wwwnc.cdc.gov/eid/article/27/8/21-0018_article. Accessed 13 Mar 2023.

Rearigh L, Stohs E, Freifeld A, Zimmer A (2020) De-escalation of empiric broad spectrum antibiotics in hematopoietic stem cell transplant recipients with febrile neutropenia. Ann Hematol 99:1917–1924

Rendo MJ, Joseph JJ, Phan LM, DeStefano CB (2022) CAR T-cell therapy for patients with multiple myeloma: current evidence and challenges. Blood Lymphat Cancer Targets Ther 12: 119–136

Rittirsch D, Flierl MA, Ward PA (2008) Harmful molecular mechanisms in sepsis. Nat Rev Immunol 8:776–787

Rohlke F, Stollman N (2012) Fecal microbiota transplantation in relapsing *Clostridium difficile* infection. Ther Adv Gastroenterol 5:403–420

Rosenfeldt V, Benfeldt E, Valerius NH, Pærregaard A, Michaelsen KF (2004) Effect of probiotics on gastrointestinal symptoms and small intestinal permeability in children with atopic dermatitis. J Pediatr 145:612–616

Rothhammer V, Mascanfroni ID, Bunse L, Takenaka MC, Kenison JE, Mayo L, Chao C-C, Patel B, Yan R, Blain M, Alvarez JI, Kébir H, Anandasabapathy N, Izquierdo G, Jung S, Obholzer N, Pochet N, Clish CB, Prinz M, Prat A et al (2016) Type I interferons and microbial metabolites of tryptophan modulate astrocyte activity and central nervous system inflammation via the aryl hydrocarbon receptor. Nat Med 22:586–597

Routy B, Le Chatelier E, Derosa L, Duong CPM, Alou MT, Daillère R, Fluckiger A, Messaoudene M, Rauber C, Roberti MP, Fidelle M, Flament C, Poirier-Colame V, Opolon P, Klein C, Iribarren K, Mondragón L, Jacquelot N, Qu B, Ferrere G et al (2018) Gut microbiome influences efficacy of PD-1-based immunotherapy against epithelial tumors. Science (New York, NY) 359:91–97

Safadi JM, Quinton AMG, Lennox BR, Burnet PWJ, Minichino A (2022) Gut dysbiosis in severe mental illness and chronic fatigue: a novel trans-diagnostic construct? A systematic review and meta-analysis. Mol Psychiatry 27:141–153

Santos SS, Brunialti MKC, Rigato O, Machado FR, Silva E, Salomao R (2012) Generation of nitric oxide and reactive oxygen species by neutrophils and monocytes from septic patients and association with outcomes. Shock 38:18–23

Scott SA, Fu J, Chang PV (2020) Microbial tryptophan metabolites regulate gut barrier function via the aryl hydrocarbon receptor. Proc Natl Acad Sci 117:19376–19387

Selby PJ, Lopes N, Mundy J, Crofts M, Millar JL, McElwain TJ (1987) Cyclophosphamide priming reduces intestinal damage in man following high dose melphalan chemotherapy. Br J Cancer 55: 531–533

Shen Q-D, Zhu H-Y, Wang L, Fan L, Liang J-H, Cao L, Wu W, Xia Y, Li J-Y, Xu W (2018) Gemcitabine-oxaliplatin plus rituximab (R-GemOx) as first-line treatment in elderly patients with diffuse large B-cell lymphoma: a single-arm, open-label, phase 2 trial. Lancet Haematol 5: e261–e269

Sheykhhasan M, Manoochehri H, Dama P (2022) Use of CAR T-cell for acute lymphoblastic leukemia (ALL) treatment: a review study. Cancer Gene Ther 29:1080–1096

Silva JPB, Navegantes-Lima KC, Oliveira ALB, Rodrigues DVS, Gaspar SLF, Monteiro VVS, Moura DP, Monteiro MC (2019) Protective mechanisms of butyrate on inflammatory bowel disease. Curr Pharm Des 24:4154–4166

Singh AP, Aijaz S (2016) Enteropathogenic E. coli: breaking the intestinal tight junction barrier. F1000Research 4:231

Smits WK, Lyras D, Lacy DB, Wilcox MH, Kuijper EJ (2016) Clostridium difficile infection. Nat Rev Dis Prim 2:1–20

Snyder M, Pasikhova Y, Baluch A (2017) Early antimicrobial De-escalation and stewardship in adult hematopoietic stem cell transplantation recipients: retrospective review. Open Forum Infect Dis 4:ofx226

Song Y, Gyarmati P (2019) Bacterial translocation in acute lymphocytic leukemia. PLoS One 14: e0214526

Stojanovska V, McQuade RM, Fraser S, Prakash M, Gondalia S, Stavely R, Palombo E, Apostolopoulos V, Sakkal S, Nurgali K (2018) Oxaliplatin-induced changes in microbiota, TLR4+ cells and enhanced HMGB1 expression in the murine colon. PLoS One 13:e0198359

Sun Q, Ji Y-C, Wang Z-L, She X, He Y, Ai Q, Li L-Q (2021) Sodium butyrate alleviates intestinal inflammation in mice with necrotizing enterocolitis. Mediat Inflamm 2021:1–12

Sundström GM, Wahlin A, Nordin-Andersson I, Suhr OB (1998) Intestinal permeability in patients with acute myeloid leukemia. Eur J Haematol 61:250–254

Taplitz RA, Kennedy EB, Flowers CR (2018) Antimicrobial prophylaxis for adult patients with cancer-related immunosuppression: ASCO and IDSA clinical practice guideline update summary. J Oncol Pract 14:692–695

Thomsen M, Vitetta L (2018) Adjunctive treatments for the prevention of chemotherapy- and radiotherapy-induced Mucositis. Integr Cancer Ther 17:1027–1047

Torres VBL, Azevedo LCP, Silva UVA, Caruso P, Torelly AP, Silva E, Carvalho FB, Vianna A, Souza PCP, Godoy MMG, Azevedo JRA, Spector N, Bozza FA, Salluh JIF, Soares M (2015) Sepsis-associated outcomes in critically ill patients with malignancies. Ann Am Thorac Soc 12: 1185–1192

Tsirigotis P, Triantafyllou K, Girkas K, Giannopoulou V, Ioannidou E, Chondropoulos S, Kalli T, Papaxoinis G, Pappa V, Papageorgiou E, Economopoulos T, Ladas SD, Dervenoulas J (2008) Keratinocyte growth factor is effective in the prevention of intestinal mucositis in patients with hematological malignancies treated with high-dose chemotherapy and autologous hematopoietic SCT: a video-capsule endoscopy study. Bone Marrow Transplant 42:337–343

Tyszka M, Biliński J, Basak GW (2021) Advances in intestinal barrier preservation and restoration in the allogeneic hematopoietic cell transplantation setting. J Clin Med 10:2508

Ubeda C, Taur Y, Jenq RR, Equinda MJ, Son T, Samstein M, Viale A, Socci ND, van den Brink MRM, Kamboj M, Pamer EG (2010) Vancomycin-resistant enterococcus domination of intestinal microbiota is enabled by antibiotic treatment in mice and precedes bloodstream invasion in humans. J Clin Invest 120:4332–4341

Usami M, Kishimoto K, Ohata A, Miyoshi M, Aoyama M, Fueda Y, Kotani J (2008) Butyrate and trichostatin A attenuate nuclear factor kappaB activation and tumor necrosis factor alpha secretion and increase prostaglandin E2 secretion in human peripheral blood mononuclear cells. Nutr Res 28:321–328

van der Velden WJFM, Blijlevens NMA, Maas FMHM, Schaap NPM, Jansen JH, van der Reijden BA, Feuth T, Dolstra H, Donnelly JP (2009) NOD2 polymorphisms predict severe acute graft-versus-host and treatment-related mortality in T-cell-depleted haematopoietic stem cell transplantation. Bone Marrow Transplant 44:243–248

Vecchio AJ, Rathnayake SS, Stroud RM (2021) Structural basis for *Clostridium perfringens* enterotoxin targeting of claudins at tight junctions in mammalian gut. Proc Natl Acad Sci 118:e2024651118

Verlinden A, Mikulska M, Knelange NS, Averbuch D, Styczynski J, Infectious Diseases Working Party (IDWP) of the European Group for Blood and Marrow Transplantation Group (EBMT) (2020) Current antimicrobial practice in febrile neutropenia across Europe and Asia: the EBMT infectious disease working party survey. Bone Marrow Transplant 55:1588–1594

Vinolo MAR, Rodrigues HG, Hatanaka E, Sato FT, Sampaio SC, Curi R (2011) Suppressive effect of short-chain fatty acids on production of proinflammatory mediators by neutrophils. J Nutr Biochem 22:849–855

Voth DE, Ballard JD (2005) *Clostridium difficile* toxins: mechanism of action and role in disease. Clin Microbiol Rev 18:247–263

Wang H-B, Wang P-Y, Wang X, Wan Y-L, Liu Y-C (2012) Butyrate enhances intestinal epithelial barrier function via up-regulation of tight junction protein Claudin-1 transcription. Dig Dis Sci 57:3126–3135

Wang B, Wu G, Zhou Z, Dai Z, Sun Y, Ji Y, Li W, Wang W, Liu C, Han F, Wu Z (2015) Glutamine and intestinal barrier function. Amino Acids 47:2143–2154

Wang R, Yang X, Liu J, Zhong F, Zhang C, Chen Y, Sun T, Ji C, Ma D (2022) Gut microbiota regulates acute myeloid leukaemia via alteration of intestinal barrier function mediated by butyrate. Nat Commun 13:2522

Wei L, Wen X-S, Xian CJ (2021) Chemotherapy-induced intestinal microbiota dysbiosis impairs mucosal homeostasis by modulating toll-like receptor Signaling pathways. Int J Mol Sci 22:9474

Willems L, Porcher R, Lafaurie M, Casin I, Robin M, Xhaard A, Andreoli AL, Rodriguez-Otero P, Dhedin N, Socié G, Ribaud P, Peffault de Latour R (2012) Clostridium difficile infection after allogeneic hematopoietic stem cell transplantation: incidence, risk factors, and outcome. Biol Blood Marrow Transplant 18:1295–1301

Williams DA (2001) Inflammatory cytokines and mucosal injury. JNCI Monogr 2001:26–30

Williams MD, Braun LA, Cooper LM, Johnston J, Weiss RV, Qualy RL, Linde-Zwirble W (2004) Hospitalized cancer patients with severe sepsis: analysis of incidence, mortality, and associated costs of care. Crit Care 8:R291–R298

Worth LJ, Thursky KA, Seymour JF, Slavin MA (2007) Vancomycin-resistant enterococcus faecium infection in patients with hematologic malignancy: patients with acute myeloid leukemia are at high-risk. Eur J Haematol 79:226–233

Yamamoto ML, Schiestl RH (2014) Lymphoma caused by intestinal microbiota. Int J Environ Res Public Health 11:9038–9049

Yan H, Ajuwon KM (2017) Butyrate modifies intestinal barrier function in IPEC-J2 cells through a selective upregulation of tight junction proteins and activation of the Akt signaling pathway. PLoS One 12:e0179586

Yang H, Wildhaber B, Teitelbaum D (2003) Keratinocyte growth factor improves epithelial function after massive small bowel resection. J Parenter Enter Nutr 27:198–206

Ye H, Lv M, Zhao X, Zhao X, Huang X (2012) Plasma level of lipopolysaccharide-binding protein is indicative of acute graft-versus-host disease following allogeneic hematopoietic stem cell transplantation. Int J Hematol 95:680–688

Yuan J-H, Xie Q-S, Chen G-C, Huang C-L, Yu T, Chen Q-K, Li J-Y (2021) Impaired intestinal barrier function in type 2 diabetic patients measured by serum LPS, Zonulin, and IFABP. J Diabetes Complicat 35:107766

Zaura E, Brandt BW, Teixeira de Mattos MJ, Buijs MJ, Caspers MPM, Rashid M-U, Weintraub A, Nord CE, Savell A, Hu Y, Coates AR, Hubank M, Spratt DA, Wilson M, Keijser BJF, Crielaard W (2015) Same exposure but two radically different responses to antibiotics: resilience of the salivary microbiome versus long-term microbial shifts in feces. MBio 6:e01693-15

Zuo L, Kuo W-T, Turner JR (2020) Tight junctions as targets and effectors of mucosal immune homeostasis. Cell Mol Gastroenterol Hepatol 10:327–340

Autophagy as a Limiting Factor to the Effectiveness of Tyrosine Kinase Inhibitors in Blood Cancers

Keli Lima, Manuela Albuquerque de Melo, Brunno Gilberto Santos de Macedo, Fabiola Traina, and João Agostinho Machado-Neto

Abstract

Autophagy, a crucial cellular process involved in maintaining homeostasis through the degradation and recycling of intracellular components, has emerged as a significant player in drug resistance mechanisms in various cancers. This chapter explores the intricate interplay between autophagy and resistance to tyrosine kinase inhibitors in three types of hematological malignancies: chronic myeloid leukemia, Philadelphia-negative myeloproliferative neoplasms, and acute myeloid leukemia. In chronic myeloid leukemia, tyrosine kinase inhibitors targeting the BCR::ABL1 fusion protein, such as imatinib, induce autophagy as a survival mechanism, contributing to therapy resistance. Inhibition of autophagy potentiates the efficacy of tyrosine kinase inhibitors, suggesting a promising strategy for overcoming resistance. Similar observations are made in Philadelphia-negative myeloproliferative neoplasms, where the inhibition of the JAK2/STAT signaling pathway drives autophagy, and the suppression of autophagy enhances the cytotoxic effects of JAK2 inhibitors like ruxolitinib.

K. Lima
Department of Pharmacology, Institute of Biomedical Sciences, University of São Paulo, São Paulo, Brazil

Laboratory of Medical Investigation in Pathogenesis and Targeted Therapy in Onco-Immuno-Hematology (LIM-31), Department of Internal Medicine, Hematology Division, Faculdade de Medicina, University of São Paulo, São Paulo, Brazil

M. A. de Melo · B. G. S. de Macedo · F. Traina (✉)
Department of Medical Images, Hematology, and Oncology, Ribeirão Preto Medical School, University of São Paulo, São Paulo, Brazil
e-mail: ftraina@fmrp.usp.br

J. A. Machado-Neto (✉)
Department of Pharmacology, Institute of Biomedical Sciences, University of São Paulo, São Paulo, Brazil
e-mail: jamachadoneto@usp.br

© The Author(s), under exclusive license to Springer Nature Switzerland AG 2024
Interdisciplinary Cancer Research, https://doi.org/10.1007/16833_2024_361
Published online: 31 August 2024

In acute myeloid leukemia with FLT3 mutations, FLT3 inhibitors induce autophagy, which also promotes resistance to therapy. Combined inhibition of autophagy and FLT3 signaling shows synergistic effects in overcoming resistance, although the complex interactions with leukemia stem cells require further investigation. In summary, this chapter underscores the importance of understanding the role of autophagy in drug resistance and its potential as a therapeutic target in altered tyrosine kinase–driven blood cancers. Future research should focus on elucidating the specific mechanisms underlying autophagy modulation in different cancer types and identifying biomarkers to guide molecularly targeted treatment strategies.

Keywords

Autophagy · Combined therapy · Drug resistance · Hematological neoplasm · Tyrosine kinase inhibitors

Abbreviations

3-MA	3-methyladenine
AMPK	Adenosine monophosphate–activated protein kinase
ATG101	Autophagy-related protein 101
ATG12:	Autophagy-related protein 12
ATG13	Autophagy-related protein 13
ATG14L	Autophagy-related protein 14-like
ATG5	Autophagy-related protein 5
BCL2	B cell lymphoma 2
BCL2L1	B cell lymphoma 2-like 1
BCL-XL	B cell lymphoma-extra large
BCR::ABL1	Breakpoint cluster region-Abelson 1
BECN1	Beclin 1
FDA	Food and Drug Administration
FIP200	Focal adhesion kinase family interacting protein of 200 kDa
FLT3	Fms-like tyrosine kinase 3
GβL	Beta subunit–type G protein
ITD	Internal tandem duplication
JAK1	Janus kinase 1
JAK2	Janus kinase 2
JAK3	Janus kinase 3
KIT	Kit proto-oncogene receptor tyrosine kinase
LC3B:	Light chain microtubule-associated protein 3B
MAPK:	Mitogen-activated protein kinase
mTOR:	Mechanistic target of rapamycin
mTORC1	mTOR signaling complex 1
PCR	Polymerase chain reaction

PDGFR	Platelet-derived growth factor receptor
Ph	Philadelphia
PI3K	Phosphoinositide 3-Kinase
PI3KC3-C1	Class III phosphatidylinositol-3-phosphate kinase complex 1
PI3P	Phosphatidylinositol-3-phosphate
PRAS40	Proline-rich AKT substrate 40 kDa
RET	Rearranged during transfection
SQSTM1/p62	Phosphotyrosine-independent ligand for the Lck SH2 domain of 62 kDa
STAT	Signal transducer and activator of transcription
TYK2	Tyrosine kinase 2
ULK1	Unc-51 like autophagy activating kinase 1

1 Introduction

Autophagy is a dynamic process whose main function is the recycling of intracellular material to maintain homeostasis and involves the rearrangement of subcellular membranes that sequester long-lived, supernumerary, and damaged organelles or proteins and deliver them to the lysosomal machinery for degradation and reuse of their byproducts (Boya et al. 2013; Choi et al. 2013; Levine and Yuan 2005; Ravikumar et al. 2010). The early stages of autophagy encompass the initiation of vesicle nucleation and subsequent expansion of an isolation membrane, often referred to as a phagophore. This phagophore's edges later merge through vesicle closure, giving rise to the autophagosome—a double-membrane vesicle responsible for sequestering cytoplasmic material. Following its formation, the autophagosome fuses with a lysosome, forming the autophagolysosome. Within this compartment, the captured material, along with the inner membrane, undergoes degradation (Levine and Kroemer 2008). Autodigestion not only provides nutrients to maintain vital cellular functions but can also remove hazardous intracellular entities such as damaged or superfluous organelles, denatured, misfolded, and aggregated proteins, and invading microorganisms. Interestingly, autophagy autodigestion is now emerging as a central biological pathway that functions to promote cellular health and longevity (Levine and Kroemer 2008; Mizushima et al. 2008).

Multiple forms of autophagy exist, each with distinct mechanisms and functions. These include microautophagy, macroautophagy, and chaperone-mediated autophagy. Microautophagy and macroautophagy are capable of engulfing large structures through both selective and nonselective processes. In contrast, chaperone-mediated autophagy selectively degrades soluble proteins (Klionsky 2005; Mizushima et al. 2008). Chaperone-mediated autophagy occurs in lysosomes and involves the selective translocation of cytosolic proteins that are marked by a pentapeptide motif across the lysosomal membrane. Microautophagy, in turn, occurs when cytosolic contents are engulfed by direct invagination of lysosomal membranes into tubulovesicular structures. In macroautophagy, hereafter termed autophagy, phagophores elongate and fuse while engulfing the cytoplasm within double-membrane vesicles called autophagosomes. Autophagosomes first fuse with endosomes to form hybrid organelles called amphisomes that then fuse with acidic lysosomes where the trapped cytosolic contents are degraded (Klionsky 2005; Massey et al. 2006; Ravikumar et al. 2010). Macroautophagy can also be classified based on the process of cargo selectivity. The non-selective bulk autophagy can indiscriminately degrade intracellular content, particularly under nutritional depletion stimuli, while selective autophagy modalities termed according to the target such as mitophagy (mitochondria), aggrephagy (protein aggregates), pexophagy (peroxisomes), and intracellular bacteria (xenophagy) relies on ubiquitylation and removal of specific cargos as means to cellular quality and quantity control (Gatica et al. 2018; Klionsky 2005; Mizushima et al. 2011; Vargas et al. 2023).

The autophagic pathway responds to a multitude of environmental cues. For instance, the presence of nutrients triggers signaling pathways reliant on mechanistic target of rapamycin (mTOR), along with the class I-AKT/PI3K pathway activated by insulin and other growth factors, which function as inhibitors of autophagy. Moreover, the AKT protein exerts a negative regulatory effect on autophagy by phosphorylating beclin 1. Conversely, adenosine monophosphate–activated protein kinase (AMPK) acts as a negative regulator of mTOR while directly phosphorylating the ULK1 (Unc-51 like autophagy activating kinase 1) protein, thereby serving as a positive modulator of autophagy in response to energy depletion (Choi et al. 2013). The mTOR protein resides in the mTOR signaling complex 1 (mTORC1), which is also composed of Raptor (protein associated with mTOR regulation), GβL (beta subunit–type G protein), and PRAS40 (proline-rich AKT substrate 40 kDa), which in turn regulate the "mTOR substrate complex," which consists of the proteins ULK1, ATG13, and ATG101 and the focal adhesion kinase family interacting protein of 200 kDa (FIP200) (Choi et al. 2013). Autophagy is also regulated by the class III phosphatidylinositol-3-phosphate kinase complex 1 (PI3KC3-C1), which is composed of a beclin 1; VPS34, a class III PI3K; VPS15, a scaffolding subunit; and ATG14L. The stimulation of this complex generates phosphatidylinositol-3-phosphate (PI3P), which promotes the nucleation of the autophagosomal membrane (Choi et al. 2013). Besides ATG9 vesicle–mediated transport and transfer of phospholipid from membrane sources such as mitochondria, endoplasmic reticulum, Golgi apparatus, and even the plasma membrane itself, the autophagosomal elongation and maturation comprise the activity of two cooperative ubiquitin-type conjugation systems. The ATG5-ATG12-ATG16L system, where the ATG12 protein is conjugated to ATG5 by the enzymes ATG7 and ATG10, then forms a complex with ATG16L1, which participates in the maturation of the autophagic membrane by colocalizing the conjugation system to the elongating phagophore and a second LC3 (light chain microtubule-associated protein 3B) conjugation system, where LC3 and its homologs are modified with the cellular phospholipid, phosphatidylethanolamine (PE) (Choi et al. 2013). The precursor form of LC3 undergoes cleavage by the ATG4B protease, resulting in the generation of the LC3-I form, characterized by an exposed lipid conjugation site in the C-terminal region. Subsequently, the conjugation of phosphatidylethanolamine (PE) with LC3-I takes place through the sequential action of ATG7 and ATG3 activities. In mammals, the conversion of LC3-I (the free form) to LC3-II (the PE-conjugated form) represents a pivotal regulatory step in autophagosome formation (Choi et al. 2013).

Autophagy plays a vital role in promoting cell survival and assists malignant cells in resisting metabolic and therapeutic stress (Amaravadi et al. 2019). Autophagy has emerged as a potent cancer escape mechanism implicated in the development of drug resistance across various cancer types, including central nervous system tumors, melanoma, lung cancer, bladder cancer, and thyroid cancer. Combining autophagy inhibition with standard therapies holds promise in mitigating and potentially reversing therapy resistance in these cancers (Goodall et al. 2014; Levy et al. 2017; Liu et al. 2015; McAfee et al. 2012; Wang et al. 2017; Zou et al. 2013). Studies have demonstrated that genetic or pharmacological inhibition of autophagy increases the cytotoxicity of cancer chemotherapeutic agents (Carew et al. 2007; White 2012). Thus, one can differentially target autophagy in a context- and disease-stage-specific manner, making autophagy inhibition useful in promoting tumor regression (Mizushima et al. 2008; Mulcahy Levy et al. 2017). However, the role of autophagy in cancer is complex and some variables must be taken into consideration, such as tumor subtypes, stages of tumor progression, cellular context, or drugs that induce this process (White 2012). For example, although some studies demonstrate that autophagy can confer a survival advantage in established tumors, other data suggest that autophagy has a role in tumor suppression by promoting death in tumor cells resistant to apoptosis (Degenhardt et al. 2006). Furthermore, the removal of intracellular elements that amplify genomic instability, or even autophagic cell death itself to a lesser extent, participates in autophagy-related tumor suppression functions (Folkerts et al. 2019; Seo et al. 2022). Collectively, these findings underscore the necessity of investigating the role of autophagy across various cancer types, encompassing both hematological malignancies and those with specific mutations (Larrue et al. 2016).

2 Autophagy as a Consequence of Treatment with Tyrosine Kinase Inhibitors in Chronic Myeloid Leukemia

In 1960, Nowell and Hungerford (Nowell and Hungerford 1960) discovered the Philadelphia (Ph) chromosome, the first consistent chromosomal abnormality associated with a specific type of leukemia, which was a paradigm shift in the understanding of cancer biology. Next, Rowley (Rowley 1973) described the Ph chromosome as a result of a t(9;22) reciprocal chromosomal translocation. Finally, the involvement of the *ABL1* gene, present on chromosome 9, and *BCR*, present on chromosome 22, was described, and the fusion gene *BCR*::*ABL1* and its chimeric protein were characterized (Bartram et al. 1983; Groffen et al. 1984). The BCR::ABL1 fusion protein presents constitutive tyrosine kinase activity that is responsible for the activation of cell survival and proliferation signaling pathways, of which we highlight the MAPK and PI3K/AKT/mTOR pathways (Deininger et al. 2000). Given the central role of BCR::ABL1 in the pathogenesis of chronic myeloid leukemia, selective pharmacological inhibitors for this oncoprotein were developed (Goldman and Melo 2003). Imatinib (formerly CGP57418B or STI571) was the most promising inhibitor in preclinical studies (Druker et al. 1996) and was the first tyrosine kinase inhibitor used in clinical trials in patients with chronic myeloid leukemia (Druker et al. 2001; O'Brien et al. 2003). The initial results of clinical studies were impressive and changed the paradigms of targeted treatment in oncology. Long-term studies confirm the safety and efficacy of imatinib in chronic myeloid leukemia (Hochhaus et al. 2017). Due to the development of resistance or intolerance to imatinib, novel inhibitors of BCR::ABL1 or signaling pathways regulated by it have been developed; highlighted in the clinic are dasatinib, nilotinib, and ponatinib, and currently tyrosine kinase inhibitors for the treatment of chronic myeloid leukemia are in the third generation (Shoukier et al. 2021).

Considering that many patients present a complete molecular response for a long period, eligibility criteria for discontinuation of tyrosine kinase inhibitors, as well as for monitoring, were created (Inzoli et al. 2022). While a considerable number of patients experience remission after treatment withdrawal, current clinical and biological factors linked to improved outcomes do not provide precise stratification for eligibility or assessment of relapse risk (Inzoli et al. 2022), which indicates that studies to better understand the mechanisms of resistance and/or permanence of the neoplastic cell are still necessary.

The induction of autophagy by tyrosine kinase inhibitors has been previously documented as a survival mechanism in experimental models of chronic myeloid leukemia (Bellodi et al. 2009; Can et al. 2011). Treating 32D cells expressing the BCR::ABL1 p210 isoform and K562 (a BCR::ABL1^{+} cell line) with pharmacological inhibitors of autophagy, such as chloroquine and bafilomycin A1, enhances imatinib-induced apoptosis. Similarly, pharmacological inhibitors of autophagy further reduced the colony formation cell percentage when compared to the imatinib and dasatinib effects alone upon hematopoietic stem cells isolated from chronic myeloid leukemia patients (Bellodi et al. 2009). In K562 cells, autophagy induced by ruxolitinib was also reported, which was attributed to the regulation of genes involved in the AKT/mTOR pathway (Bagca et al. 2016). Ianniciello et al. (2017) demonstrated that autophagy plays a crucial role in the proliferation of CD34^{+} cells derived from chronic myeloid leukemia under hypoxic conditions. Interestingly, this reliance on autophagy was not observed in normal CD34^{+} cells. These findings suggest that targeting autophagy could be a selective strategy to combat malignant hematopoietic cells while sparing normal ones (Ianniciello et al. 2017). Given this body of evidence, future studies could address autophagy biomarkers in the discontinuation responses of tyrosine kinase inhibitors for chronic myeloid leukemia, since this biological processor can be pharmacologically modulated. Indeed, given the promising in vitro results regarding the reduction of chronic myeloid leukemia stem cells (Bellodi et al. 2009), a phase II clinical trial was developed (CHOICES—CHlOroquine and Imatinib Combination to Eliminate Stem cells, ClinicalTrials.gov: NCT01227135). In the CHOICES trial, the objective was to assess the combination of imatinib with hydroxychloroquine in chronic myeloid leukemia patients who had attained a substantial cytogenetic response but still exhibited detectable residual disease as evidenced by quantitative PCR. The results of the clinical trial indicate a modest improvement in quantitative PCR levels at 12 and 24 months, suggesting autophagy inhibition perhaps of clinical value in patients with chronic myeloid leukemia in the blast phase (Horne et al. 2020).

3 Autophagy as a Consequence of Treatment with JAK2 Inhibitors in Philadelphia-Negative Myeloproliferative Neoplasms

In humans, the *JAK2* gene is located on chromosome 9p24 and possesses 25 exons encoding a protein containing 1,132 amino acids. The JAK2 protein is a member of the Janus kinase (JAK) family, which acts as non-receptor tyrosine kinase. JAK2 plays an important role in cytokine receptor signaling and consists of four members, JAK1, JAK2, JAK3, and TYK2 (Gnanasambandan and Sayeski 2011; Yamaoka et al. 2004). In 2005, the presence of a recurrent mutation in the pseudokinase domain (JH2) of JAK2 was reported to drive the constitutive activation of the protein (James et al. 2005; Kralovics et al. 2005; Levine et al. 2005; Scott et al. 2005). This mutation occurs in exon 14 and is a missense mutation (g.1849 $G > T$) that generates a substitution of valine for phenylalanine at amino acid position 617 (V617F) (Chen and Mullally 2014). This discovery broke paradigms in understanding the pathogenesis of classical Philadelphia-negative myeloproliferative neoplasms. The presence of the $JAK2^{V617F}$ mutation is observed in ~95% of patients with polycythemia vera, and 50–60% with essential thrombocythemia and primary myelofibrosis (Gou et al. 2022). When activated, JAK2 phosphorylates signal transducer and activator of transcription (STAT) proteins and activates downstream pathways such as PI3K/AKT/mTOR (Chen and Mullally 2014; Wolf et al. 2013). Among the transcriptional targets of STAT proteins, the genes *BCL2* and *BCL2L1* (also known as BCL-XL protein) stand out as an additional autophagy regulation interface alongside mTOR activation (Guo et al. 2015; Li et al. 2010; Sepulveda et al. 2007; Wolf et al. 2013).

Given the central role of the JAK2/STAT signaling axis, pharmacological inhibitors of this pathway have been developed. Ruxolitinib (formerly known as INCB018424) is a selective JAK1/2 inhibitor approved by the Food and Drug Administration (FDA) for the treatment of intermediate- and high-risk primary myelofibrosis, and polycythemia vera patients with inadequate response or intolerance to hydroxyurea (Quintas-Cardama et al. 2010). Ruxolitinib is well tolerated, reduces inflammatory cytokines and splenomegaly, and improves constitutional symptoms in patients with primary myelofibrosis (Harrison et al. 2012; Pardanani et al. 2011; Verstovsek et al. 2012). Ruxolitinib also controls the hematocrit, reduces the spleen volume, and improves symptoms in patients with polycythemia vera

(Vannucchi et al. 2015). Despite the clinical benefits, ruxolitinib treatment does not reverse bone marrow fibrosis and does not lead to the elimination of the malignant clone, indicating the presence of gaps in knowledge that require further investigation.

In $JAK2^{V617F}$ experimental models, it was observed that ruxolitinib reduces the activation of mTOR and the expression of BCL-2 and BCL-XL, which was associated with the triggering of autophagy, as evidenced by the consumption of LC3BI/II and SQSTM1/p62 (Courdy et al. 2023; Ishida et al. 2018; Machado-Neto et al. 2020). Pharmacological studies with autophagy inhibitors (3-methyladenine, bafilomycin A1, hydroxychloroquine, and SAR405) highlighted that, in this context, autophagy acts as a resistance mechanism since the combination of ruxolitinib with autophagy inhibitors results in the accumulation of mitochondrial damage and increased apoptosis in $JAK2^{V617F}$ cells (Courdy et al. 2023; Machado-Neto et al. 2020). Interestingly, Machado-Neto et al. (2020) found that approximately 15% of patients with Philadelphia-negative myeloproliferative neoplasia harbor mutations in autophagy-related genes. This genetic alteration was linked to advanced age and hinted at a potential for poorer survival outcomes (Machado-Neto et al. 2020). In polycythemia vera and essential thrombocythemia patients, the impairment of autophagy in platelets can exert a negative bioenergetic effect on platelet function, promoting platelet hyperreactivity and dysfunctional clot retraction, which could be associated with the increased risk for both thrombosis and hemorrhage (Sanchez et al. 2022). Thus, understanding autophagy in the pathophysiology of myeloproliferative neoplasms and response to JAK2 inhibitors is important for improving therapeutic strategies and reducing the risk of adverse events in myeloproliferative neoplasms patients.

4 Autophagy as a Consequence of Treatment with FLT3 Inhibitors in Acute Myeloid Leukemia

The Fms-like tyrosine kinase 3 (FLT3), also known as CD135, is a member of the class III receptor tyrosine kinase family, which also includes KIT and PDGFR, and is expressed by hematopoietic progenitors. The FLT3 is composed of five domains: an extracellular immunoglobulin-like domain, a transmembrane domain, a juxtamembrane domain, and two tyrosine kinase domains (TKD1/2) (Rosnet et al. 1993; Stirewalt and Radich 2003). The activation of the FLT3 receptor with its ligand (FLT3L) plays an important role in the proliferation and differentiation of hemopoietic progenitor cells (Gilliland and Griffin 2002). The FLT3 kinase is expressed in blast cells in the majority of patients with acute myeloid leukemia, and the presence of FLT3 mutations has been detected in around 30% of these patients (Kiyoi et al. 1998; Nakao et al. 1996). Studies carried out by independent research groups demonstrate that the presence of FLT3-internal tandem duplication (ITD) confers an unfavorable prognosis, independent of karyotype, and is associated with lower overall and event-free survival (Abu-Duhier et al. 2000; Cloos et al. 2006; Kiyoi et al. 1999; Mrozek and Bloomfield 2006; Rombouts et al. 2000). The treatment of acute myeloid leukemia with FLT3-ITD mutations is currently considered a challenge (Kindler et al. 2010). Although selective FLT3 inhibitors are currently in clinical use, clinical trials with these drugs now show limited efficacy with transient responses (Hospital et al. 2017; Wander et al. 2014). Therefore, mechanisms related to the acquisition of resistance to these drugs deserve to be investigated.

The connections between FLT3 and autophagy in acute myeloid leukemia have been progressively uncovered (Du et al. 2021). According to Mulcahy Levy and colleagues, inhibition of autophagy can overcome resistance to tyrosine kinase inhibitors in vitro studies using human cell lines and primary leukemic cells (Levy et al. 2017). Heydt and collaborators have also observed that the use of SAR405, a class III PI3K inhibitor, was able to overcome resistance to quizartinib in MV4–11 and MOLM-14 cells (Heydt et al. 2018).

The involvement of autophagy in the post-translational degradation of FLT3-ITD was also studied. In 2016, Larrue et al. showed that proteasome inhibitors like bortezomib can trigger autophagy, leading to the degradation of FLT3-ITD proteins in leukemic cells (Larrue et al. 2016). Rudat and colleagues also discovered that RET, a highly activated tyrosine kinase receptor in acute myeloid leukemia, can drive the activation of mTOR and then inhibit autophagic degradation of the FLT3 protein (Rudat et al. 2018). Hence, the suppression of RET by pharmacological inhibitors (vandetanib or danusertib), in combination with a selective inhibitor of FLT3 (crenolanib), synergistically diminishes the survival and proliferation of FLT3-mutated cells. According to this study, the induction of autophagy by proteasome inhibitors or RET inhibitors, when combined with FLT3 inhibitors, presents a potential therapeutic avenue. This combined approach also helps prevent acquired resistance following treatments with tyrosine kinase inhibitors in acute myeloid leukemia patients with FLT3 mutation (Rudat et al. 2018).

In cell lines and primary cells from patients with acute myeloid leukemia that present the FLT3-ITD and $FLT3^{D815Y}$ mutations, the pharmacological inhibition of autophagy (chloroquine) overcomes the resistance to the multi-kinase inhibitor sorafenib (Xu et al. 2022). An analysis of integration with phospho-proteomics highlights autophagy as a pivotal pathway modulated by FLT3 inhibition in MV4-11 cells (Koschade et al. 2022). Genetic (shATG7, shULK1, and shBECN1) or pharmacological (ROC-235, Lys05, PIK-III, and SAR-405) inhibition of autophagy appears to synergize effects with FLT3 inhibitors (quizartinib, crenolanib, and gilteritinib) in cell lines, primary cells from patients with acute myeloid leukemia, and in vivo murine models, reinforcing the role of autophagy in resistance to FLT3 inhibitors in acute myeloid leukemia with FLT3 mutation (Dupont et al. 2022; Koschade et al. 2022). It has also been demonstrated that inhibition of BTK (CG-806) alters autophagic flux and acts as a key event in overcoming FLT3 inhibitor resistance in experimental models of acute myeloid leukemia with mutation in FLT3 (Zhang et al. 2023). On the other hand, inhibiting autophagy diminished quiescence and depleted the repopulating potential of FLT3-ITD leukemia stem cells. Surprisingly, treatment with FLT3 kinase inhibitor (quizartinib) reduced mitochondrial respiration and countered the effects of autophagy inhibition (Lys05) on leukemia stem cell attrition. These findings unveil unforeseen antagonistic interactions between combined oncogenic tyrosine kinase inhibition and autophagy inhibition in leukemia stem cells, underscoring the context-dependency and necessity for further investigation before a clinical application for this therapeutic strategy (Qiu et al. 2022).

5 Conclusion

Integrating enhanced clinical studies with comprehensive molecular and cellular analyses represents the optimal approach to deepen our understanding of the biological mechanisms underlying autophagy in various cancers. This strategy allows for the development of a more informed and rational basis for determining when and how to manipulate autophagy during cancer therapy (Levy et al. 2017).

Here, the impact of mutations in three oncogenes, *BCR*::*ABL1*, *JAK2*, and *FLT3*, relevant for the physiopathology of blood cancers, on the regulation of autophagy, is presented. In general, the constitutive activation that mimics the presence of growth factors, which impact the STAT/BCL-2 and PI3K/AKT/mTOR pathways, seems to be a common point in the modulation of autophagy (Fig. 1). Pharmacological inhibition of these oncogenic pathways through tyrosine kinase inhibitors triggers autophagy and acts as a resistance mechanism, simulating deprivation of nutrient growth factors.

In certain scenarios, genetic or pharmacological inhibition of autophagy can, indeed, reduce cell proliferation and viability. Moreover, most of these studies demonstrate a potentiating effect, including synergistic interactions with tyrosine kinase inhibitors in chronic myeloid leukemia, myeloproliferative neoplasms, and acute myeloid leukemia experimental models. A summary of the main tyrosine kinase and autophagy inhibitors used in the preclinical studies is presented in Fig. 2.

Another point worth highlighting is the combination of tyrosine kinase and autophagy inhibitors on leukemic stem cells. As previously presented, in chronic myeloid leukemia stem cells, the inhibition of autophagy seemingly leads to the eradication of neoplastic clones (Bellodi et al. 2009), whereas in FLT3-mutated acute myeloid leukemia, this combination may result in antagonistic effects on leukemic stem cells (Qiu et al. 2022). Thus, future studies are still necessary to define the best role of autophagy inhibitors in context and specific cellular subpopulations. A summary of preclinical studies using tyrosine kinase and autophagy inhibitors in hematological neoplasm with constitutive active tyrosine kinase signaling is presented in Table 1.

In conclusion, autophagy plays a significant role in various blood cancers, including chronic myeloid leukemia, Philadelphia-negative myeloproliferative neoplasms, and acute myeloid leukemia. In these hematological neoplasms, constitutive activation of signaling pathways such as JAK2/STAT and PI3K/AKT/mTOR influences autophagy modulation, and tyrosine kinase inhibition is commonly

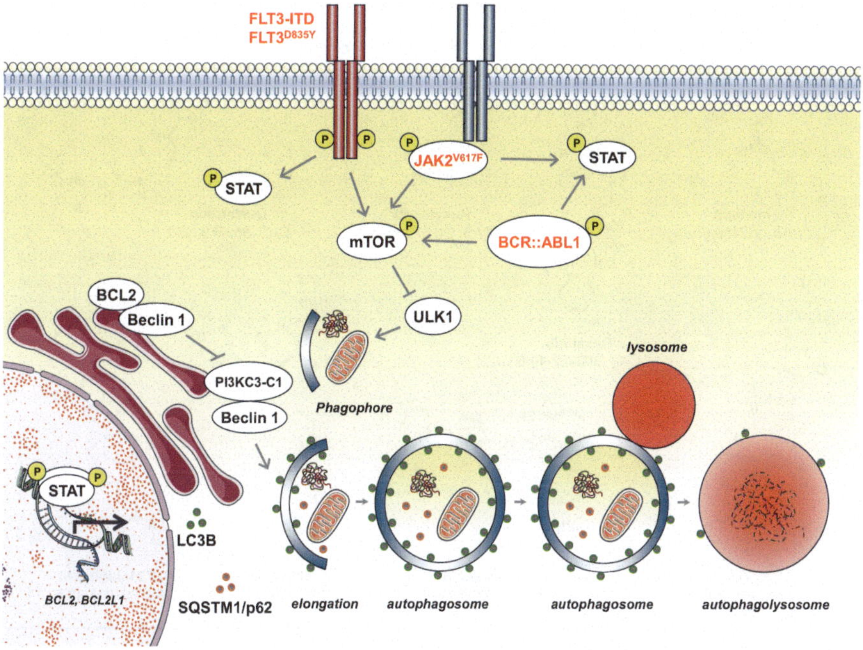

Fig. 1 A potential model for modulation of autophagy by mutations in tyrosine kinases, BCR::ABL1, JAK2, and FLT3, in blood cancer cells. The constitutive activation by mutations in *BCR::ABL1*, *JAK2*, and *FLT3* mimics the availability of growth factors and leads to STAT phosphorylation and activation of its target genes, including *BCL2* and *BCL2L1* (also known as BCL-XL protein). BCL-2 and/or BCL-XL bind to and inhibit beclin 1, thereby inhibiting the initiation of autophagy. Activation of these oncoproteins also induces mTOR activation which also inhibits autophagy. Thus, the use of tyrosine kinase inhibitors in this context could disarm the constitutive activation of these signaling pathways, similar to a situation of starvation, and activate autophagy as a survival mechanism. Abbreviations: *P* phosphorylation. (Figure created using Servier Medical Art (http://www.servier.com))

employed as a treatment strategy. However, resistance to the tyrosine kinase inhibitors poses a significant challenge, and autophagy emerges as a potentially exploitable resistance mechanism. Combining autophagy inhibitors with tyrosine kinase inhibitors shows promise in preclinical studies, suggesting a potential strategy to overcome resistance and improve treatment outcomes. Additionally, the role of autophagy in leukemic stem cells differs across cancer types, demanding deeper investigation to promote directed therapeutic strategies. Thus, future research should focus on understanding the precise mechanisms underlying autophagy modulation in different blood cancers and identifying biomarkers to guide molecularly targeted treatment approaches.

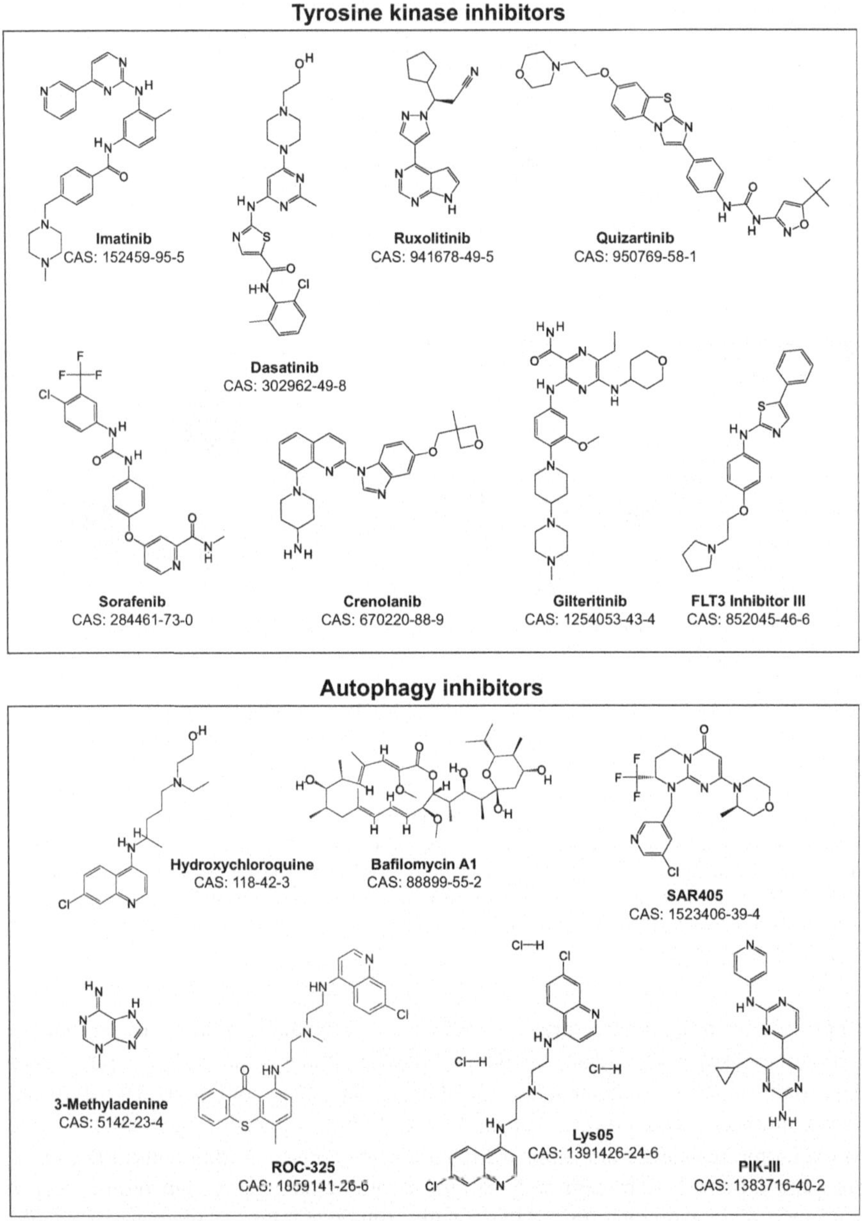

Fig. 2 Representation of the chemical structures of tyrosine kinase and autophagy inhibitors used for chronic myeloid leukemia, Philadelphia-negative myeloproliferative neoplasm, and acute myeloid leukemia experimental models. *CAS* Chemical Abstracts Service

Table 1 Summary of preclinical studies using TKI and autophagy inhibitors in tyrosine kinase–driven blood cancers

Disease	Experimental models	Molecular alteration	TKI	TKI target	Autophagy significance	Autophagy inhibitor	Outcome	Reference
CML	K-562, 32D-p210BCR::ABL1, primary CML cells	BCR::ABL1	Imatinib and dasatinib	ABL1	Resistance to therapy	Chloroquine and bafilomycin A1	Pharmacological inhibition of autophagy improves imatinib-induced apoptosis	Bellodi et al. (2009)
CML	K-562 cells	BCR::ABL1	Imatinib	ABL1	Not evaluated	Not used	Imatinib induces autophagy	Can et al. (2011)
CML	K-562 cells	BCR::ABL1	Ruxolitinib	JAK1/2	Cell death induction	Not used	Ruxolitinib induces autophagic cell death	Bagca et al. (2016)
CML	K-562 and primary CML ($CD34^+$) cells	BCR::ABL1	Not used	N/A	Cell survival under hypoxia	shRNA targeting ATG7	CML cells deficient for autophagy have impaired proliferation and increased apoptosis	Ianniciello et al. (2017)
CML	Clinical trial (62 CML patients)	BCR::ABL1	Imatinib	ABL1	Persistence of leukemia stem cells	Hydroxychloroquine	Modest improvement in quantitative PCR levels at 12 and 24 months, suggesting autophagy inhibition maybe of clinical value	Horne et al. 2020

(continued)

Table 1 (continued)

Disease	Experimental models	Molecular alteration	TKI	TKI target	Autophagy significance	Autophagy inhibitor	Outcome	Reference
							in chronic phase CML	
MPN	HEL and PVTL-2 cells	$JAK2^{V617F}$	Ruxolitinib	JAK1/2	Resistance to therapy	Chloroquine	Pharmacological inhibition of autophagy improves ruxolitinib-induced apoptosis	Ishida et al. (2018)
MPN	SET-2 cells	$JAK2^{V617F}$	Ruxolitinib	JAK1/2	Resistance to therapy	3-MA, bafilomycin A1, and hydroxychloroquine	Pharmacological inhibition of autophagy improves ruxolitinib-induced apoptosis	Machado-Neto et al. (2020)
MPN	HEL, SET-2, and primary MPN ($CD34^+$) cells	$JAK2^{V617F}$	Ruxolitinib	JAK1/2	Resistance to therapy	Bafilomycin A1, chloroquine, SAR405	Pharmacological inhibition of autophagy improves ruxolitinib-induced apoptosis	Courdy et al. (2023)
AML	MV4–11 and MOLM-14 cells	FLT3-ITD	FLT3 inhibitor III and quizartinib	FLT3	Cell proliferation support	Bafilomycin A1 and SAR405	Autophagy sustains the FLT3-ITD-dependent proliferation and	Heydt et al. (2018).

							autophagy inhibition overcomes acquired resistance to FLT3 inhibitors	
AML	MOLM-14, Ba/F3-FLT3-ITD, Ba/F3-FLT3^{D835Y}, Ba/F3-FLT3ITD$^{+/D835Y}$, and primary AML cells	FLT3-ITD and FLT3^{D835Y}	Sorafenib	Multikinase inhibitor	Resistance to therapy	Chloroquine	Autophagy inhibition overcomes FLT3 inhibitor resistance in FLT3 mutated AML	Xu et al. (2022)
AML	MV4–11 and MOLM-14 cells	FLT3-ITD	Quizartinib, crenolanib, and gilteritinib	FLT3	Resistance to therapy	ROC-235 and Lys05	Autophagy inhibition overcomes primary treatment resistance against FLT3-targeted therapy and improve its efficacy	Koschade et al. (2022)
AML	MV4–11 and MOLM-14 cells	FLT3-ITD	Gilteritinib	FLT3	Cell survival support	PIK-III and SAR-405	Autophagy inhibition reduces expansion and enhances apoptosis in FLT3-ITD AML cells and relapse rate in vivo	Dupont et al. (2022)

(continued)

Table 1 (continued)

Disease	Experimental models	Molecular alteration	TKI	TKI target	Autophagy significance	Autophagy inhibitor	Outcome	Reference
AML	MV4–11, MOLM-13, and primary AML cells	FLT3-ITD	Quizartinib	FLT3	Cell survival support	Lys05	Antagonistic interactions between combined oncogenic tyrosine kinase inhibition and autophagy inhibition in leukemia stem cells	Qiu et al. (2022)
AML	MOLM-14, Ba/F3-FLT3-ITD, Ba/F3-FLT3^{D835Y}, Ba/F3-FLT3ITD$^{+/D835Y}$ and primary AML cells	FLT3-ITD and FLT3^{D835Y}	Quizartinib and CG-806	FLT3 and BKT	Resistance to therapy	Chloroquine	Autophagy inhibition overcomes FLT3 inhibitor resistance in FLT3 mutated AML	Zhang et al. (2023)

Abbreviations: *CML* chronic myeloid leukemia, *MPN* myeloproliferative neoplasm, *AML* acute myeloid leukemia, *ITD* internal tandem duplication, *TKI* tyrosine kinase inhibitor. Gene and proteins are described according to the HUGO Gene Nomenclature Committee (HGNC)

Acknowledgments K.L. received a fellowship from the São Paulo Research Foundation (FAPESP) (grant 2020/12842-0). M.A.M. received a fellowship from FAPESP (grant 2021/11112-0). B.G.S.M. (grant 2022/03871-1). This study was supported by grants 2019/23864-7, 2021/11606-3, and 2013/08135-2 from FAPESP, and grants 465539/2014-9, 409401/2021-8, and 309614/2022-8 from the National Council for Scientific and Technological Development (CNPq). This study was financed in part by the Coordenação de Aperfeiçoamento de Pessoal de Nível Superior, Brasil (CAPES), Finance Code 001.

Compliance with Ethical Standards The authors declare no conflict of interest.

References

Abu-Duhier FM, Goodeve AC, Wilson GA, Gari MA, Peake IR, Rees DC, Vandenberghe EA, Winship PR, Reilly JT (2000) FLT3 internal tandem duplication mutations in adult acute myeloid leukaemia define a high-risk group. Br J Haematol 111:190–195

Amaravadi RK, Kimmelman AC, Debnath J (2019) Targeting Autophagy in Cancer: Recent Advances and Future Directions. Cancer Discov 9:1167–1181

Bagca BG, Ozalp O, Kurt CC, Mutlu Z, Saydam G, Gunduz C, Avci CB (2016) Ruxolitinib induces autophagy in chronic myeloid leukemia cells. Tumour Biol 37:1573–1579

Bartram CR, de Klein A, Hagemeijer A, van Agthoven T, Geurts van Kessel A, Bootsma D, Grosveld G, Ferguson-Smith MA, Davies T, Stone M et al (1983) Translocation of c-abl oncogene correlates with the presence of a Philadelphia chromosome in chronic myelocytic leukaemia. Nature 306:277–280

Bellodi C, Lidonnici MR, Hamilton A, Helgason GV, Soliera AR, Ronchetti M, Galavotti S, Young KW, Selmi T, Yacobi R, Van Etten RA, Donato N, Hunter A, Dinsdale D, Tirro E, Vigneri P, Nicotera P, Dyer MJ, Holyoake T, Salomoni P, Calabretta B (2009) Targeting autophagy potentiates tyrosine kinase inhibitor-induced cell death in Philadelphia chromosome-positive cells, including primary CML stem cells. J Clin Invest 119:1109–1123

Boya P, Reggiori F, Codogno P (2013) Emerging regulation and functions of autophagy. Nat Cell Biol 15:713–720

Can G, Ekiz HA, Baran Y (2011) Imatinib induces autophagy through BECLIN-1 and ATG5 genes in chronic myeloid leukemia cells. Hematology 16:95–99

Carew JS, Nawrocki ST, Kahue CN, Zhang H, Yang C, Chung L, Houghton JA, Huang P, Giles FJ, Cleveland JL (2007) Targeting autophagy augments the anticancer activity of the histone deacetylase inhibitor SAHA to overcome Bcr-Abl-mediated drug resistance. Blood 110:313–322

Chen E, Mullally A (2014) How does JAK2V617F contribute to the pathogenesis of myeloproliferative neoplasms? Hematology Am Soc Hematol Educ Program 2014:268–276

Choi AM, Ryter SW, Levine B (2013) Autophagy in human health and disease. N Engl J Med 368:651–662

Cloos J, Goemans BF, Hess CJ, van Oostveen JW, Waisfisz Q, Corthals S, de Lange D, Boeckx N, Hahlen K, Reinhardt D, Creutzig U, Schuurhuis GJ, Zwaan C, M., Kaspers, G.J. (2006) Stability and prognostic influence of FLT3 mutations in paired initial and relapsed AML samples. Leukemia 20:1217–1220

Courdy C, Platteeuw L, Ducau C, De Araujo I, Boet E, Sahal A, Saland E, Edmond V, Tavitian S, Bertoli S, Cougoul P, Granat F, Poillet L, Marty C, Plo I, Sarry JE, Manenti S, Mansat-De Mas V, Joffre C (2023) Targeting PP2A-dependent autophagy enhances sensitivity to ruxolitinib in JAK2(V617F) myeloproliferative neoplasms. Blood Cancer J 13:106

Degenhardt K, Mathew R, Beaudoin B, Bray K, Anderson D, Chen G, Mukherjee C, Shi Y, Gelinas C, Fan Y, Nelson DA, Jin S, White E (2006) Autophagy promotes tumor cell survival and restricts necrosis, inflammation, and tumorigenesis. Cancer Cell 10:51–64

Deininger MW, Goldman JM, Melo JV (2000) The molecular biology of chronic myeloid leukemia. Blood 96:3343–3356

Druker BJ, Tamura S, Buchdunger E, Ohno S, Segal GM, Fanning S, Zimmermann J, Lydon NB (1996) Effects of a selective inhibitor of the Abl tyrosine kinase on the growth of Bcr-Abl positive cells. Nat Med 2:561–566

Druker BJ, Talpaz M, Resta DJ, Peng B, Buchdunger E, Ford JM, Lydon NB, Kantarjian H, Capdeville R, Ohno-Jones S, Sawyers CL (2001) Efficacy and safety of a specific inhibitor of the BCR-ABL tyrosine kinase in chronic myeloid leukemia. N Engl J Med 344:1031–1037

Du W, Xu A, Huang Y, Cao J, Zhu H, Yang B, Shao X, He Q, Ying M (2021) The role of autophagy in targeted therapy for acute myeloid leukemia. Autophagy 17:2665–2679

Dupont M, Huart M, Lauvinerie C, Bidet A, Guitart AV, Villacreces A, Vigon I, Desplat V, El Habhab A, Pigneux A, Ivanovic Z, Brunet De la Grange P, Dumas PY, Pasquet JM (2022) Autophagy targeting and hematological mobilization in FLT3-ITD acute myeloid leukemia decrease repopulating capacity and relapse by inducing apoptosis of committed leukemic cells. Cancers (Basel) 14

Folkerts H, Hilgendorf S, Vellenga E, Bremer E, Wiersma VR (2019) The multifaceted role of autophagy in cancer and the microenvironment. Med Res Rev 39:517–560

Gatica D, Lahiri V, Klionsky DJ (2018) Cargo recognition and degradation by selective autophagy. Nat Cell Biol 20:233–242

Gilliland DG, Griffin JD (2002) The roles of FLT3 in hematopoiesis and leukemia. Blood 100: 1532–1542

Gnanasambandan K, Sayeski PP (2011) A structure-function perspective of Jak2 mutations and implications for alternate drug design strategies: the road not taken. Curr Med Chem 18: 4659–4673

Goldman JM, Melo JV (2003) Chronic myeloid leukemia--advances in biology and new approaches to treatment. N Engl J Med 349:1451–1464

Goodall ML, Wang T, Martin KR, Kortus MG, Kauffman AL, Trent JM, Gately S, MacKeigan JP (2014) Development of potent autophagy inhibitors that sensitize oncogenic BRAF V600E mutant melanoma tumor cells to vemurafenib. Autophagy 10:1120–1136

Gou P, Zhang W, Giraudier S (2022) Insights into the potential mechanisms of JAK2V617F somatic mutation contributing distinct phenotypes in myeloproliferative neoplasms. Int J Mol Sci 23

Groffen J, Stephenson JR, Heisterkamp N, de Klein A, Bartram CR, Grosveld G (1984) Philadelphia chromosomal breakpoints are clustered within a limited region, bcr, on chromosome 22. Cell 36:93–99

Guo J, Roberts L, Chen Z, Merta PJ, Glaser KB, Shah OJ (2015) JAK2V617F drives Mcl-1 expression and sensitizes hematologic cell lines to dual inhibition of JAK2 and Bcl-xL. PLoS One 10:e0114363

Harrison C, Kiladjian JJ, Al-Ali HK, Gisslinger H, Waltzman R, Stalbovskaya V, McQuitty M, Hunter DS, Levy R, Knoops L, Cervantes F, Vannucchi AM, Barbui T, Barosi G (2012) JAK inhibition with ruxolitinib versus best available therapy for myelofibrosis. N Engl J Med 366: 787–798

Heydt Q, Larrue C, Saland E, Bertoli S, Sarry JE, Besson A, Manenti S, Joffre C, Mansat-De Mas V (2018) Oncogenic FLT3-ITD supports autophagy via ATF4 in acute myeloid leukemia. Oncogene 37:787–797

Hochhaus A, Larson RA, Guilhot F, Radich JP, Branford S, Hughes TP, Baccarani M, Deininger MW, Cervantes F, Fujihara S, Ortmann CE, Menssen HD, Kantarjian H, O'Brien SG, Druker BJ (2017) Long-term outcomes of Imatinib treatment for chronic myeloid leukemia. N Engl J Med 376:917–927

Horne GA, Stobo J, Kelly C, Mukhopadhyay A, Latif AL, Dixon-Hughes J, McMahon L, Cony-Makhoul P, Byrne J, Smith G, Koschmieder S, BrUmmendorf TH, Schafhausen P, Gallipoli P, Thomson F, Cong W, Clark RE, Milojkovic D, Helgason GV, Foroni L, Nicolini FE, Holyoake TL, Copland M (2020) A randomised phase II trial of hydroxychloroquine and imatinib versus imatinib alone for patients with chronic myeloid leukaemia in major cytogenetic response with residual disease. Leukemia 34:1775–1786

Hospital MA, Green AS, Maciel TT, Moura IC, Leung AY, Bouscary D, Tamburini J (2017) FLT3 inhibitors: clinical potential in acute myeloid leukemia. Onco Targets Ther 10:607–615

Ianniciello A, Dumas PY, Drullion C, Guitart A, Villacreces A, Peytour Y, Chevaleyre J, Brunet de la Grange P, Vigon I, Desplat V, Priault M, Sbarba PD, Ivanovic Z, Mahon FX, Pasquet JM (2017) Chronic myeloid leukemia progenitor cells require autophagy when leaving hypoxia-induced quiescence. Oncotarget 8:96984–96992

Inzoli E, Aroldi A, Piazza R, Gambacorti-Passerini C (2022) Tyrosine Kinase Inhibitor discontinuation in chronic myeloid leukemia: eligibility criteria and predictors of success. Am J Hematol 97:1075–1085

Ishida S, Akiyama H, Umezawa Y, Okada K, Nogami A, Oshikawa G, Nagao T, Miura O (2018) Mechanisms for mTORC1 activation and synergistic induction of apoptosis by ruxolitinib and BH3 mimetics or autophagy inhibitors in JAK2-V617F-expressing leukemic cells including newly established PVTL-2. Oncotarget 9:26834–26851

James C, Ugo V, Le Couedic JP, Staerk J, Delhommeau F, Lacout C, Garcon L, Raslova H, Berger R, Bennaceur-Griscelli A, Villeval JL, Constantinescu SN, Casadevall N, Vainchenker W (2005) A unique clonal JAK2 mutation leading to constitutive signalling causes polycythaemia vera. Nature 434:1144–1148

Kindler T, Lipka DB, Fischer T (2010) FLT3 as a therapeutic target in AML: still challenging after all these years. Blood 116:5089–5102

Kiyoi H, Towatari M, Yokota S, Hamaguchi M, Ohno R, Saito H, Naoe T (1998) Internal tandem duplication of the FLT3 gene is a novel modality of elongation mutation which causes constitutive activation of the product. Leukemia 12:1333–1337

Kiyoi H, Naoe T, Nakano Y, Yokota S, Minami S, Miyawaki S, Asou N, Kuriyama K, Jinnai I, Shimazaki C, Akiyama H, Saito K, Oh H, Motoji T, Omoto E, Saito H, Ohno R, Ueda R (1999) Prognostic implication of FLT3 and N-RAS gene mutations in acute myeloid leukemia. Blood 93:3074–3080

Klionsky DJ (2005) The molecular machinery of autophagy: unanswered questions. J Cell Sci 118:7–18

Koschade SE, Klann K, Shaid S, Vick B, Stratmann JA, Tholken M, Meyer LM, Nguyen TD, Campe J, Moser LM, Hock S, Baker F, Meyer CT, Wempe F, Serve H, Ullrich E, Jeremias I, Munch C, Brandts CH (2022) Translatome proteomics identifies autophagy as a resistance mechanism to on-target FLT3 inhibitors in acute myeloid leukemia. Leukemia 36:2396–2407

Kralovics R, Passamonti F, Buser AS, Teo SS, Tiedt R, Passweg JR, Tichelli A, Cazzola M, Skoda RC (2005) A gain-of-function mutation of JAK2 in myeloproliferative disorders. N Engl J Med 352:1779–1790

Larrue C, Saland E, Boutzen H, Vergez F, David M, Joffre C, Hospital MA, Tamburini J, Delabesse E, Manenti S, Sarry JE, Recher C (2016) Proteasome inhibitors induce FLT3-ITD degradation through autophagy in AML cells. Blood 127:882–892

Levine B, Kroemer G (2008) Autophagy in the pathogenesis of disease. Cell 132:27–42

Levine B, Yuan J (2005) Autophagy in cell death: an innocent convict? J Clin Invest 115:2679–2688

Levine RL, Wadleigh M, Cools J, Ebert BL, Wernig G, Huntly BJ, Boggon TJ, Wlodarska I, Clark JJ, Moore S, Adelsperger J, Koo S, Lee JC, Gabriel S, Mercher T, D'Andrea A, Frohling S, Dohner K, Marynen P, Vandenberghe P, Mesa RA, Tefferi A, Griffin JD, Eck MJ, Sellers WR, Meyerson M, Golub TR, Lee SJ, Gilliland DG (2005) Activating mutation in the tyrosine kinase JAK2 in polycythemia vera, essential thrombocythemia, and myeloid metaplasia with myelofibrosis. Cancer Cell 7:387–397

Levy JMM, Towers CG, Thorburn A (2017) Targeting autophagy in cancer. Nat Rev Cancer 17: 528–542

Li G, Miskimen KL, Wang Z, Xie XY, Brenzovich J, Ryan JJ, Tse W, Moriggl R, Bunting KD (2010) STAT5 requires the N-domain for suppression of miR15/16, induction of bcl-2, and survival signaling in myeloproliferative disease. Blood 115:1416–1424

Liu JT, Li WC, Gao S, Wang F, Li XQ, Yu HQ, Fan LL, Wei W, Wang H, Sun GP (2015) Autophagy inhibition overcomes the antagonistic effect between Gefitinib and cisplatin in epidermal growth factor receptor mutant non--small-cell lung cancer cells. Clin Lung Cancer 16:e55–e66

Machado-Neto JA, Coelho-Silva JL, Santos FPS, Scheucher PS, Campregher PV, Hamerschlak N, Rego EM, Traina F (2020) Autophagy inhibition potentiates ruxolitinib-induced apoptosis in JAK2(V617F) cells. Investig New Drugs 38:733–745

Massey AC, Zhang C, Cuervo AM (2006) Chaperone-mediated autophagy in aging and disease. Curr Top Dev Biol 73:205–235

McAfee Q, Zhang Z, Samanta A, Levi SM, Ma XH, Piao S, Lynch JP, Uehara T, Sepulveda AR, Davis LE, Winkler JD, Amaravadi RK (2012) Autophagy inhibitor Lys05 has single-agent antitumor activity and reproduces the phenotype of a genetic autophagy deficiency. Proc Natl Acad Sci USA 109:8253–8258

Mizushima N, Levine B, Cuervo AM, Klionsky DJ (2008) Autophagy fights disease through cellular self-digestion. Nature 451:1069–1075

Mizushima N, Yoshimori T, Ohsumi Y (2011) The role of Atg proteins in autophagosome formation. Annu Rev Cell Dev Biol 27:107–132

Mrozek K, Bloomfield CD (2006) Chromosome aberrations, gene mutations and expression changes, and prognosis in adult acute myeloid leukemia. Hematol Am Soc Hematol Educ Program 2006:169–177

Mulcahy Levy JM, Zahedi S, Griesinger AM, Morin A, Davies KD, Aisner DL, Kleinschmidt-DeMasters BK, Fitzwalter BE, Goodall ML, Thorburn J, Amani V, Donson AM, Birks DK, Mirsky DM, Hankinson TC, Handler MH, Green AL, Vibhakar R, Foreman NK, Thorburn A (2017) Autophagy inhibition overcomes multiple mechanisms of resistance to BRAF inhibition in brain tumors. elife 6

Nakao M, Yokota S, Iwai T, Kaneko H, Horiike S, Kashima K, Sonoda Y, Fujimoto T, Misawa S (1996) Internal tandem duplication of the flt3 gene found in acute myeloid leukemia. Leukemia 10:1911–1918

Nowell P, Hungerford D (1960) A minute chromosome in human chronic granulocytic leukemia. Science 132:1497

O'Brien SG, Guilhot F, Larson RA, Gathmann I, Baccarani M, Cervantes F, Cornelissen JJ, Fischer T, Hochhaus A, Hughes T, Lechner K, Nielsen JL, Rousselot P, Reiffers J, Saglio G, Shepherd J, Simonsson B, Gratwohl A, Goldman JM, Kantarjian H, Taylor K, Verhoef G, Bolton AE, Capdeville R, Druker BJ (2003) Imatinib compared with interferon and low-dose cytarabine for newly diagnosed chronic-phase chronic myeloid leukemia. N Engl J Med 348: 994–1004

Pardanani A, Vannucchi AM, Passamonti F, Cervantes F, Barbui T, Tefferi A (2011) JAK inhibitor therapy for myelofibrosis: critical assessment of value and limitations. Leukemia 25:218–225

Qiu S, Kumar H, Yan C, Li H, Paterson AJ, Anderson NR, He J, Yang J, Xie M, Crossman DK, Lu R, Welner RS, Bhatia R (2022) Autophagy inhibition impairs leukemia stem cell function in FLT3-ITD AML but has antagonistic interactions with tyrosine kinase inhibition. Leukemia 36: 2621–2633

Quintas-Cardama A, Vaddi K, Liu P, Manshouri T, Li J, Scherle PA, Caulder E, Wen X, Li Y, Waeltz P, Rupar M, Burn T, Lo Y, Kelley J, Covington M, Shepard S, Rodgers JD, Haley P, Kantarjian H, Fridman JS, Verstovsek S (2010) Preclinical characterization of the selective JAK1/2 inhibitor INCB018424: therapeutic implications for the treatment of myeloproliferative neoplasms. Blood 115:3109–3117

Ravikumar B, Sarkar S, Davies JE, Futter M, Garcia-Arencibia M, Green-Thompson ZW, Jimenez-Sanchez M, Korolchuk VI, Lichtenberg M, Luo S, Massey DC, Menzies FM, Moreau K, Narayanan U, Renna M, Siddiqi FH, Underwood BR, Winslow AR, Rubinsztein DC (2010) Regulation of mammalian autophagy in physiology and pathophysiology. Physiol Rev 90: 1383–1435

Rombouts WJ, Blokland I, Lowenberg B, Ploemacher RE (2000) Biological characteristics and prognosis of adult acute myeloid leukemia with internal tandem duplications in the Flt3 gene. Leukemia 14:675–683

Rosnet O, Schiff C, Pebusque MJ, Marchetto S, Tonnelle C, Toiron Y, Birg F, Birnbaum D (1993) Human FLT3/FLK2 gene: cDNA cloning and expression in hematopoietic cells. Blood 82: 1110–1119

Rowley JD (1973) Letter: A new consistent chromosomal abnormality in chronic myelogenous leukaemia identified by quinacrine fluorescence and Giemsa staining. Nature 243:290–293

Rudat S, Pfaus A, Cheng YY, Holtmann J, Ellegast JM, Buhler C, Marcantonio DD, Martinez E, Gollner S, Wickenhauser C, Muller-Tidow C, Lutz C, Bullinger L, Milsom MD, Sykes SM, Frohling S, Scholl C (2018) RET-mediated autophagy suppression as targetable co-dependence in acute myeloid leukemia. Leukemia 32:2189–2202

Sanchez GR, Esparza O, McMahon B, Hernandez G, Le D, Nemkov T, D'Alessandro A, Thorburn A, López JA, Davizon-Castillo P (2022) Dysregulated autophagy in platelets from JAK2V617F-positive polycythemia vera and essential thrombocythemia underlies bioenergetic failure as a possible mechanism for thrombohemorrhagic complications. Blood 140(Supplement 1):9590–9591

Scott LM, Campbell PJ, Baxter EJ, Todd T, Stephens P, Edkins S, Wooster R, Stratton MR, Futreal PA, Green AR (2005) The V617F JAK2 mutation is uncommon in cancers and in myeloid malignancies other than the classic myeloproliferative disorders. Blood 106:2920–2921

Seo W, Silwal P, Song IC, Jo EK (2022) The dual role of autophagy in acute myeloid leukemia. J Hematol Oncol 15:51

Sepulveda P, Encabo A, Carbonell-Uberos F, Minana MD (2007) BCL-2 expression is mainly regulated by JAK/STAT3 pathway in human CD34+ hematopoietic cells. Cell Death Differ 14: 378–380

Shoukier M, Kubiak M, Cortes J (2021) Review of new-generation tyrosine kinase inhibitors for chronic myeloid leukemia. Curr Oncol Rep 23:91

Stirewalt DL, Radich JP (2003) The role of FLT3 in haematopoietic malignancies. Nat Rev Cancer 3:650–665

Vannucchi AM, Kiladjian JJ, Griesshammer M, Masszi T, Durrant S, Passamonti F, Harrison CN, Pane F, Zachee P, Mesa R, He S, Jones MM, Garrett W, Li J, Pirron U, Habr D, Verstovsek S (2015) Ruxolitinib versus standard therapy for the treatment of polycythemia vera. N Engl J Med 372:426–435

Vargas JNS, Hamasaki M, Kawabata T, Youle RJ, Yoshimori T (2023) The mechanisms and roles of selective autophagy in mammals. Nat Rev Mol Cell Biol 24:167–185

Verstovsek S, Mesa RA, Gotlib J, Levy RS, Gupta V, DiPersio JF, Catalano JV, Deininger M, Miller C, Silver RT, Talpaz M, Winton EF, Harvey JH Jr, Arcasoy MO, Hexner E, Lyons RM, Paquette R, Raza A, Vaddi K, Erickson-Viitanen S, Koumenis IL, Sun W, Sandor V, Kantarjian HM (2012) A double-blind, placebo-controlled trial of ruxolitinib for myelofibrosis. N Engl J Med 366:799–807

Wander SA, Levis MJ, Fathi AT (2014) The evolving role of FLT3 inhibitors in acute myeloid leukemia: quizartinib and beyond. Ther Adv Hematol 5:65–77

Wang W, Kang H, Zhao Y, Min I, Wyrwas B, Moore M, Teng L, Zarnegar R, Jiang X, Fahey TJ 3rd (2017) Targeting autophagy sensitizes BRAF-mutant thyroid cancer to Vemurafenib. J Clin Endocrinol Metab 102:634–643

White E (2012) Deconvoluting the context-dependent role for autophagy in cancer. Nat Rev Cancer 12:401–410

Wolf A, Eulenfeld R, Gabler K, Rolvering C, Haan S, Behrmann I, Denecke B, Haan C, Schaper F (2013) JAK2-V617F-induced MAPK activity is regulated by PI3K and acts synergistically with PI3K on the proliferation of JAK2-V617F-positive cells. JAKSTAT 2:e24574

Xu D, Chen Y, Yang Y, Yin Z, Huang C, Wang Q, Jiang L, Jiang X, Yin C, Liu Q, Yu G (2022) Autophagy activation mediates resistance to FLT3 inhibitors in acute myeloid leukemia with FLT3-ITD mutation. J Transl Med 20:300

Yamaoka K, Saharinen P, Pesu M, Holt VE 3rd, Silvennoinen O, O'Shea JJ (2004) The Janus kinases (Jaks). Genome Biol 5:253

Zhang W, Yu G, Zhang H, Basyal M, Ly C, Yuan B, Ruvolo V, Piya S, Bhattacharya S, Zhang Q, Borthakur G, Battula V, Konopleva M, Rice WG, Andreeff M (2023) Concomitant targeting of FLT3 and BTK overcomes FLT3 inhibitor resistance in acute myeloid leukemia through the inhibition of autophagy. Haematologica 108:1500–1514

Zou Y, Ling YH, Sironi J, Schwartz EL, Perez-Soler R, Piperdi B (2013) The autophagy inhibitor chloroquine overcomes the innate resistance of wild-type EGFR non-small-cell lung cancer cells to erlotinib. J Thorac Oncol 8:693–702

Immunogenetic Factors in the Pathogenesis of Myeloproliferative Neoplasms

Velizar Shivarov

Abstract

This chapter summarizes the findings of recent reports on immunogenetic studies in myeloproliferative neoplasms (MPNs) without *BCR::ABL1* translocation. The focus is on three entities, namely essential thrombocythemia (ET), polycythemia vera (PV), and primary myelofibrosis (PMF), which are driven by JAK2 V617F or CALR mutations. We mainly discuss the identified potential protective human leukocyte antigen class I or II (HLA-I or -II) alleles in MPNs with either type of mutations. We further mention the potential role of polymorphisms in NKG2D ligand genes that has been investigated in the settings of MPNs. Finally, we describe the known mechanisms of immune surveillance evasion in MPNs. In sum, more in-depth understanding of the role of various immunogenetic factors in MPNs will help optimizing the therapeutic approaches in those diseases.

Keywords

Chronic eosinophilic leukemia · Chronic myeloid leukemia · Chronic neutrophilic leukemia · Myeloproliferative neoplasms · Primary myelofibrosis

1 Introduction

The latest edition of the World Health Organization (WHO) classification of hematolymphoid tumors recognizes the defined in the previous editions category of myeloproliferative neoplasms (MPNs) comprising the following entities: chronic myeloid leukemia (CML), polycythemia vera (PV), essential thrombocythemia (ET), primary myelofibrosis (PMF), chronic neutrophilic leukemia (CNL), chronic

V. Shivarov (✉)
Department of Experimental Research, Medical University Pleven, Pleven, Bulgaria
e-mail: vshivarov@abv.bg

© The Author(s), under exclusive license to Springer Nature Switzerland AG 2024

Interdisciplinary Cancer Research, https://doi.org/10.1007/16833_2024_350
Published online: 4 September 2024

eosinophilic leukemia (CEL), juvenile myelomonocytic leukemia (JMML), and myeloproliferative neoplasm, not otherwise specified (MPN, NOS) (Khoury et al. 2022). The remainder of the text will focus on the three main *BCR::ABL1*-negative myeloproliferative neoplasms, namely PV, ET, and PMF, and the term "MPN" will be used to designate those entities unless explicitly specified otherwise. At a genetic level, the pathogenesis of the vast majority of those cases is driven by hotspot mutations in just three genes *JAK2*, *MPL*, and *CALR* (Khoury et al. 2022; Luque Paz et al. 2023; Grinfeld et al. 2018).

JAK2 V617F mutation was initially reported in 2005 and was subsequently identified in more than 95% PV patients in up to 60% ET or PMF (Grinfeld et al. 2018). This mutation leads to constitutional activation of the cytoplasmic tyrosine kinase JAK2, which signal downstream of the receptors for the major hematopoietic factors such as thrombopoietin, erythropoietin, and granulocyte-colony-stimulating factor, causing increased myeloproliferation. *MPL* gene encodes the cell surface receptor for thrombopoietin (TpoR). A number of hotspot missense mutations in its exon 10 (mostly M515L and M515K) are reported in less than 5% of the MPN cases with ET or PMF. These mutations cause constitutive signaling through TpoR and thereby skewing of hematopoiesis toward megakaryocyte lineage (Grinfeld et al. 2017). CALR mutations are reported in up to 1/3 of ET and PMF patients. These are exclusively insertions/deletions in the last exon 9 of the gene (most commonly a 52 base pair deletion (type 1 mutations) or five base pair insertion (type 2 mutations)) causing a 1 base pair shift in the reading frame and translation of a neomorphic C-terminus of the protein (Grinfeld et al. 2017). The main mechanism of action of those mutations is as an extracellular chaperone for TpoR causing its constitutive activation but additional cell intrinsic effects are not excluded (Grinfeld et al. 2017; How et al. 2019; Shivarov et al. 2014).

A subset of cases may not have mutation if either of those genes and are designated as triple-negative cases (Grinfeld et al. 2018). Additional frequently co-mutated genes are *TET2*, *ASXL1*, and *DNMT3A* and some splicing factors, epigenetic modifiers, and signaling pathways genes such as *SRSF2*, *SF3B1*, *U2AF1*, *ZRSR2*, *EZH2*, *STAG2*, *IDH1*, *IDH2*, *CBL*, *KRAS*, *NRAS*, and *TP53* (Khoury et al. 2022; Luque Paz et al. 2023; Grinfeld et al. 2018). Additional genetic mutations are more characteristic for PMF cases or are acquired in the late stages of evolution of PV and ET (Luque Paz et al. 2023). However, in general, MPNs remain tumors with low mutational burden and therefore are considered less immunogenic (Luque Paz et al. 2023; Grinfeld et al. 2018). On the other hand, PV and ET can have slow evolution over many years and decades until they evolve into myelofibrotic stage or aggressive blast phase (Luque Paz et al. 2023). Furthermore, recently, it became evident that in some cases, clonal hematopoiesis of indeterminate potential (CHIP) can evolve into an overt MPN at an advanced age with a lag of many years since the initial identification of the mutated hematopoietic clone (Luque Paz et al. 2023). Another common feature of MPNs is that they are considered hematopoietic stem cell (HSC) disease as the driver mutations are commonly identified in the HSC or early hematopoietic progenitor compartment leading to a hematopoiesis with a bias toward myeloid lineage (Mead and Mullally 2017; Williams et al. 2020). Finally, a common feature of the clinically overt MPNs is the hyperinflammatory state in the body, which is considered to be triggered by the neoplastic clone but can also contribute to sustained myeloproliferation (Fisher et al. 2021; Hermouet et al. 2015).

This unique interplay between multiple cancer hallmarks in the different stages of development of MPNs raises the logical question about the mechanisms of evasion of immune surveillance in those malignancies as well as about the genetic and non-genetic factors, which may influence it. The latter is particularly important as they can be exploited for therapeutic purposes such as the development of semi-generic neoantigen-based vaccines either alone or in combination with other immunotherapeutic or targeted therapies (Handlos Grauslund et al. 2021; Braun and Zeiser 2020).

Below, we will summarize the current knowledge regarding the immunogenetic factors, which may play a role in the pathogenesis of MPNs, and discuss the potential clinical implications and directions for further research.

2 Immunogenicity of MPNs

As mentioned above, MPNs have a low mutational burden which theoretically makes them less immunogenic. However, the limited diversity in terms of main driver mutations allows for an easy assessment of the potential immunogenicity with standard immunological assays for T and B cell responses. Most studies use prediction of T cell neoepitopes with several prediction algorithms. JAK2 V617F was initially identified as an unsuitable target for the development of antigen-based vaccines because of the low frequency of the HLA-I alleles, which are predicted to present peptides derived from it (Warren and Holt 2010). However, subsequently Holmström et al. (2017) demonstrated the induction of specific CD8+ T cells HLA-A2 positive healthy individual by the VLNYGVCFC peptide. It might be a rare event that JAK2 V617F induce immune response in MPN patients as another study using in silico prediction of HLA binding and RNA sequencing did not identify JAK2 V617F–derived neoepitopes in an MPN cohort of mixed genetic background (Schischlik et al. 2019). On the other hand, CALR mutations are expectedly more immunogenic based on the long neomorphic C-terminus (Holmström et al. 2020). Indeed, several groups independently showed T cell response against CALRmut neoantigens. For example, Holmström et al. (2019) identified specific T cells in a large proportion of JAK2 V617F+ MPN patients or healthy individuals after stimulation with long (more than 20 amino acid residues) and short (9- and 10-mer) peptides. Strikingly, the proportion of healthy individuals displaying T cell responses against CALR mutant peptides was identical between younger (median age 19 years) and elderly (median age 57 years) individuals (Holmström et al. 2019). The same group was also able to demonstrate spontaneous HLA-DR restricted CD4+ T cell responses to CALR neoantigen in CALRmut+ MPN patients (Holmström et al., 2016, 2018). Using tetramer staining and flow cytometry, Tubb et al. (2018) were able to demonstrate the presence of CALRmut neoantigen–specific CD8+ T cells in healthy donors. They additionally cloned putative-specific T cell receptors (TCRs) and generated TCR-engineered

T cells, which however failed to recognize CALRmut+ MPN cells. Similarly to those studies, Bozkus et al. (2019) demonstrated that pulled CALRmut proteins with length of 14–15 AA could elicit naïve T cell response in healthy donors who were presumably neoantigen naïve. They further narrowed down the list of possible neoantigens and used 9-mer peptides to demonstrate specific CD8+ T cell response. One of the peptides RMMRTKMRM they tested elicited strong response in healthy subjects who harbored HLA-I alleles, which were predicted to have high binding affinity to that neoantigen. Our findings supported this observation as we were able to demonstrate T cell response to RMMRTKMRM in unstimulated peripheral blood mononuclear cells (PBMCs) from at least one healthy individual, and we considered redundant testing of additional subjects for CALRmut neoantigen–specific T cell responses. Taken together, all these findings show that neomorphic C-terminus of CALRmut is not only antigenic but also sufficiently immunogenic at least in people with intact immune system. The evidence for immunogenicity of CALRmut led to a clinical testing of a novel CALRmut peptide vaccine. Phase I trial with repetitive vaccination with 36-mer peptide RMRRMRRTRRKMRRKMSPARPRTSCREACLQGWTEA failed to demonstrate significant clinical and hematological responses in the enrolled ten patients (Handlos Grauslund et al. 2021). The study, however, demonstrated the induction of strong T cell responses in the majority, but in most patients, the responses were almost exclusively of CD4+ T cell (Handlos Grauslund et al. 2021), which is consistent with the ex vivo studies discussed above (Holmström et al. 2016, 2018; Cimen Bozkus et al. 2019). In our recent unpublished study, we used ELISPOT assays with 9-mer peptides which were considered likely neoantigens presented via HLA-I molecules presumably inducing CD8+ T cell responses. We were not able to demonstrate such CD8+ T cell responses in our CALRmut patients, which was consistent with the data from Bozkus et al. (2019). More interestingly, CALRmut may induce humoral immune response as Kyllesbech et al. (2023) showed that healthy individuals may have detectable IgG antibody levels against CALRmut. Therefore, acquisition of CALR mutation may be a relatively frequent event, but mutated cells would very rarely propagate the clone to an overt MPN because of efficient immune surveillance mechanisms. Finally, HLA ligandome analysis in JAK2 V617F+ and CALRmut+ MPNs did not reveal any frequently presented tumor-associated JAK2 V617F–, CALRmut-, or MPLmut-derived peptides (Bilich 2022), which can be interpreted as either poor immunogenicity or effective HLA-mediated restriction of early oncogenesis.

3 JAK2 V617F+ MPNs and HLA Polymorphisms

Led by the observations regarding immunogenicity of the major types of MPN-related mutations, we questioned whether HLA-genetic background may play a role in immune surveillance in those diseases by restricting the development of mutated clones (Ivanova et al. 2020). To address this question, several years ago we undertook a population genetic approach and genotyped at high resolution 139 JAK2 V617F+ MPN patients and 622 healthy controls with a homogeneous genetic background (Bulgarian population) (Ivanova et al. 2020). We showed that neither HLA-I homozygosity nor HLA evolutionary diversity (HED) (Chowell et al. 2019) at any locus was associated with the presence of JAK2 V617F mutation. Furthermore, we estimated the ability of the HLA-I genotype (so-called patient harmonic mean best rank (PHBR)) of any subject to present neoantigens from JAK2 V617F using the bioinformatics approach developed previously by a US group (Marty et al. 2017). We did not identify any significant differences in the estimated PHBRs between MPN patients and healthy controls. We additionally searched for specific protective alleles for the development of JAK2 V617F MPN. We identified three alleles (HLA-A*02:01, HLA-B*35:01, and HLA-C*15:02) that were significantly depleted among MPN patients as compared to the controls (Ivanova et al. 2020). The most interesting hit was HLA-B*35:01 allele as it was predicted to bind a 9-mer JAK2 V617F neoantigen (LVLNYGVCF) (Ivanova et al. 2020). We also used the molecular dynamics approach to demonstrate that HLA-B*35:01 can bind LVLNYGVCF. Additional bioinformatics analyses showed that LVLNYGVCF can likely be derived from proteasomal degradation of the native JAK2 V617F protein and can be efficiently transported to the TAP system, thus fulfilling all the prerequisites for a neoantigen (Ivanova et al. 2020). Those analyses collectively suggested that *HLA-B**35:01 may be a protective allele for the development of JAK2 V617F–driven MPNs by the presentation of the neoantigen LVLNYGVCF (Ivanova et al. 2020). We subsequently expanded our analysis to the role of *HLA-II* genes in JAK2 V617F+ MPNs (Shivarov et al. 2023a). Using an identical approach, we did not observe any significant differences in the HED for *HLA-II* loci between MPN patients and healthy controls (Shivarov et al. 2023a). However, two HLA-II alleles (DQB1*04:02 and DPB1*03:01) were statistically less frequent among MPN patients (Shivarov et al. 2023a).

4 CALR Mutation (CALRmut) + MPNs and HLA Polymorphisms

After the failure of the neoepitope-based vaccine for CALRmut MPNs, Gigoux et al. searched for plausible explanation (Gigoux et al. 2022). Based on their analysis of HLA-I allele frequencies, patients with six HLA-I alleles (HLA-A*11:01, HLA-B*08:01, HLA-B*44:02, HLA-C*07:01, HLA-C*07:02, and HLA-C*06:02) were underrepresented among CALRmut+ vs. JAK2 V617F+ MPN patients (Gigoux et al. 2022). Using a more stringent approach, we recently identified HLA-C*06:02 allele as the only potentially protective allele for CALRmut+ MPN in the Bulgarian population (unpublished) (Shivarov et al. 2023b). Taken together, our and other groups' data suggest that there specific HLA-I alleles can restrict differentially JAK2 V617F– and CALRmut-driven oncogenesis (Shivarov et al. 2023b).

5 Non-HLA Immunogenetic Systems and MPNs

Tumor cells with low expression of classical MHC molecules can still be rejected through active immune surveillance through NKG2D-mediated recognition (Lanier 2015; Nausch and Cerwenka 2008). In support of the major contribution of this mechanism, the expression of NKG2D ligands by cancer cells has been shown to be of prognostic relevance in a number of caner types (Dhar and Wu 2018). The major class of NKG2D ligands consists of major histocompatibility complex (MHC) class I –related molecules (MIC). The *MICA* and *MICB* loci in humans are highly polymorphic albeit to a lesser extent if compared to HLA-I loci. To complement our observations on the role of HLA-I polymorphisms in JAK2 V617F+ MPNs, we studied the distribution of *MICA* and *MICB* polymorphisms in the same cohort of Bulgarian patients (Ivanova et al. 2023). Notably, we identified *MICA*008:001* and *MICA*016* polymorphisms as significantly less frequent among JAK2 V617F+ MPN patients (Ivanova et al. 2023). No *MICB*-depleted alleles were identified in MPNs vs. healthy controls (Ivanova et al. 2023).

6 Mechanisms of Immune Escape in MPNs

The findings of potentially protective alleles for the development of MPNs suggesting active immune surveillance in their development justifies further interest in the potential mechanisms of its escape. Some initial studies suggested that *HLA-I* and *–II* genes might be significantly downregulated in peripheral blood cells from MPN patients (Skov et al. 2013; Cardoso et al. 2018). We analyzed publicly available datasets of gene expression studies in JAK2 V617F+ CD34+ cells and showed that bone marrow CD34+ cells from PV and ET patients had significant downregulation of a number of genes including *HLA-A, HLA-B*, and *TAPBP* (Ivanova et al. 2020). Peripheral blood CD34+ cells from myelofibrosis patients had only one downregulated gene *TAPBP* (Ivanova et al. 2020). We also observed downregulation of *HLA-II* genes *HLA-DRB1*, *HLA-DRA*, and *HLA-DRB4* in JAK2 V617F bone marrow CD34+ from ET and PV patients compared to normal CD34+ cells (Shivarov et al. 2023a). Furthermore, *CD74* and *CIITA* genes were also significantly downregulated in JAK2V617F CD34+ cells. However, JAK2 V617F + peripheral blood CD34+ cells from PMF patients upregulated several HLA-II pathway genes (*HLA-DPB1, HLA-DPB1, HLA-DRA, HLA-DRB1, HLA-DMA, HLA-DMB, and HLA-DOA*) (Shivarov et al. 2023a). Notably, treatment with interferon-alfa and ruxolitinib may upregulate MHC I genes expression at least transiently (Ivanova et al. 2020; Skov et al. 2017), while ruxolitinib-persistent cells showed downregulation of a number of *HLA-I* and *–II* genes such as *HLA-A, HLA-B, HLA-C, HLA-E, HLA-G, CALR, TAP1, CD74, HLA-DQA1, HLA-DPA1, HLA-DRA, HLA-DQB1, HLA- DPB1, and HLA—DMB* (Ivanova et al. 2020; Shivarov et al. 2023a).

Some reports proposed that JAK2 V617F+ but not CALRmut+ MPNs can escape immune surveillance through direct upregulation of immune checkpoint ligand

PD-L1 (Prestipino et al. 2018; Holmstrom et al. 2018; Milosevic Feenstra et al. 2022). Prestipino et al. (2018) demonstrated such JAK2 V617F-mediated upregulation of PD-L1 in bone marrow cells. Our analyses did not show significant upregulation of PD-L1 in CD34+ cells from MPN patients (Ivanova et al. 2020). However, we found that ruxolitinib may increase *PD-L1* expression and downregulate expression of NKG2D ligands such as *ULBP2, RAET1E,* and *RAET1G* in ex vivo long-term treated JAK2 V617F+ cell line (Ivanova et al. 2020, 2023). Finally, an additional mechanism of immune escape in JAK2 V617F + MPNs might be higher levels of soluble sMICA molecules in the patients' sera presumably because of increased shedding in MPNs (Ivanova et al. 2023).

Along with its intrinsic antigenic and immunogenic properties, CALR exon 9 mutations in MPN may directly contribute to immune escape by a number of not mutually exclusive mechanisms. Firstly, mutated CALR protein may slightly decrease expression on HLA-I molecules and thereby limit antigen presentation (Arshad and Cresswell 2018). Besides, the redistribution of CALR mutants because of the loss of ER-retention signal (Eder-Azanza et al. 2014) and differential binding of Ca^{2+} ions (Shivarov et al. 2014; Pietra et al. 2016; Ibarra et al. 2022) may directly affect the unfolded protein response (Ibarra et al. 2022; Mansier et al. 2019; Nam et al. 2019) and generation of neoantigens in the ER with glycoproteins being specifically sensitive to this effect (Schürch et al. 2022). Finally, mutated CALR proteins can be secreted (Liu et al. 2020) and work as a roque cytokine binding and activating TpoR in trans but also as a decoy to CALR receptor on professional phagocytic cells, thereby shifting the balance to suppression of phagocytosis of the transformed cells (Liu et al. 2020). Analogously, it is possible that mutated CALR proteins may diminish NK cell activation through NKp46 receptor (Sen Santara et al. 2023). Our unpublished analyses suggested that at the transcriptional level, the presence of CALRmut does not impair the HLA-I pathway (Shivarov et al. 2023b). Notably, based on the scRNA-Seq data, expression of *TAP1* and *TAP2* genes in both wild-type and mutant cells was low, which may point to a lower potential of HSC and progenitor cells to present TAP-dependent antigens and therefore being less immunogenic in general (Shivarov et al. 2023b). However, in our scRNA-Seq analysis, CALRmut+ patients did not show increased expression of *PD-L1* by MPN-SCs, which was consistent with a previous report as mentioned above (Shivarov et al. 2023b; Milosevic Feenstra et al. 2022).

Finally, several studies suggest a distinct role for NK cells in immune surveillance in MPNs and its modulation by immunotherapy and targeted therapy (Riley et al. 2015, 2016; Costa et al. 2022; Schönberg et al. 2015), suggesting further avenues for immunogenetic studies in MPNs.

7 Conclusion

In the last decade, several studies suggested that immunogenetic factors such as specific HLA-I alleles may restrict the development of MPNs in a driver mutation-dependent manner. Furthermore, some of the possible mechanisms for the evasion of active immune surveillance by MPN-SCs were elucidated, but many details remain insufficiently investigated. These observations have important practical implications as they point to the intrinsic limitations of currently available therapeutic approaches and suggest opportunities for their optimization or even the development of new therapeutic strategies.

Acknowledgements Our work on immunogenetics of MPNs was partially supported by the Bulgarian National Science Fund project KP-06-H41/2.

Disclosures The author declares no potential conflicts of interest.

References

Arshad N, Cresswell P (2018) Tumor-associated calreticulin variants functionally compromise the peptide loading complex and impair its recruitment of MHC-I. J Biol Chem 293(25):9555–9569
Bilich T (2022) Characterization of T-cell epitopes for antigen-specific immunotherapy concepts targeting myeloproliferative neoplasms and multiple myeloma. Universität Tübingen
Braun LM, Zeiser R (2020) Immunotherapy in myeloproliferative diseases. Cells 9(6):1559
Cardoso EM, Esgalhado AJ, Patrão L, Santos M, Neves VP, Martinez J et al (2018) Distinctive CD8+ T cell and MHC class I signatures in polycythemia vera patients. Ann Hematol 97: 1563–1575
Chowell D, Krishna C, Pierini F, Makarov V, Rizvi NA, Kuo F et al (2019) Evolutionary divergence of HLA class I genotype impacts efficacy of cancer immunotherapy. Nat Med 25(11):1715–1720. https://doi.org/10.1038/s41591-019-0639-4
Cimen Bozkus C, Roudko V, Finnigan JP, Mascarenhas J, Hoffman R, Iancu-Rubin C et al (2019) Immune checkpoint blockade enhances shared neoantigen-induced T-cell immunity directed against mutated calreticulin in myeloproliferative neoplasms. Cancer Discov 9(9):1192–1207
Costa AFDO, Marani LO, Bianco TM, Arantes AQ, Lopes IA, Pereira-Martins DA, Palma LC, Scheucher PS, Schiavinato JLDS, Binelli LS, Silva CA (2022) Altered distribution and function of NK-cell subsets lead to impaired tumor surveillance in JAK2V617F myeloproliferative neoplasms. Front Immunol 13:768592
Dhar P, Wu JD (2018) NKG2D and its ligands in cancer. Curr Opin Immunol 51:55–61
Eder-Azanza L, Navarro D, Aranaz P, Novo F, Cross N, Vizmanos J (2014) Bioinformatic analyses of CALR mutations in myeloproliferative neoplasms support a role in signaling. Leukemia 28(10):2106–2109
Fisher DAC, Fowles JS, Zhou A, Oh ST (2021) Inflammatory pathophysiology as a contributor to myeloproliferative neoplasms. Front Immunol 12:683401
Gigoux M, Holmström MO, Zappasodi R, Park JJ, Pourpe S, Bozkus CC et al (2022) Calreticulin mutant myeloproliferative neoplasms induce MHC-I skewing, which can be overcome by an optimized peptide cancer vaccine. Sci Transl Med 14(649):eaba4380
Grinfeld J, Nangalia J, Green AR (2017) Molecular determinants of pathogenesis and clinical phenotype in myeloproliferative neoplasms. Haematologica 102(1):7
Grinfeld J, Nangalia J, Baxter EJ, Wedge DC, Angelopoulos N, Cantrill R et al (2018) Classification and personalized prognosis in myeloproliferative neoplasms. N Engl J Med 379(15): 1416–1430
Handlos Grauslund J, Holmström MO, Jørgensen NG, Klausen U, Weis-Banke SE, El Fassi D et al (2021) Therapeutic cancer vaccination with a peptide derived from the calreticulin exon 9 mutations induces strong cellular immune responses in patients with CALR-mutant chronic myeloproliferative neoplasms. Front Oncol 11:637420

Hermouet S, Bigot-Corbel E, Gardie B (2015) Pathogenesis of myeloproliferative neoplasms: role and mechanisms of chronic inflammation. Mediat Inflamm 2015:1

Holmström M, Riley C, Svane I, Hasselbalch H, Andersen M (2016) The CALR exon 9 mutations are shared neoantigens in patients with CALR mutant chronic myeloproliferative neoplasms. Leukemia 30(12):2413–2416

Holmstrom MO, Hjortso MD, Ahmad SM, Met O, Martinenaite E, Riley C et al (2017) The JAK2V617F mutation is a target for specific T cells in the JAK2V617F-positive myeloproliferative neoplasms. Leukemia 31(2):495–498. https://doi.org/10.1038/leu.2016.290

Holmström M, Martinenaite E, Ahmad S, Met Ö, Friese C, Kjær L et al (2018) The calreticulin (CALR) exon 9 mutations are promising targets for cancer immune therapy. Leukemia 32(2): 429–437

Holmstrom MO, Riley CH, Skov V, Svane IM, Hasselbalch HC, Andersen MH (2018) Spontaneous T-cell responses against the immune check point programmed-death-ligand 1 (PD-L1) in patients with chronic myeloproliferative neoplasms correlate with disease stage and clinical response. Onco Targets Ther 7(6):e1433521. https://doi.org/10.1080/2162402X.2018.1433521

Holmström MO, Ahmad SM, Klausen U, Bendtsen SK, Martinenaite E, Riley CH et al (2019) High frequencies of circulating memory T cells specific for calreticulin exon 9 mutations in healthy individuals. Blood Cancer J 9(2):8

Holmström MO, Cordua S, Skov V, Kjær L, Pallisgaard N, Ellervik C et al (2020) Evidence of immune elimination, immuno-editing and immune escape in patients with hematological cancer. Cancer Immunol Immunother 69:315–324

How J, Hobbs GS, Mullally A (2019) Mutant calreticulin in myeloproliferative neoplasms. Blood J Am Soc Hematol 134(25):2242–2248

Ibarra J, Elbanna YA, Kurylowicz K, Ciboddo M, Greenbaum HS, Arellano NS et al (2022) Type I but not type II calreticulin mutations activate the IRE1α/XBP1 pathway of the unfolded protein response to drive myeloproliferative neoplasms. Blood Cancer Discov 3(4):298–315

Ivanova M, Tsvetkova G, Lukanov T, Stoimenov A, Hadjiev E, Shivarov V (2020) Probable HLA-mediated immunoediting of JAK2 V617F-driven oncogenesis. Exp Hematol 92:75–88

Ivanova M, Tsvetkova G, Lessichkova S, Gesheva N, Hadjiev E, Shivarov V (2023) Exploration of the role of NKG2D ligands MICA and MICB in JAK2 V617F-positive myeloproliferative neoplasms. HLA 102(2):168–178

Khoury JD, Solary E, Abla O, Akkari Y, Alaggio R, Apperley JF et al (2022) The 5th edition of the World Health Organization classification of haematolymphoid tumours: myeloid and histiocytic/dendritic neoplasms. Leukemia 36(7):1703–1719

Kyllesbech C, Trier N, Mughal F, Hansen P, Holmström M, El Fassi D et al (2023) Antibodies to calnexin and mutated calreticulin are common in human sera. Curr Res Transl Med 71(2): 103380

Lanier LL (2015) NKG2D receptor and its ligands in host defense. Cancer Immunol Res 3(6): 575–582

Liu P, Zhao L, Loos F, Marty C, Xie W, Martins I et al (2020) Immunosuppression by mutated calreticulin released from malignant cells. Mol Cell 77(4):748–760

Luque Paz D, Kralovics R, Skoda RC (2023) Genetic basis and molecular profiling in myeloproliferative neoplasms. Blood 141(16):1909–1921

Mansier O, Prouzet-Mauléon V, Jégou G, Barroso K, Raymundo DP, Chauveau A et al (2019) The expression of myeloproliferative neoplasm-associated calreticulin variants depends on the functionality of ER-associated degradation. Cancers 11(12):1921

Marty R, Kaabinejadian S, Rossell D, Slifker MJ, van de Haar J, Engin HB et al (2017) MHC-I genotype restricts the oncogenic mutational landscape. Cell 171(6):1272–83 e15. https://doi.org/10.1016/j.cell.2017.09.050

Mead AJ, Mullally A (2017) Myeloproliferative neoplasm stem cells. Blood J Am Soc Hematol. 129(12):1607–1616

Milosevic Feenstra JD, Jäger R, Schischlik F, Ivanov D, Eisenwort G, Rumi E et al (2022) PD-L1 overexpression correlates with JAK2-V617F mutational burden and is associated with 9p uniparental disomy in myeloproliferative neoplasms. Am J Hematol 97(4):390–400

Nam AS, Kim K-T, Chaligne R, Izzo F, Ang C, Taylor J et al (2019) Somatic mutations and cell identity linked by genotyping of transcriptomes. Nature 571(7765):355–360

Nausch N, Cerwenka A (2008) NKG2D ligands in tumor immunity. Oncogene 27(45):5944–5958

Pietra D, Rumi E, Ferretti V, Di Buduo C, Milanesi C, Cavalloni C et al (2016) Differential clinical effects of different mutation subtypes in CALR-mutant myeloproliferative neoplasms. Leukemia 30(2):431–438

Prestipino A, Emhardt AJ, Aumann K, O'Sullivan D, Gorantla SP, Duquesne S et al (2018) Oncogenic JAK2(V617F) causes PD-L1 expression, mediating immune escape in myeloproliferative neoplasms. Sci Transl Med 10(429). https://doi.org/10.1126/scitranslmed.aam7729

Riley CH, Hansen M, Brimnes MK, Hasselbalch HC, Bjerrum OW, Straten PT et al (2015) Expansion of circulating CD56bright natural killer cells in patients with JAK2-positive chronic myeloproliferative neoplasms during treatment with interferon-α. Eur J Haematol 94(3): 227–234

Riley CH, Brimnes MK, Hansen M, Jensen MK, Hasselbalch HC, Kjaer L et al (2016) Interferon-α induces marked alterations in circulating regulatory T cells, NK cell subsets, and dendritic cells in patients with JAK 2V617F-positive essential thrombocythemia and polycythemia vera. Eur J Haematol 97(1):83–92

Schischlik F, Jager R, Rosebrock F, Hug E, Schuster M, Holly R et al (2019) Mutational landscape of the transcriptome offers putative targets for immunotherapy of myeloproliferative neoplasms. Blood 134(2):199–210. https://doi.org/10.1182/blood.2019000519

Schönberg K, Rudolph J, Vonnahme M, Parampalli Yajnanarayana S, Cornez I, Hejazi M et al (2015) JAK inhibition impairs NK cell function in myeloproliferative neoplasms. Cancer Res 75(11):2187–2199

Schürch PM, Malinovska L, Hleihil M, Losa M, Hofstetter MC, Wildschut MH et al (2022) Calreticulin mutations affect its chaperone function and perturb the glycoproteome. Cell Rep 41(8):111689

Sen Santara S, Lee D-J, Crespo Â, Hu JJ, Walker C, Ma X et al (2023) The NK cell receptor NKp46 recognizes ecto-calreticulin on ER-stressed cells. Nature 616(7956):348–356

Shivarov V, Ivanova M, Tiu R (2014) Mutated calreticulin retains structurally disordered C terminus that cannot bind Ca2+: some mechanistic and therapeutic implications. Blood Cancer J 4(2):e185-e

Shivarov V, Tsvetkova G, Hadjiev E, Ivanova M (2023a) The relevance of HLA class II genes in JAK2 V617F-positive myeloproliferative neoplasms. Hum Immunol 84(3):199–207

Shivarov V, Tsvetkova G, Micheva I, Lesichkova S, Petrova J, Ivanova A et al (2023b) P1001: modulation of mutant calreticulin-driven oncogenesis in myeloproliferative neoplasms by classical HLA genes. HemaSphere 7(S3):e4399399

Skov V, Riley CH, Thomassen M, Larsen TS, Jensen MK, Bjerrum OW et al (2013) Whole blood transcriptional profiling reveals significant down-regulation of human leukocyte antigen class I and II genes in essential thrombocythemia, polycythemia vera and myelofibrosis. Leuk Lymphoma 54(10):2269–2273

Skov V, Riley CH, Thomassen M, Kjær L, Stauffer Larsen T, Bjerrum OW et al (2017) The impact of interferon-alpha2 on HLA genes in patients with polycythemia vera and related neoplasms. Leuk Lymphoma 58(8):1914–1921

Tubb VM, Schrikkema DS, Croft NP, Purcell AW, Linnemann C, Freriks MR et al (2018) Isolation of T cell receptors targeting recurrent neoantigens in hematological malignancies. J Immunother Cancer 6(1):1–14

Warren RL, Holt RA (2010) A census of predicted mutational epitopes suitable for immunologic cancer control. Hum Immunol 71(3):245–254

Williams N, Lee J, Moore L, Baxter EJ, Hewinson J, Dawson KJ et al. (2020) Phylogenetic reconstruction of myeloproliferative neoplasm reveals very early origins and lifelong evolution. BioRxiv. 2020.11.09.374710

Germline Predisposition in the Field of Myeloproliferative Neoplasms

Emanuela Sant'Antonio, Oscar Borsani, and Elisa Rumi

Abstract

Philadelphia-negative myeloproliferative neoplasms (MPN) are a group of clonal malignancies of the myeloid lineage, characterized by a deregulated cellular proliferation driven by somatically acquired driver mutations that ultimately result in a hyperactivated JAK/STAT signaling. MPN have traditionally been described as sporadic, late-onset cancers. However, it has been increasingly recognized that a subset of these neoplasms has early onset and/or aggregates within family, thus suggesting that they may occur on the background of germline mutations, both inherited or occurring de novo. Several lines of evidence support the role of germline genetic factors in the pathogenesis of MPN. The inherited predisposition to MPN occurs either because of rare but highly penetrant gene variants (i.e., true familial MPN with Mendelian inheritance, which consists in malignant disorder that occurs in two or more family members and that mimics sporadic cases as regards rate of vascular events and risk of disease progression) or because of more common but weakly penetrant risk factors. Some of the most important predisposing genetic variations include the *JAK2* 46/1 haplotype and the rs2736100 variant of the telomerase reverse transcriptase (*TERT*) gene. On the other hand, germline copy number variations – such as duplication of autophagy-related 2B (*ATG2B*) and GSK3B-interacting protein (*GSKIP*) genes – and mutations of the retinoblastoma binding protein 6 (*RBBP6*) gene have been identified as highly penetrant predisposition loci for familial myeloid neoplasms, including MPN. Taken together, recent discoveries support the notion that MPN

E. Sant'Antonio (✉)
Division of Hematology, Azienda USL Toscana Nord Ovest, Lucca, Italy
e-mail: santantonio.emanuela@hsr.it

O. Borsani · E. Rumi
Department of Molecular Medicine, University of Pavia, Pavia, Italy

Division of Hematology, Fondazione IRCCS Policlinico San Matteo, Pavia, Italy

© The Author(s), under exclusive license to Springer Nature Switzerland AG 2022

Interdisciplinary Cancer Research, https://doi.org/10.1007/16833_2022_3
Published online: 1 December 2022

might occur on a background of germline mutations more frequently than previously thought; moreover, the spectrum of known tumor predisposition syndromes can be broader than anticipated. Finally, environmental and lifestyle influences on susceptibility to MPN have also been reported.

Keywords

Germline mutations · Myeloproliferative neoplasms · Predisposition · Signaling

1 Introduction

According to the World Health Organization (WHO) classification, the term "myeloid" applies to all cells belonging to the erythroid, granulocytic (neutrophil, eosinophil, basophil), megakaryocytic, monocytic/macrophage, and mast cell lineages. Myeloid neoplasms (MN) are a composite group of disorders caused by the clonal proliferation of cells belonging to the myeloid lineage and characterized by hyperproliferation and/or hindered differentiation of abnormal myeloid stem and progenitor cells.

Myeloproliferative neoplasms (MPN) are a group of clonal malignancies of the myeloid lineage characterized by a deregulated cellular proliferation, with no significant loss of differentiation capacity, due to a hyperactivated JAK/STAT signaling (Cazzola and Kralovics 2014). According to the last WHO classification (Arber et al. 2016), this category comprises four main disorders, i.e., chronic myeloid leukemia, BCR-ABL1-positive (CML), polycythemia vera (PV), essential thrombocythemia (ET), and primary myelofibrosis (PMF), in its overt and prefibrotic subtype, and three rarer diseases, i.e., chronic eosinophilic leukemia, not otherwise specified (CEL, NOS), chronic neutrophilic leukemia (CNL), and MPN-unclassifiable (MPN-U). MPN have been traditionally classified according to the presence or absence of the Philadelphia chromosome into Philadelphia-positive (i.e., CML) and Philadelphia-negative (i.e., PV, ET, PMF) MPN: the present work refers to the latter group.

Philadelphia-negative MPN are classified according to the predominant cell lineages involved by the hyperproliferative phenotype, which is mainly, but not exclusively, linked to a single myeloid lineage: elevated platelet (PLT) count (i.e., thrombocytosis) in ET; elevated red blood cell (RBC) count (i.e., erythrocytosis) in PV; and finally, megakaryocytic hyperplasia and BM fibrosis in PMF. From a molecular point of view, most cases of Philadelphia-negative MPN are caused by a somatically acquired, gain-of-function mutation in the hematopoietic stem cell and progenitor cell compartment, activating JAK/STAT signaling pathway and inducing an increased sensitivity to low levels of various cytokines, such as erythropoietin (EPO), insulin-like growth factor-1 (IGF-1), IL-3, and GM-CSF, or even an autonomous cell proliferation, in the absence of their relative growth factor (Vainchenker and Kralovics 2017). Therefore, such mutations act as disease-initiating events, being deeply involved in disease pathogenesis, and are hence termed “phenotypic driver mutations.” They are usually located in one of three genes, i.e., *JAK2*, *CALR*, and *MPL*, occurring mostly in a mutually exclusive manner (Mora et al. 2021).

In addition, sporadic alterations in epigenetic modifying genes, spliceosome genes, and cytokine signaling regulation genes have been reported: these somatic events are not specific nor limited to the MPN genetic landscape and occur together with the abovementioned driver events. These mutations may affect genes involved in DNA methylation (*TET2*, *DNMT3A*, *IDH1/2*), chromatin structure (*EZH2*, *ASXL1*), or mRNA splicing (*SF3B1*, *SRSF2*, *U2AF1*) (Lundberg et al. 2014; Schischlik and Kralovics 2017).

While the range of these rare mutations detected in MPN is expanding, their clinical and prognostic significance is still under active evaluation. However, few of them have been already included in validated prognostic scoring system as MIPSS70 (Guglielmelli et al. 2018) which is currently used for selecting appropriate candidates for allotransplantation in the setting of PMF.

2 Evidence Suggesting a Germline Predisposition to Myeloproliferative Neoplasms

MN have traditionally been described as sporadic, late-onset cancers, diagnosed at median age > 60 years. However, it has been increasingly recognized that a subset of these neoplasms has early onset and/or aggregates within family, suggesting they arise from germline mutations, both inherited or occurring de novo. The relevance of MN arising in the context of predisposing germline mutations is highlighted by their inclusion in the 2016 revision to the WHO classification of MN as the new category “Myeloid neoplasms with germline predisposition.” The 2016 WHO classification (Arber et al. 2016), indeed, recognizes four subtypes, based on the presence or absence of accompanying clinical features: (I) MN with germline predisposition without a pre-existing disorder or organ dysfunction; (II) MN with germline predisposition and pre-existing platelet disorders; (III) MN with germline predisposition associated with other organ disfunction; and (IV) MN with germline predisposition associated with inherited bone marrow failure syndromes and telomere biology disorders. In familial neoplasms, mutations are present in the heterozygous condition, most commonly resulting in loss of functional alleles and subsequent haploinsufficiency. Though the biological mechanisms that drive progression to an overt hematologic malignancy are the topic of active research, they usually involve different cooperating events in bone marrow stem or progenitor cells, often involving the same genes previously affected by germline mutations (Klco and Mullighan 2021). It should be noted, however, that these entities mostly comprise familial cases of acute leukemia (AL) and/or myelodysplastic syndrome (MDS), while germline predisposition to MPN is still not officially recognized by the 2016 WHO classification.

There are, however, several lines of evidence that support the role of germline genetic factors in the pathogenesis of MPN. The observation of familial clustering of

MPN was the first clinical clue that pointed to the hypothesis that germline factors played an important role in MPN pathogenesis (Bellanne-Chantelot et al. 2006; Rumi et al. 2007). This clustering, indeed, is an observation that exceeds that expected by chance in the general population. Patterns of inheritance are markedly heterogeneous, suggesting there are probably a variety of different germline mutations driving the predisposition. Indeed, in a significant fraction of families, MPN are inherited in an autosomal dominant fashion (Kralovics et al. 2003; Rumi et al. 2006), whereas other families follow a recessive pattern and/or the disease shows a weaker penetrance (Rumi et al. 2007).

"True" familial MPN describes a malignant disorder that occurs in two or more family members and that mimics sporadic cases as regards rate of vascular events and risk of disease progression. This entity is characterized by clonal hematopoiesis, incomplete penetrance, and, often, multilineage proliferation, as opposed to those cases of hereditary thrombocytosis (Han et al. 2021), in which hematopoiesis is polyclonal and risk of thromboembolic events is thought to be rare or absent. The term *familial* is used to refer to a mostly unknown germline lesion that predisposes to the acquisition of MPN driver mutations; interestingly, driver mutations and MPN phenotypes usually vary between family members (Rumi and Cazzola 2017), supporting the notion that the germline predisposition creates a fertile ground for the subsequent acquisition of driver and/or subclonal mutations that modulate disease presentation and individual patients' outcome. Indeed, whatever the inherited genetic factors are, they cannot be the sole cause of clonal proliferation, and development of MPN requires the acquisition of additional somatic mutations. We may speculate that the germline genetic event could increase genomic instability either perturbing DNA replication or altering cellular mechanisms of DNA repair, so causing a somewhat weaker predisposition than that seen with inherited mutations in true tumor suppressor genes (Jones and Cross 2013).

Based on a large Swedish study, the risk of developing an MPN in first-degree relatives of affected patients is 5- to 7-fold higher than that in the general population; this risk rises up to 12-fold if we consider the risk of developing ET in first-degree relatives of ET patients (Landgren et al. 2008). Moreover, two Italian studies analyzed a large population of MPN patients and found a prevalence of inherited disease of 7–11% (Rumi et al. 2007; Olcaydu et al. 2011). As regards patients' characteristics, literature data are consistent in reporting that the age of onset is lower in familial MPN as compared with sporadic cases, and there is some evidence for anticipation between generations. The notion of familial aggregation has been recently confirmed and extended to other subtypes of myeloid neoplasms, thanks to two large, population-based studies that focused on the whole group of hematological neoplasms (Sud et al. 2018, 2019).

Moreover, some MPN patients have been shown to be "biclonal," which means that there are at least two independent sources of malignant blood cells which cannot be traced back to a common ancestral cell, as demonstrated by different pattern of X-chromosome inactivation (Beer et al. 2009; Hussein et al. 2009). An inherited genetic predisposition in such individuals has been postulated based on the notion that two independent clonal outgrowths in a single patient had a very low chance of spontaneous occurrence (Harutyunyan and Kralovics 2012).

From a genetic standpoint, several works have already reported on common polymorphisms that predispose to MPN. They will be addressed in much details in a dedicated section.

3 Genetic Predisposition to Myeloproliferative Neoplasms

There are two main patterns of inherited predisposition to MPN: (I) "true" familial MPN with Mendelian inheritance, with the underlying risk alleles presumed to be rare and relatively highly penetrant (Bellanne-Chantelot et al. 2006; Rumi et al. 2007; Malak et al. 2012), and (II) a more general predisposition, driven by weakly penetrant, common risk factors. Known predisposing germline genetic variations and associated MPN risks include *JAK2* 46/1 haplotype (up to fourfold) (Jones et al. 2009; Kilpivaara et al. 2009; Olcaydu et al. 2009), telomerase reverse transcriptase (*TERT*) promoter variant (rs2736100) (twofold) (Jager et al. 2014), duplication of autophagy-related 2B (*ATG2B*) and GSK3B-interacting protein (*GSKIP*) (high risk of progression from ET to myelofibrosis or leukemia) (Saliba et al. 2015), and germline retinoblastoma binding protein 6 (*RBBP6*) mutations (nearly 5% of familial MPN cases) (Harutyunyan et al. 2016).

3.1 *JAK2* 46/1 Haplotype

Few years after its discovery, it was noted that the *JAK2*V617F mutation is not acquired in a random fashion, but rather preferentially occurs on a specific haplotype of *JAK2* referred to as "46/1" or "GGCC" (Jones et al. 2009; Olcaydu et al. 2009). This haplotype is located in the 3′-portion of the gene and includes *JAK2* and two other genes, insulin-like 6 (*INSL6*) and insulin-like 4 (*INSL4*), that are not normally transcribed in the hematopoietic system. It has a quite high frequency in Europe, with roughly 50% of individuals carrying at least one allele. Its association with MPN has been confirmed also in Japanese (Ohyashiki et al. 2012) and Chinese cases (Zhang et al. 2012), indicating that the underlying mechanism is not limited to Caucasians and probably has a relatively old evolutionary basis. This single genetic feature covers roughly half of the population attributable risk of developing an MPN, i.e., the proportion of the risk that may be explained by inherited factors. Moreover, *JAK2* 46/1 is one of the strongest predisposition factors identified to date that is associated with the development of a molecularly defined neoplasm; in detail, individuals carrying a 46/1 allele have an odds ratio of developing MPN being threefold to fourfold higher than noncarriers (Kilpivaara et al. 2009). However, even though 46/1 appears to be the most important common risk factor for MPN development, its penetrance is, indeed, quite low so that it cannot be used to predict disease development. Moreover, it has been inconsistently associated with *JAK2*-unmutated disease (Jones et al. 2010; Patnaik et al. 2010).

Furthermore, the *JAK2* 46/1 haplotype does not significantly contribute to familial MPN, in which there is no strong enrichment of this haplotype compared to sporadic MPN (Olcaydu et al. 2011).

Its role, however, seems to be wider than expected: indeed, an excess of *JAK2* 46/1 has been reported in patients with normal karyotype acute myeloid leukemia, with a possible negative prognostic significance (Andrikovics et al. 2010; Nahajevszky et al. 2011). This is an interesting finding especially in the context of recent population-based investigations that addressed and evaluated the risk of developing a myeloid or a lymphoid disease in first-degree relatives of adults with hematological neoplasms (Goldin et al. 2012; Sud et al. 2018, 2019) and suggests that the biological consequences of a single predisposition locus could influence a wide range of hematopoietic cell types, giving a broader contribution to the pathogenesis of leukemia.

However, how a particular inherited background influences the acquisition of the V617F mutations is still to be settled. Two general hypotheses have been proposed to explain this association: the first, termed "hypermutability," considers 46/1 as more genetically unstable, acquiring V617F at a faster rate than other haplotypes. Indeed, there have been reports of patients in which the *JAK2*V617F mutation has been acquired at least twice (Lambert et al. 2009; Olcaydu et al. 2009) and cases in which *JAK2*V617F and exon 12 mutations co-occurred in a biclonal disease (Li et al. 2008), a finding that has a very low chance of spontaneous occurrence, given the low incidence of MPN. However, we may argue that these individuals have a general and higher propensity to acquire further mutations, which is not limited to the *JAK2* genes, as testified by the occurrence of several cases that are positive for both *JAK2*V617F and *BCR-ABL* (Hussein et al. 2007; Hassan et al. 2015; Bader and Dreiling 2019; Ali et al. 2021). On the other hand, according to the second hypothesis, called "fertile ground," the *JAK2*V617F arises on all haplotypes at an equal rate, but there is an additional factor associated with the 46/1 haplotype that provides a selective advantage to the mutated clone; in this regard, a higher number of hematopoietic colonies has been reported in individuals carrying the 46/1 haplotype (Jones et al. 2009).

Despite these hypotheses and clinical observations, the biological consequences of the 46/1 haplotype are still far from being settled, and we may speculate that elements of both the hypermutability theory and the fertile ground hypotheses may co-occur and contribute to the acquisition of *JAK2* and other mutations.

4 Telomere Length and Risk of MPN

Telomere length (TL) has been investigated as a marker of susceptibility for the development of neoplasia, and, indeed, it has been found to be associated with risk of suffering from several cancer types, including hematological neoplasms (Weischer et al. 2013; The Telomeres Mendelian Randomization Collaboration et al. 2017). TL is, at least in part, genetically determined, and 11 relevant single nucleotide polymorphisms (SNPs) have been identified, through genome-wide association studies

(Mangino et al. 2012; Codd et al. 2013; The Telomeres Mendelian Randomization Collaboration et al. 2017). Its measurement has been traditionally done through flow-FISH analysis or qPCR, but more recently a genetic score has been defined to assess their cumulative influence. It was called "teloscore" and has been successfully used as proxy of genetically determined leukocyte TL. Of note, the associations between leukocyte TL and cancer risk have not consistently shown the same trend if TL was directly assessed through qPCR or through the teloscore. In detail, as regards MPN, results have been partially inconsistent: five studies have found an association between short telomeres and increased risk of development of the disease, and all of them relied on direct TL measurement (Bernard et al. 2009; Elena et al. 2011; Spanoudakis et al. 2011; Ruella et al. 2013; Dahlstrom et al. 2015; Alimam et al. 2018).

On the other hand, a recent study used the genetic score to test TL role in a large European case-control study and observed that genetically determined longer telomeres were significantly associated with an increased risk of developing MPN (Giaccherini et al. 2020).

In further detail, the authors replicated the well-known association between the *TERT*-rs2736100-C allele and an increased risk of developing MPN that have been previously and consistently reported (Jager et al. 2014; Oddsson et al. 2014; Tapper et al. 2015; Trifa et al. 2016b) and observed for the first time a potential association between higher risk of MPN and the *OBFC1*-rs9420907-C allele and the *PXK*-rs6772228-T SNP (Giaccherini et al. 2020). From a biological standpoint, *TERT* encodes for the telomerase reverse transcriptase, and *OBFC1* codes for a component of the CTC1-STN1-TEN1 (CST) complex, which protect TL through telomere capping and by regulating telomerase access to telomeres, while the role of the protein product of *PXK* gene in regulating TL is less clear.

Though results on the topic of TL and risk of MPN have been partially inconsistent, we should acknowledge that the latter approach, based on the genetic risk score, has the advantage of being independent from potential bias derived from study design and from environmental factors that are associated with TL variation. Moreover, sample size of the study from Giaccherini et al. was almost fourfold that of the previously mentioned studies.

5 Germline Mutations in Genes of DNA Repair Pathway

Recently, whole-exome sequencing of five families with parent-child or sibling pairs affected by *JAK2*V617F-mutated MPN revealed the presence of rare germline variants in known tumor predisposition genes of the DNA repair pathway, especially hereditary breast and ovarian cancers, as *ATM*, *CHEK2*, and the highly penetrant *BRCA1* and *BRCA2* genes (Elbracht et al. 2021). In this study, two patients (affected by ET and PV) experienced progression to AL, thus suggesting a high risk of leukemic transformation in these familial cases.

While an association of germline variants in *ATM* or *CHEK2* and *JAK2*V617F MPN has already been reported (Hinds et al. 2016), this work showed for the first time an association between germline variants in *BRCA1/2* and MPN, thus expanding the spectrum of tumor predisposition syndromes.

6 Other Predisposition Alleles

In the last few years, other predisposition alleles have been reported to be linked to MPN development. In details, variant in *MECOM*, *HBS1L-MYB*, SH2B3, *TET2*, *ATM*, *CHEK2*, *PINT*, and *GF11B* genes was identified in MPN patients (Tapper et al. 2015). Combined effect of *JAK2/TER/MECOM* variants was estimated to have a population association rate of 51% for *JAK2*-negative MPN (Tapper et al. 2015). The same germline variants identify individuals with a predisposition not only to MPN but also to *JAK2*V617F clonal hematopoiesis, a more common phenomenon that may foreshadow the development of an overt neoplasm (Hinds et al. 2016).

It has been demonstrated that an excess of the A3669G (rs6198) variant of the glucocorticoid receptor β (*GRβ*) gene characterized PV but not ET cases. Expression of higher levels of this dominant negative isoform of GRβ probably contributes to increased erythrocytosis by blocking glucocorticoids stimulation (Varricchio et al. 2011). The same variant has also been associated with blast-phase transformation in *JAK2*V617F-positive PMF; indeed, the risk allele G/G correlated with a higher number of circulating CD34+ cells, with splenomegaly and with a higher leukocyte count at diagnosis (Poletto et al. 2012).

Evidences suggest that polymorphisms in genes involved in oxidative stress may influence the risk of developing Philadelphia-negative MPN. In particular, a protective role has been described for the *CAT*-262 C > T and *GPX1* Pro198Leu polymorphisms, while a risk effect has been reported for the *GSTP1* Ile105Val variant. From a biological point of view, these variants could produce different general, non-specific, effects at multiple levels, thus resulting in imbalances in the oxidative stress status and, finally, promoting carcinogenesis (Trifa et al. 2016a).

Apoptosis, a programmed cell death process, plays a key role in regulating the balance between cellular proliferation and cell death especially in tissues with a high cellular turnover, such as hematopoietic system. Thus, it's not surprising that aberrations in apoptotic processes could promote carcinogenesis by allowing accumulation of non-functional or damaged cells. Accordingly, it has been demonstrated that apoptosis is often deregulated in hematologic malignancies (Zaman et al. 2014). Cysteine aspartate proteases, commonly known as caspases, are a family of intracellular proteins that play a critical role in both intrinsic and extrinsic pathways of apoptosis. The C653T (rs1132312) variant of the *CASP9* gene, whose function is important in regulating megakaryocyte turnover in MPN, has been related to a significantly increased risk of ET (Azevedo et al. 2019).

Nowadays, there is a bulk of evidence that polymorphisms in genes encoding for key enzymes involved in the phase II of biotransformation of xenobiotics (e.g., *GSTM1*, *GSTP1*, *GSTT1*, *NQO1*) could play an important role in the pathogenesis of malignant neoplasms. These genes and their protein products are involved in detoxification of a variety of exogenous and endogenous genotoxicants that can otherwise damage cell genome and promote carcinogenesis. In a recent study, the large deletion rs71748309 of *GSTT1* gene and the rs1800566 (609C > T) of *NQO1* gene have been associated with the risk of development of Philadelphia-negative MPN (Ovsepyan et al. 2019).

It has been well described that proinflammatory cytokines are overall increased in MPN; however, a clear relationship between their role in MPN pathogenesis is still lacking. Tumor necrosis factor alpha (*TNF-α*), the most important and studied pro-inflammatory cytokine, is involved in several infectious, autoimmune, and inflammatory processes. Polymorphisms in the promoter region of the *TNF-α* gene (i.e., TNF-238 G > A – rs rs361525 and TNF-308 G > A – rs1800629) have been linked to different hematological malignancies. It has been hypothesized that these polymorphisms could act by promoting *TNF-α* gene transcription, thereby promoting inflammation and inflammatory-driven cancer progression. The TNF-308A variant has been associated with an increased susceptibility to prostate cancer, breast cancer, and non-Hodgkin's lymphoma. A recent study also confirmed the association between the TNF-308A allele and susceptibility to MPN development, especially to PV and PMF (Macedo et al. 2016).

Finally, recent evidences support the hypothesis that inherited factors may have a broader effect on susceptibility to developing malignancy; indeed, epidemiological studies have consistently shown that MPN patients have an increased risk of developing both hematological cancers and solid tumors, as compared to the general population (Barbui et al. 2019; Marchetti et al. 2020). In detail, MPN patients have a significantly increased risk of developing a lymphoid neoplasm, estimated to be threefold to more than fivefold higher than the general population (Marchetti et al. 2018; Holst et al. 2020; Rumi et al. 2020), and a slightly increased risk of developing solid tumors (1.2- to 1.6-fold) (Frederiksen et al. 2011). These associations could in part be accounted for by prior leukemogenic treatments which promote further malignancy, or for other hematological malignancies possibly the presence of *JAK2*V617F itself, which has been linked to increased risk of secondary cancer and DNA damage. Finally, it has been suggested that the alleles that drive MPN predisposition might be clustered in definite groups of individuals, as supported by the observations that the incidence of MPN is heterogenous; e.g., MPN are tenfold more frequent in Ashkenazi Jews as compared with other populations in Northern Israel (Chaiter et al. 1992).

7 Conclusion

The increased accessibility and a wider use of new sequencing technologies have allowed for a much deeper knowledge about the genetic background of sporadic MPN which now go beyond the classical driver mutations and comprise new somatically acquired mutations. The *JAK2* 46/1 haplotype is now well recognized as a strong risk factor that predisposes to acquisition of the *JAK2*V617F mutations: this was an important finding that opened the way to the uncovering and understanding of the genetic predisposition to MPN. It is now clear that germline factors play an important role in the pathogenesis of MPN, especially in patients with a young age at diagnosis and/or with other relatives affected with MPN. It will be interesting to see how this inherited predisposition influences initiation, clinical phenotype, course and transformation driven by acquired mutations. Thus, further candidate and systematic studies are needed to identify more risk variants with variable frequencies and degrees of penetrance.

We hope that next-generation sequencing will clarify the genetic and molecular basis of familial MPN, by allowing the discovery of further rare, highly penetrant variants. Indeed, identification of an underlying hereditary cancer predisposition is of major clinical relevance for both the individual patients and their families because it will make possible to act by preventive care and to enhance early recognition and treatment of emerging diseases. Indeed, novel therapeutic strategies, such as PARP-inhibitors, could be proposed and evaluated according to recent observations that link *BRCA1/2* germline variants to Philadelphia-negative MPN.

Taken together, these data suggest that MPN are more frequently caused by germline mutations than previously assumed. Indeed, the spectrum of known tumor predisposition syndromes could be broader than anticipated.

However, one should keep in mind that also both environmental and lifestyle influences have an influence on MPN development. Studies have mainly focused on risk linked to exposures to different genotoxic products (e.g., chemicals) through lifestyle choices or occupation. Despite the vast heterogeneity in defining such hard-to-measure risks, there are recurrent data that suggest an association between benzene exposure and development of PV.

Many other studies are needed to deciphering the modality by which rare and common germline factors interact with somatically acquired mutations and environmental factors to contribute to the overall risk of MPN development.

Acknowledgements None.

Compliance with Ethical Standards The authors declare that there is no conflict of interest.

References

Ali EAH, Al-Akiki S, Yassin MA (2021) A case report of BCR-ABL-JAK2-positive chronic myeloid leukemia with complete hematological and major molecular response to Dasatinib. Case Rep Oncol 14:690–694

Alimam S, McLornan DP, Jiang J, Radia D, Mufti GJ, Harrison CN (2018) Shortened telomeres in essential thrombocythemia: clinicopathological and treatment correlations. Haematologica 103: e234–e236

Andrikovics H, Nahajevszky S, Koszarska M, Meggyesi N, Bors A, Halm G, Lueff S, Lovas N, Matrai Z, Csomor J, Rasonyi R, Egyed M, Varkonyi J, Mikala G, Sipos A, Kozma A, Adam E, Fekete S, Masszi T, Tordai A (2010) JAK2 46/1 haplotype analysis in myeloproliferative neoplasms and acute myeloid leukemia. Leukemia 24:1809–1813

Arber DA, Orazi A, Hasserjian R, Thiele J, Borowitz MJ, Le Beau MM, Bloomfield CD, Cazzola M, Vardiman JW (2016) The 2016 revision to the World Health Organization classification of myeloid neoplasms and acute leukemia. Blood 127:2391–2405

Azevedo AP, Silva SN, Reichert A, Lima F, Júnior E, Rueff J (2019) The role of caspase genes polymorphisms in genetic susceptibility to Philadelphia-negative myeloproliferative neoplasms in a Portuguese population. Pathol Oncol Res 25:961–969

Bader G, Dreiling B (2019) Concurrent JAK2-positive myeloproliferative disorder and chronic myelogenous leukemia: a novel entity? A case report with review of the literature. J Investig Med High Impact Case Rep 7:2324709619832322

Barbui T, Ghirardi A, Masciulli A, Carobbio A, Palandri F, Vianelli N, De Stefano V, Betti S, Di Veroli A, Iurlo A, Cattaneo D, Delaini F, Bonifacio M, Scaffidi L, Patriarca A, Rumi E, Casetti IC, Stephenson C, Guglielmelli P, Elli EM, Palova M, Bertolotti L, Erez D, Gomez M, Wille K, Perez-Encinas M, Lunghi F, Angona A, Fox ML, Beggiato E, Benevolo G, Carli G, Cacciola R, McMullin MF, Tieghi A, Recasens V, Marchetti M, Griesshammer M, Alvarez-Larran A, Vannucchi AM, Finazzi G (2019) Second cancer in Philadelphia negative myeloproliferative neoplasms (MPN-K). A nested case-control study. Leukemia 33:1996–2005

Beer PA, Jones AV, Bench AJ, Goday-Fernandez A, Boyd EM, Vaghela KJ, Erber WN, Odeh B, Wright C, McMullin MF, Cullis J, Huntly BJ, Harrison CN, Cross NC, Green AR (2009) Clonal diversity in the myeloproliferative neoplasms: independent origins of genetically distinct clones. Br J Haematol 144:904–908

Bellanne-Chantelot C, Chaumarel I, Labopin M, Bellanger F, Barbu V, De Toma C, Delhommeau F, Casadevall N, Vainchenker W, Thomas G, Najman A (2006) Genetic and clinical implications of the Val617Phe JAK2 mutation in 72 families with myeloproliferative disorders. Blood 108:346–352

Bernard L, Belisle C, Mollica L, Provost S, Roy DC, Gilliland DG, Levine RL, Busque L (2009) Telomere length is severely and similarly reduced in JAK2V617F-positive and -negative myeloproliferative neoplasms. Leukemia 23:287–291

Cazzola M, Kralovics R (2014) From Janus kinase 2 to calreticulin: the clinically relevant genomic landscape of myeloproliferative neoplasms. Blood 123:3714–3719

Chaiter Y, Brenner B, Aghai E, Tatarsky I (1992) High incidence of myeloproliferative disorders in Ashkenazi Jews in northern Israel. Leuk Lymphoma 7:251–255

Codd V, Nelson CP, Albrecht E, Mangino M, Deelen J, Buxton JL, Hottenga JJ, Fischer K, Esko T, Surakka I, Broer L, Nyholt DR, Mateo Leach I, Salo P, Hagg S, Matthews MK, Palmen J, Norata GD, O'Reilly PF, Saleheen D, Amin N, Balmforth AJ, Beekman M, de Boer RA, Bohringer S, Braund PS, Burton PR, de Craen AJ, Denniff M, Dong Y, Douroudis K, Dubinina E, Eriksson JG, Garlaschelli K, Guo D, Hartikainen AL, Henders AK, Houwing-Duistermaat JJ, Kananen L, Karssen LC, Kettunen J, Klopp N, Lagou V, van Leeuwen EM, Madden PA, Magi R, Magnusson PK, Mannisto S, McCarthy MI, Medland SE, Mihailov E, Montgomery GW, Oostra BA, Palotie A, Peters A, Pollard H, Pouta A, Prokopenko I, Ripatti S, Salomaa V, Suchiman HE, Valdes AM, Verweij N, Vinuela A, Wang X, Wichmann HE, Widen E, Willemsen G, Wright MJ, Xia K, Xiao X, van Veldhuisen DJ, Catapano AL, Tobin MD, Hall AS, Blakemore AI, van Gilst WH, Zhu H, Consortium CA, Erdmann J, Reilly MP, Kathiresan S, Schunkert H, Talmud PJ, Pedersen NL, Perola M, Ouwehand W, Kaprio J, Martin NG, van Duijn CM, Hovatta I, Gieger C, Metspalu A, Boomsma DI, Jarvelin MR, Slagboom PE, Thompson JR, Spector TD, van der Harst P, Samani NJ (2013) Identification of seven loci affecting mean telomere length and their association with disease. Nat Genet 45:422–427, 427e421–422

Dahlstrom J, Zhang X, Ghaderi M, Hultcrantz M, Bjorkholm M, Xu D (2015) Dysregulation of shelterin factors coupled with telomere shortening in Philadelphia chromosome negative myeloproliferative neoplasms. Haematologica 100:e402–e405

Elbracht M, Meyer R, Kricheldorf K, Gezer D, Thomas E, Betz B, Kurth I, Teichmann LL, Brummendorf TH, Germing U, Isfort S, Koschmieder S (2021) Germline variants in DNA repair genes, including BRCA1/2, may cause familial myeloproliferative neoplasms. Blood Adv 5:3373–3376

Elena C, Rumi E, Portolan M, Della Porta MG, Pascutto C, Passamonti F (2011) Flow-FISH evaluation of telomere length in Philadelphia-negative myeloproliferative neoplasms. Haematologica 96:1236–1238

Frederiksen H, Farkas DK, Christiansen CF, Hasselbalch HC, Sorensen HT (2011) Chronic myeloproliferative neoplasms and subsequent cancer risk: a Danish population-based cohort study. Blood 118:6515–6520

Giaccherini M, Macauda A, Sgherza N, Sainz J, Gemignani F, Maldonado JMS, Jurado M, Tavano F, Mazur G, Jerez A, Góra-Tybor J, Gołos A, Mohedo FH, Lopez JM, Várkonyi J, Spadano R, Butrym A, Canzian F, Campa D (2020) Genetic polymorphisms associated with telomere length and risk of developing myeloproliferative neoplasms. Blood Cancer J 10:89

Goldin LR, Kristinsson SY, Liang XS, Derolf AR, Landgren O, Bjorkholm M (2012) Familial aggregation of acute myeloid leukemia and myelodysplastic syndromes. J Clin Oncol 30:179–183

Guglielmelli P, Lasho TL, Rotunno G, Mudireddy M, Mannarelli C, Nicolosi M, Pacilli A, Pardanani A, Rumi E, Rosti V, Hanson CA, Mannelli F, Ketterling RP, Gangat N, Rambaldi A, Passamonti F, Barosi G, Barbui T, Cazzola M, Vannucchi AM, Tefferi A (2018) MIPSS70: mutation-enhanced international prognostic score system for transplantation-age patients with primary myelofibrosis. J Clin Oncol 36:310–318

Han EY, Catherwood M, McMullin MF (2021) Hereditary thrombocytosis: the genetic landscape. Br J Haematol 194:1098–1105

Harutyunyan AS, Kralovics R (2012) Role of germline genetic factors in MPN pathogenesis. Hematol Oncol Clin North Am 26:1037–1051

Harutyunyan AS, Giambruno R, Krendl C, Stukalov A, Klampfl T, Berg T, Chen D, Milosevic Feenstra JD, Jager R, Gisslinger B, Gisslinger H, Rumi E, Passamonti F, Pietra D, Muller AC, Parapatics K, Breitwieser FP, Herrmann R, Colinge J, Bennett KL, Superti-Furga G, Cazzola M, Hammond E, Kralovics R (2016) Germline RBBP6 mutations in familial myeloproliferative neoplasms. Blood 127:362–365

Hassan A, Dogara LG, Babadoko AA, Awwalu S, Mamman AI (2015) Coexistence of JAK2 and BCR-ABL mutation in patient with myeloproliferative neoplasm. Niger Med J 56:74–76

Hinds DA, Barnholt KE, Mesa RA, Kiefer AK, Do CB, Eriksson N, Mountain JL, Francke U, Tung JY, Nguyen HM, Zhang H, Gojenola L, Zehnder JL, Gotlib J (2016) Germ line variants predispose to both JAK2 V617F clonal hematopoiesis and myeloproliferative neoplasms. Blood 128:1121–1128

Holst JM, Plesner TL, Pedersen MB, Frederiksen H, Moller MB, Clausen MR, Hansen MC, Hamilton-Dutoit SJ, Norgaard P, Johansen P, Ramm Eberlein T, Mortensen BK, Mathiasen G, Ovlisen A, Wang R, Wang C, Zhang W, Beier Ommen H, Stentoft J, Ludvigsen M, Tam W, Chan WC, Inghirami G, d'Amore F (2020) Myeloproliferative and lymphoproliferative malignancies occurring in the same patient: a nationwide discovery cohort. Haematologica 105:2432–2439

Hussein K, Bock O, Seegers A, Flasshove M, Henneke F, Buesche G, Kreipe HH (2007) Myelofibrosis evolving during imatinib treatment of a chronic myeloproliferative disease with coexisting BCR-ABL translocation and JAK2V617F mutation. Blood 109:4106–4107

Hussein K, Bock O, Theophile K, Schlue J, Ballmaier M, Kroger N, Gohring G, Busche G, Kreipe H (2009) Biclonal expansion and heterogeneous lineage involvement in a case of chronic myeloproliferative disease with concurrent MPLW515L/JAK2V617F mutation. Blood 113: 1391–1392

Jager R, Harutyunyan AS, Rumi E, Pietra D, Berg T, Olcaydu D, Houlston RS, Cazzola M, Kralovics R (2014) Common germline variation at the TERT locus contributes to familial clustering of myeloproliferative neoplasms. Am J Hematol 89:1107–1110

Jones AV, Cross NC (2013) Inherited predisposition to myeloproliferative neoplasms. Ther Adv Hematol 4:237–253

Jones AV, Chase A, Silver RT, Oscier D, Zoi K, Wang YL, Cario H, Pahl HL, Collins A, Reiter A, Grand F, Cross NC (2009) JAK2 haplotype is a major risk factor for the development of myeloproliferative neoplasms. Nat Genet 41:446–449

Jones AV, Campbell PJ, Beer PA, Schnittger S, Vannucchi AM, Zoi K, Percy MJ, McMullin MF, Scott LM, Tapper W, Silver RT, Oscier D, Harrison CN, Grallert H, Kisialiou A, Strike P, Chase

AJ, Green AR, Cross NC (2010) The JAK2 46/1 haplotype predisposes to MPL-mutated myeloproliferative neoplasms. Blood 115:4517–4523

Kilpivaara O, Mukherjee S, Schram AM, Wadleigh M, Mullally A, Ebert BL, Bass A, Marubayashi S, Heguy A, Garcia-Manero G, Kantarjian H, Offit K, Stone RM, Gilliland DG, Klein RJ, Levine RL (2009) A germline JAK2 SNP is associated with predisposition to the development of JAK2(V617F)-positive myeloproliferative neoplasms. Nat Genet 41:455–459

Klco JM, Mullighan CG (2021) Advances in germline predisposition to acute leukaemias and myeloid neoplasms. Nat Rev Cancer 21:122–137

Kralovics R, Stockton DW, Prchal JT (2003) Clonal hematopoiesis in familial polycythemia vera suggests the involvement of multiple mutational events in the early pathogenesis of the disease. Blood 102:3793–3796

Lambert JR, Everington T, Linch DC, Gale RE (2009) In essential thrombocythemia, multiple JAK2-V617F clones are present in most mutant-positive patients: a new disease paradigm. Blood 114:3018–3023

Landgren O, Goldin LR, Kristinsson SY, Helgadottir EA, Samuelsson J, Bjorkholm M (2008) Increased risks of polycythemia vera, essential thrombocythemia, and myelofibrosis among 24,577 first-degree relatives of 11,039 patients with myeloproliferative neoplasms in Sweden. Blood 112:2199–2204

Li S, Kralovics R, De Libero G, Theocharides A, Gisslinger H, Skoda RC (2008) Clonal heterogeneity in polycythemia vera patients with JAK2 exon12 and JAK2-V617F mutations. Blood 111: 3863–3866

Lundberg P, Karow A, Nienhold R, Looser R, Hao-Shen H, Nissen I, Girsberger S, Lehmann T, Passweg J, Stern M, Beisel C, Kralovics R, Skoda RC (2014) Clonal evolution and clinical correlates of somatic mutations in myeloproliferative neoplasms. Blood 123:2220–2228

Macedo LC, de Cesare QF, Pagliari-E-Silva S, Pagnano KB, Rodrigues C, de Alencar JB, Sell AM, Visentainer JE (2016) Association of TNF polymorphisms with JAK2 (V617F) myeloproliferative neoplasms in Brazilian patients. Blood Cells Mol Dis 57:54–57

Malak S, Labopin M, Saint-Martin C, Bellanne-Chantelot C, Najman A, French Group of Familial Myeloproliferative D (2012) Long term follow up of 93 families with myeloproliferative neoplasms: life expectancy and implications of JAK2V617F in the occurrence of complications. Blood Cells Mol Dis 49:170–176

Mangino M, Hwang SJ, Spector TD, Hunt SC, Kimura M, Fitzpatrick AL, Christiansen L, Petersen I, Elbers CC, Harris T, Chen W, Srinivasan SR, Kark JD, Benetos A, El Shamieh S, Visvikis-Siest S, Christensen K, Berenson GS, Valdes AM, Vinuela A, Garcia M, Arnett DK, Broeckel U, Province MA, Pankow JS, Kammerer C, Liu Y, Nalls M, Tishkoff S, Thomas F, Ziv E, Psaty BM, Bis JC, Rotter JI, Taylor KD, Smith E, Schork NJ, Levy D, Aviv A (2012) Genome-wide meta-analysis points to CTC1 and ZNF676 as genes regulating telomere homeostasis in humans. Hum Mol Genet 21:5385–5394

Marchetti M, Carobbio A, Capitoni E, Barbui T (2018) Lymphoproliferative disorders in patients with chronic myeloproliferative neoplasms: a systematic review. Am J Hematol 93:698–703

Marchetti M, Ghirardi A, Masciulli A, Carobbio A, Palandri F, Vianelli N, Rossi E, Betti S, Di Veroli A, Iurlo A, Cattaneo D, Finazzi G, Bonifacio M, Scaffidi L, Patriarca A, Rumi E, Casetti IC, Stephenson C, Guglielmelli P, Elli EM, Palova M, Rapezzi D, Erez D, Gomez M, Wille K, Perez-Encinas M, Lunghi F, Angona A, Fox ML, Beggiato E, Benevolo G, Carli G, Cacciola R, McMullin MF, Tieghi A, Recasens V, Isfort S, Pane F, De Stefano V, Griesshammer M, Alvarez-Larran A, Vannucchi AM, Rambaldi A, Barbui T (2020) Second cancers in MPN: survival analysis from an international study. Am J Hematol 95:295–301

Mora B, Siracusa C, Rumi E, Maffioli M, Casetti IC, Barraco D, Merli M, Rossi M, Ubezio M, Accetta R, Libera L, Pietra D, Trotti C, Uccella S, Pallotti F, Casalone R, Bertu L, Arcaini L, Della Porta MG, Passamonti F (2021) Platelet count predicts driver mutations' co-occurrence in low JAK2 mutated essential thrombocythemia and myelofibrosis. Leukemia 35:1490–1493

Nahajevszky S, Andrikovics H, Batai A, Adam E, Bors A, Csomor J, Gopcsa L, Koszarska M, Kozma A, Lovas N, Lueff S, Matrai Z, Meggyesi N, Sinko J, Sipos A, Varkonyi A, Fekete S,

Tordai A, Masszi T (2011) The prognostic impact of germline 46/1 haplotype of Janus kinase 2 in cytogenetically normal acute myeloid leukemia. Haematologica 96:1613–1618

Oddsson A, Kristinsson SY, Helgason H, Gudbjartsson DF, Masson G, Sigurdsson A, Jonasdottir A, Jonasdottir A, Steingrimsdottir H, Vidarsson B, Reykdal S, Eyjolfsson GI, Olafsson I, Onundarson PT, Runarsson G, Sigurdardottir O, Kong A, Rafnar T, Sulem P, Thorsteinsdottir U, Stefansson K (2014) The germline sequence variant rs2736100_C in TERT associates with myeloproliferative neoplasms. Leukemia 28:1371–1374

Ohyashiki JH, Yoneta M, Hisatomi H, Iwabuchi T, Umezu T, Ohyashiki K (2012) The C allele of JAK2 rs4495487 is an additional candidate locus that contributes to myeloproliferative neoplasm predisposition in the Japanese population. BMC Med Genet 13:6

Olcaydu D, Harutyunyan A, Jager R, Berg T, Gisslinger B, Pabinger I, Gisslinger H, Kralovics R (2009) A common JAK2 haplotype confers susceptibility to myeloproliferative neoplasms. Nat Genet 41:450–454

Olcaydu D, Rumi E, Harutyunyan A, Passamonti F, Pietra D, Pascutto C, Berg T, Jager R, Hammond E, Cazzola M, Kralovics R (2011) The role of the JAK2 GGCC haplotype and the TET2 gene in familial myeloproliferative neoplasms. Haematologica 96:367–374

Ovsepyan VA, Tregubova EV, Luchinin AS, Minaeva NV (2019) Gene polymorphism of xenobiotic biotransformation enzymes in patients with classical Ph-negative myeloproliferative neoplasms. Bull Exp Biol Med 167:767–770

Patnaik MM, Lasho TL, Finke CM, Gangat N, Caramazza D, Siragusa S, Hanson CA, Pardanani A, Tefferi A (2010) MPL mutation effect on JAK2 46/1 haplotype frequency in JAK2V617F-negative myeloproliferative neoplasms. Leukemia 24:859–860

Poletto V, Rosti V, Villani L, Catarsi P, Carolei A, Campanelli R, Massa M, Martinetti M, Viarengo G, Malovini A, Migliaccio AR, Barosi G (2012) A3669G polymorphism of glucocorticoid receptor is a susceptibility allele for primary myelofibrosis and contributes to phenotypic diversity and blast transformation. Blood 120:3112–3117

Ruella M, Salmoiraghi S, Risso A, Carobbio A, Buttiglieri S, Spatola T, Sivera P, Ricca I, Barbui T, Tarella C, Rambaldi A (2013) Telomere shortening in Ph-negative chronic myeloproliferative neoplasms: a biological marker of polycythemia vera and myelofibrosis, regardless of hydroxycarbamide therapy. Exp Hematol 41:627–634

Rumi E, Cazzola M (2017) Advances in understanding the pathogenesis of familial myeloproliferative neoplasms. Br J Haematol 178:689–698

Rumi E, Passamonti F, Pietra D, Della Porta MG, Arcaini L, Boggi S, Elena C, Boveri E, Pascutto C, Lazzarino M, Cazzola M (2006) JAK2 (V617F) as an acquired somatic mutation and a secondary genetic event associated with disease progression in familial myeloproliferative disorders. Cancer 107:2206–2211

Rumi E, Passamonti F, Della Porta MG, Elena C, Arcaini L, Vanelli L, Del Curto C, Pietra D, Boveri E, Pascutto C, Cazzola M, Lazzarino M (2007) Familial chronic myeloproliferative disorders: clinical phenotype and evidence of disease anticipation. J Clin Oncol 25:5630–5635

Rumi E, Barate C, Benevolo G, Maffioli M, Ricco A, Sant'Antonio E (2020) Myeloproliferative and lymphoproliferative disorders: state of the art. Hematol Oncol 38:121–128

Saliba J, Saint-Martin C, Di Stefano A, Lenglet G, Marty C, Keren B, Pasquier F, Valle VD, Secardin L, Leroy G, Mahfoudhi E, Grosjean S, Droin N, Diop M, Dessen P, Charrier S, Palazzo A, Merlevede J, Meniane JC, Delaunay-Darivon C, Fuseau P, Isnard F, Casadevall N, Solary E, Debili N, Bernard OA, Raslova H, Najman A, Vainchenker W, Bellanne-Chantelot C, Plo I (2015) Germline duplication of ATG2B and GSKIP predisposes to familial myeloid malignancies. Nat Genet 47:1131–1140

Schischlik F, Kralovics R (2017) Mutations in myeloproliferative neoplasms – their significance and clinical use. Expert Rev Hematol 10:961–973

Spanoudakis E, Bazdiara I, Pantelidou D, Kotsianidis I, Papadopoulos V, Margaritis D, Xanthopoulidis G, Moustakidis E, Mantzourani S, Bourikas G, Tsatalas C (2011) Dynamics of telomere's length and telomerase activity in Philadelphia chromosome negative myeloproliferative neoplasms. Leuk Res 35:459–464

Sud A, Chattopadhyay S, Thomsen H, Sundquist K, Sundquist J, Houlston RS, Hemminki K (2018) Familial risks of acute myeloid leukemia, myelodysplastic syndromes, and myeloproliferative neoplasms. Blood 132:973–976

Sud A, Chattopadhyay S, Thomsen H, Sundquist K, Sundquist J, Houlston RS, Hemminki K (2019) Analysis of 153 115 patients with hematological malignancies refines the spectrum of familial risk. Blood 134:960–969

Tapper W, Jones AV, Kralovics R, Harutyunyan AS, Zoi K, Leung W, Godfrey AL, Guglielmelli P, Callaway A, Ward D, Aranaz P, White HE, Waghorn K, Lin F, Chase A, Baxter EJ, Maclean C, Nangalia J, Chen E, Evans P, Short M, Jack A, Wallis L, Oscier D, Duncombe AS, Schuh A, Mead AJ, Griffiths M, Ewing J, Gale RE, Schnittger S, Haferlach T, Stegelmann F, Dohner K, Grallert H, Strauch K, Tanaka T, Bandinelli S, Giannopoulos A, Pieri L, Mannarelli C, Gisslinger H, Barosi G, Cazzola M, Reiter A, Harrison C, Campbell P, Green AR, Vannucchi A, Cross NC (2015) Genetic variation at MECOM, TERT, JAK2 and HBS1L-MYB predisposes to myeloproliferative neoplasms. Nat Commun 6:6691

The Telomeres Mendelian Randomization Collaboration, Haycock PC, Burgess S, Nounu A, Zheng J, Okoli GN, Bowden J, Wade KH, Timpson NJ, Evans DM, Willeit P, Aviv A, Gaunt TR, Hemani G, Mangino M, Ellis HP, Kurian KM, Pooley KA, Eeles RA, Lee JE, Fang S, Chen WV, Law MH, Bowdler LM, Iles MM, Yang Q, Worrall BB, Markus HS, Hung RJ, Amos CI, Spurdle AB, Thompson DJ, O'Mara TA, Wolpin B, Amundadottir L, Stolzenberg-Solomon R, Trichopoulou A, Onland-Moret NC, Lund E, Duell EJ, Canzian F, Severi G, Overvad K, Gunter MJ, Tumino R, Svenson U, van Rij A, Baas AF, Bown MJ, Samani NJ, van t'Hof FNG, Tromp G, Jones GT, Kuivaniemi H, Elmore JR, Johansson M, McKay J, Scelo G, Carreras-Torres R, Gaborieau V, Brennan P, Bracci PM, Neale RE, Olson SH, Gallinger S, Li D, Petersen GM, Risch HA, Klein AP, Han J, Abnet CC, Freedman ND, Taylor PR, Maris JM, Aben KK, Kiemeney LA, Vermeulen SH, Wiencke JK, Walsh KM, Wrensch M, Rice T, Turnbull C, Litchfield K, Paternoster L, Standl M, Abecasis GR, JP SG, Li Y, Mijatovic V, Sapkota Y, Low SK, Zondervan KT, Montgomery GW, Nyholt DR, van Heel DA, Hunt K, Arking DE, Ashar FN, Sotoodehnia N, Woo D, Rosand J, Comeau ME, Brown WM, Silverman EK, Hokanson JE, Cho MH, Hui J, Ferreira MA, Thompson PJ, Morrison AC, Felix JF, Smith NL, Christiano AM, Petukhova L, Betz RC, Fan X, Zhang X, Zhu C, Langefeld CD, Thompson SD, Wang F, Lin X, Schwartz DA, Fingerlin T, Rotter JI, Cotch MF, Jensen RA, Munz M, Dommisch H, Schaefer AS, Han F, Ollila HM, Hillary RP, Albagha O, Ralston SH, Zeng C, Zheng W, Shu XO, Reis A, Uebe S, Huffmeier U, Kawamura Y, Otowa T, Sasaki T, Hibberd ML, Davila S, Xie G, Siminovitch K, Bei JX, Zeng YX, Forsti A, Chen B, Landi S, Franke A, Fischer A, Ellinghaus D, Flores C, Noth I, Ma SF, Foo JN, Liu J, Kim JW, Cox DG, Delattre O, Mirabeau O, Skibola CF, Tang CS, Garcia-Barcelo M, Chang KP, Su WH, Chang YS, Martin NG, Gordon S, Wade TD, Lee C, Kubo M, Cha PC, Nakamura Y, Levy D, Kimura M, Hwang SJ, Hunt S, Spector T, Soranzo N, Manichaikul AW, Barr RG, Kahali B, Speliotes E, Yerges-Armstrong LM, Cheng CY, Jonas JB, Wong TY, Fogh I, Lin K, Powell JF, Rice K, Relton CL, Martin RM, Davey Smith G (2017) Association between telomere length and risk of cancer and non-neoplastic diseases: a Mendelian Randomization Study. JAMA Oncol 3:636–651

Trifa AP, Bănescu C, Dima D, Bojan AS, Tevet M, Moldovan VG, Vesa Ş, Murat M, Pop IV, Skrypnyk C, Popp RA (2016a) Among a panel of polymorphisms in genes related to oxidative stress, CAT-262 C>T, GPX1 Pro198Leu and GSTP1 Ile105Val influence the risk of developing BCR-ABL negative myeloproliferative neoplasms. Hematology 21:520–525

Trifa AP, Banescu C, Tevet M, Bojan A, Dima D, Urian L, Torok-Vistai T, Popov VM, Zdrenghea M, Petrov L, Vasilache A, Murat M, Georgescu D, Popescu M, Patrinoiu O, Balea M, Costache R, Coles E, Saguna C, Berbec N, Vladareanu AM, Mihaila RG, Bumbea H, Cucuianu A, Popp RA (2016b) TERT rs2736100 A>C SNP and JAK2 46/1 haplotype significantly contribute to the occurrence of JAK2 V617F and CALR mutated myeloproliferative neoplasms – a multicentric study on 529 patients. Br J Haematol 174:218–226

Vainchenker W, Kralovics R (2017) Genetic basis and molecular pathophysiology of classical myeloproliferative neoplasms. Blood 129:667–679

Varricchio L, Masselli E, Alfani E, Battistini A, Migliaccio G, Vannucchi AM, Zhang W, Rondelli D, Godbold J, Ghinassi B, Whitsett C, Hoffman R, Migliaccio AR (2011) The dominant negative β isoform of the glucocorticoid receptor is uniquely expressed in erythroid cells expanded from polycythemia vera patients. Blood 118:425–436

Weischer M, Nordestgaard BG, Cawthon RM, Freiberg JJ, Tybjaerg-Hansen A, Bojesen SE (2013) Short telomere length, cancer survival, and cancer risk in 47102 individuals. J Natl Cancer Inst 105:459–468

Zaman S, Wang R, Gandhi V (2014) Targeting the apoptosis pathway in hematologic malignancies. Leuk Lymphoma 55:1980–1992

Zhang X, Hu T, Wu Z, Kang Z, Liu W, Guan M (2012) The JAK2 46/1 haplotype is a risk factor for myeloproliferative neoplasms in Chinese patients. Int J Hematol 96:611–616

Enteropathy-Associated T-Cell Lymphoma: Epidemiology, Natural History, and Management in the Current Era

Rajeev Sandal, Ashish Chauhan, Aditya Jandial, Kundan Mishra, Pulkit Rastogi, Pravesh Dhiman, and Asheesh Kumar

Abstract

Celiac disease (CeD) is a chronic autoimmune enteropathy which is seen in genetically predisposed individuals (HLA DQ2 and HLA DQ8 positive) as a result of intolerance to dietary gluten. The mortality in CeD is due to refractory celiac disease (RCD) and enteropathy-associated T-cell lymphoma (EATL). EATL is a rare entity accounting only up to less than 5% of the primary gastrointestinal lymphomas and is found in the older age group (60–70 years) with most of the cases reported from Europe. A strong association has been found between CeD and EATL, in particular RCD2. Diagnosis of EATL is confirmed by histopathology and immunophenotyping of representative biopsies from

R. Sandal (✉)
Department of Radiotherapy and Oncology, IGMC Shimla, Shimla, Himachal Pradesh, India
e-mail: drrajeevsandal@yahoo.com

A. Chauhan
Department of Gastroenterology, IGMC Shimla, Shimla, Himachal Pradesh, India

A. Jandial
Department of Clinical Haematology and Medical Oncology, Post Graduate Institute of Medical Education and Research, Chandigarh, India

K. Mishra
Department of Internal Medicine Army Hospital Research and Referral, New Delhi, India

P. Rastogi
Department of Haematology, Post Graduate Institute of Medical Education and Research, Chandigarh, India

P. Dhiman
Department of Medical Oncology and Clinical Haematology, AIIMS Bilaspur, Bilaspur, Himachal Pradesh, India

A. Kumar
Department of Nephrology, AIIMS, Vijaypur, Jammu and Kashmir, India

© The Author(s), under exclusive license to Springer Nature Switzerland AG 2023
Interdisciplinary Cancer Research, https://doi.org/10.1007/16833_2022_114
Published online: 31 January 2023

intestine. Staging is done with computed tomography (CT) or FDG-PET/CT imaging. Most of patients are diagnosed in advanced stage and not suitable for chemotherapy at time of diagnosis. Treatment is surgery and chemotherapy followed by autologous stem cell therapy. Anthracycline-based CHOP regimen is most commonly used, and brentuximab vedotin can be added in CD30 positive patients. It has aggressive course and very poor outcome with all the currently available therapies for peripheral T-cell Lymphoma. Target therapies or recruiting more patients in clinical trials may give some better outcome in future. Early detection of celiac disease by implementation of common surveillance may improve outcome.

Keywords

Brentuximab vedotin · Celiac disease · CHOP · Enteropathy-associated T-cell lymphoma (EATL) · Monomorphic epitheliotropic intestinal T-cell lymphoma · Peripheral T-cell lymphoma

Abbreviations

BV plus CHP	Brentuximab vedotin plus cyclophosphamide, doxorubicin, prednisone
CeD	Celiac disease
CHOEP	CHOP plus etoposide
CHOP	Cyclophosphamide, doxorubicin, vincristine, and prednisone
EATL	Enteropathy associated T-cell lymphoma
EPOCH	Etoposide, prednisolone, vincristine, cyclophosphamide, and doxorubicin
GFD	Gluten-free diet
Hyper-CVAD	Cyclophosphamide, vincristine, doxorubicin, and dexamethasone
IEL	Intraepithelial lymphocyte
MEITL	Monomorphic epitheliotropic intestinal T-cell lymphoma
RCD	Refractory celiac disease

1 Introduction

Celiac disease (CeD) is a multisystem disorder characterized by chronic autoimmune enteropathy seen in genetically predisposed individuals (HLA DQ2 and HLA DQ8 positive) as a result of intolerance to dietary gluten (Green and Cellier 2007; Biagi and Corazza 2010; Lindfors et al. 2019). The only available treatment for CeD is a strict gluten-free diet (GFD) which leads to clinical improvement and histological healing in the majority of the patients. Prospective data has revealed that despite strict GFD, few patients persist to have villous atrophy, and those patients who persist to have villous atrophy even after 1 year of strict GFD, after exclusion of the rest of the causes of villous atrophy, are classified as refractory celiac disease (RCD), and presently 5% of CeD patients have refractory celiac disease. CeD has been shown to have an increased mortality compared to the general population (Biagi and Corazza 2010). Despite better care and early diagnosis, mortality in CeD has largely remained unchanged. The mortality in CeD is due to refractory celiac disease and enteropathy-associated T-cell lymphoma (EATL) (Corrao et al. 2001). Based on the type of intraepithelial lymphocytes (IELs), RCD can be of two types: RCD type 1 and RCD type 2. RCD type 1 is characterized by persistent villous atrophy after 1 year of strict GFD, but with normal intraepithelial lymphocyte (IEL) phenotype (surface CD3 and CD8 positive). RCD type 2 is characterized by the abnormal clonal expansion of IELs, lacking CD3, CD8, and T-cell receptor markers, but exhibiting surface CD3 marker phenotype. The RCD type 2 frequently (50% within 5 years) evolves into EATL, which carries a poor prognosis. The link between CeD and EATL is based on common HLA alleles, gluten sensitivity in patients of EATL, and lesser incidence of EATL in those on GFD. CeD is a very common disease with varying prevalence across the globe, but global prevalence is around 1% (Green and Cellier 2007; Malamut and Cellier 2015). For the sake of this review, we will focus on EATL (Delabie et al. 2011). Incidence of EATL has increased over the years, with a large Surveillance, Epidemiology, and End Results (SEER) database showing the incidence of 0.006/100000, with incidence increasing to 0.05/100000 in those aged above 60 (Gough et al. 1962). EATL has two types: type 1 EATL that has retained the present reference for EATL, a disease related to CeD, and type 2 EATL, now known as monomorphic epitheliotropic intestinal T-cell lymphoma (MEITL), which is unrelated to CeD.

2 Epidemiology

EATL is a rare entity accounting only up to less than 5% of the primary gastrointestinal lymphomas and is found in the older age group (60–70 years) with most of the cases reported from Europe (Isaacson and Wright 1978). The strongest association of EATL has been with CeD, and studies have reported EATL in 38–100% of celiac patients. The incidence of EATL in CeD is 0.22 to 0.19/100000 and prevalence is <5%. In a Dutch study of 116 cases of EATL between 2000 and 2006, the incidence was found to be 0.10/100000, with the most common site being the proximal small intestine, and the incidence was found to be 2.08/100000 in those above 50 years, emphasizing EATL as an old age disease (Tse et al. 2012). EATL is the most common type of intestinal lymphoma but still is a rare disease. Type 1 EATL or what is now known as EATL only has mostly been seen in North European ancestry, and type 2 EATL has been reported mostly from Asia. EATL classification has been revised with type 1 EATL being classified as EATL and type 2 EATL now being renamed as monomorphic epitheliotropic intestinal T-cell lymphoma (Cellier et al. 2000). As per a multicenter study from Asia, with 38 patients in 19 years who were all type 2, the most common site of involvement was small bowel (63%), predominantly male population with a mean age of 59 years and overall survival of 7 months (Swerdlow et al. 2016). Most common presentations were perforation, abdominal pain, and obstruction. In another study with diverse population groups with 65 EATL patients, only 4% were above 65 years of age; most had bone marrow involvement, and almost all had intermediate-/high-grade disease.

EATL is an aggressive form of peripheral T-cell lymphoma which is derived from small intestinal IELs and carries a sinister prognosis. An association with CeD was described first in 1962, and histological features were described a decade later (Berman et al. 1998; McBride et al. 2010). Risk factors associated with EATL are poor adherence to gluten, delayed diagnosis (>10 years), advanced age at diagnosis (>50 years), and HLA DQ2 homozygosity (Di Sabatino et al. 2012).

3 Pathophysiology

A strong association has been found between CeD and EATL, in particular RCD. It was in a European study where the HLA DQA1*0501, DQB1*0201 genotype was found in 90% of both patients with CeD and EATL, pointing toward an association. Among RCD, it's RCD with aberrant IELs which progresses to EATL. EATL may be diagnosed 3.9–10 years after CeD, or they may be diagnosed concurrently or CeD may be diagnosed postmortem. The association of RCD with clonal T-cell proliferation and eventually EATL was shown in a seminal paper by Cellier, where out of 19 analyzed patients, aberrant T-cell phenotype with intracytoplasmic CD3 and absent surface T-cell markers as CD3, CD4, and CD8 was seen in 84% ($n = 16$) of the patients and TCRγ gene rearrangement was seen in 76% of patients and eventually 3 patients progressed to EATL (Varghese et al. 2013). RCD is considered a strong risk factor for EATL. IL15 is the mediator considered to have a role in epithelial damage caused due to its overexpression, leading to the emergence of clonal T cells. CD4 T cells and IL15 promote the expansion of CD8 T cells and are responsible for loss of immunoregulatory control over these T cells in RCD (Amiot et al. 2012). The genes involved in the pathogenesis of EATL have been explained under genetics section.

4 Diagnosis

4.1 Clinical Presentation

Patients having a preceding history of CeD and sudden worsening of preexisting signs and symptoms of celiac disease like steatorrhea, flatulence, abdomen distension, weight loss, severe anemia, neurologic disorders, and osteopenia/osteoporosis in spite of GFD should raise suspicion EATL. History of childhood CeD is unusual, and most such patients are diagnosed with CeD at old age (more than 39 years) or at the time of presentation with EATL (Cellier et al. 2000; Tse et al. 2012).

Common presenting symptoms reported in various studies are abdominal pain (65–100%), diarrhea (40–70%), systemic B symptoms (30–80%), fatigue (38%), infection (23%), lymphadenopathy (15%), hepatosplenomegaly (6%), and pruritus (3%), respectively. The signs of severe malabsorption and poor nutritional intake are seen in a major proportion of patients at the time of diagnosis and are associated with a poor outcome (Delabie et al. 2011; Swerdlow et al. 2016).

Patients may present with gastrointestinal complications like small intestinal (jejunal) obstruction and perforation (25–50%) and/or lower gastrointestinal bleeding due to common location of primary site (Swerdlow et al. 2016).

4.2 Common Location and Spread of EATL

Proximal jejunum is the most common site for EATL, followed by other parts of the small intestine, stomach, or colon. Some case reports of involvement of the liver, spleen, and lymph nodes have been described in the literature, but involvement of the lung (5–16%), skin (5%), or soft tissues is very rare (Tse et al. 2012).

One series has described small intestine involvement in 80–90% of cases, followed by mesenteric/intra-abdominal lymph nodes (35%), large intestine (16%), para-aortic/iliac lymph nodes (11%), stomach (8%), inguinal/femoral lymph nodes (6%), and other sites (≤5%), respectively. Nearly 3–18% of patients have marrow involvement at the time of diagnosis. Rarely central nervous system involvement can be seen in EATL (Berman et al. 1998; Delabie et al. 2011). Splenic atrophy and mesenteric lymph node cavitation have been reported in various case reports, significance of which is unknown (McBride et al. 2010; Di Sabatino et al. 2012). Infrequently hemophagocytic syndrome may be initial presentation or may also be diagnosed during the course of illness (16–40% of patients) in EATL (Amiot et al. 2012; Varghese et al. 2013).

4.3 Staging

Lugano classification for the staging of lymphomas (derived from Ann Arbor staging with Cotswolds modifications) is used for EATL as similar to other lymphomas (Cheson et al. 2014).This staging may or may not include modification given by Rohatiner et al. for staging gastrointestinal lymphoma (Rohatiner et al. 1994). The majority of the patients (43–90%) are diagnosed at an advanced stage (usually stage IV disease) (Delabie et al. 2011).

4.4 Endoscopy and Imaging

Endoscopy is required to inspect the suspected sites of EATL involvement within the gastrointestinal tract and for taking biopsy for histopathological confirmation (Kim et al. 2014). Newer techniques like double-balloon enteroscopy and wireless capsule endoscopy are commonly used nowadays. With the wireless capsule endoscopy, we can see the ulceration and the extent of small bowel involvement, although it is contraindicated in patients with suspected strictures (Mallant et al. 2007; Matsumoto et al. 2010; Di Sabatino et al. 2012).

On the other hand, double-balloon enteroscopy can facilitate the definitive diagnosis of EATL by allowing endoscopic biopsy and for histopathological examination. It can recognize even tiny small bowel lesions like multiple polyps, fine granular mucosa, and swollen folds and has role in diagnosing early-stage EATL (Matsumoto et al. 2010; Di Sabatino et al. 2012; Ferretti et al. 2020). CT enterography is the most common modality for staging EATL and confirmation of extraintestinal manifestations like mesenteric lymph node cavitation and splenic atrophy. It is more accurate in detecting bowel wall thickening and ulceration than small bowel enema in EATL (Mallant et al. 2007; Di Sabatino et al. 2012).

FDG-PET scan can be helpful in detecting early-stage EATL due to its high sensitivity, but inflammation and active peristaltic movement may reduce its specificity (Hadithi et al. 2006). Magnetic resonance (MR) enterography is also very sensitive in detecting early EATL which is localized to the epithelial layer of the bowel wall or is multifocal. Both PET and MR imaging can be useful in assessing the response to therapy on follow-up (Van Weyenberg et al. 2011; Di Sabatino et al. 2012).

4.5 Histology

Most EATL patients initially present as surgical emergency in the form of intestinal obstruction, perforation, or GI bleeding, and EATL is diagnosed on laparotomy. On macroscopic inspection of the intestine, there may be single or multiple circumferentially oriented jejunal ulcers with or without any tumor mass. It may appear as transparietal infiltration, with ulcerations and intestinal wall induration (Gale et al. 2000).

On microscopic examination, the malignant cells are a mixture of variably sized tumor cells invading the intestinal wall. These atypical tumor cells have round or angulated vesicular nucleus with prominent nucleoli, high mitotic index, and moderate to abundant amount of pale-staining cytoplasm. Tumor cells are often associated with infiltration of eosinophils, histiocytes, and small lymphocytes. In the early course of disease, the lesions may show a few scattered atypical cells with numerous reactive macrophages and without any appreciable mass (Chander et al. 2018).

The adjacent normal mucosa has numerous macrophages, intraepithelial T cells, and eosinophils. The intraepithelial lymphocytes (IELs) in normal uninvolved area may be clonal, and they may represent the precursor lesions to EATL. Adjacent healthy mucosa may or may not show signs of celiac disease depending on the segment analyzed, CeD associated villous atrophy is most commonly seen in the proximal small intestine, and it may be absent in distal jejunum or ileum (Fig. 1) (Chott et al. 1992).

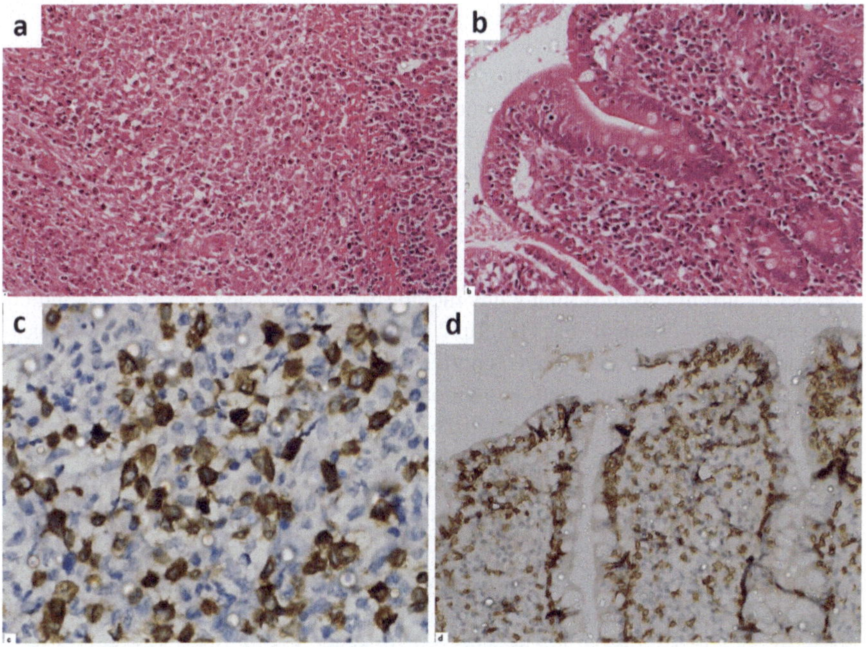

Fig. 1 Histopathology of (**a**) stricturous lesion of jejunum showing infiltration by large-sized atypical lymphoid cells in an inflammatory background of eosinophils, lymphocytes, and neutrophils (H&E, 400×). (**b**) Adjacent uninvolved portion of the jejunum shows increased intraepithelial lymphocytes (IELs) in the villi (H&E, 400×). (**c**, **d**) CD3 immunostain highlighting atypical lymphoid cells and IELs (hematoxylin counterstain & DAB chromogen, 400×)

4.6 Immunophenotyping

Tumor cells express surface CD7+ and the mucosal lymphoid antigen (CD103) in a typical case of EATL. Tumor cells lack the expression of the surface T-cell receptor/CD3 complex but are positive for intracellular CD3 (Murray et al. 1995). Other markers like CD4 (11%), CD5 (17%), and CD 8 (43%) are expressed in only a minority of cases. Previous studies have reported CD30 positivity in up to 37% EATL cases. Many cases occasionally express cytotoxic T cell-associated proteins like granzyme B, T-cell intracellular antigen-1 (TIA-1), and perforin. Aberrant expression of cytolytic natural killer (NK) cell lineage receptors can be seen in some tumor cells, but these cells have T-cell receptor rearrangements and are clearly of T-cell lineage. In situ hybridization for Epstein-Barr virus is negative in EATL (Murray et al. 1995; Schmitt-Gräff et al. 1996).

4.7 Genetics

In EATL, the T-cell receptor beta and gamma genes are clonally rearranged (Perfetti et al. 2012). Many patients have been shown to have a DQA1*0501 or DQB1*0201 HLA haplotype similar to CeD (Arber et al. 2016). Nearly 50% to 60% of the patients have complex segmental gene amplifications of the 9q31.3 region as performed by conventional cytogenetics or fluorescence in situ hybridization (FISH) (Tomita et al. 2020). This genetic change is rarely seen in any other peripheral T-cell lymphoma except monomorphic epitheliotropic intestinal T-cell lymphoma (MEITL). Monomorphic epitheliotropic intestinal T-cell lymphoma (previously referred to as type 2 EATL) can have similar genetic changes (Obermann et al. 2004; Arber et al. 2016). FISH with specific probes for this region is useful in confirming the diagnosis, but the sensitivity and specificity of this genetic abnormality is not clear (Obermann et al. 2004).

Other recurrent chromosomal gains include 1q32 (26%) and 5q35 (21%), which are common in classical EATL; 7q22 (31%), 8q24 (18%), and some cases ofdeletion16q12 have also been reported. Chromosomal 8q24 gains involving *MYC* gene have been reported in monomorphic epitheliotropic intestinal T-cell lymphoma (Ondrejka and Jagadeesh 2016).

On whole exome DNA sequencing of EATL, one-third of patients reveal mutation of *SETD2* gene. This gene encodes a histone methyltransferase which has a role in epigenetic regulation of gene expression. There are also mutations in various components of the JAK/STAT signaling pathway. The JAK/STAT pathway mutations are more common in classic EATL, whereas *SETD2* mutations are more frequent in monomorphic epitheliotropic intestinal T-cell lymphoma. These observations also confirmed the presence of 9q gains in both types of tumor. Other mutations are *KRAS*, *TP53*, or *TERT* genes which are seen in a small subset of patients (Di Sabatino et al. 2012; Ondrejka and Jagadeesh 2016; Roberti et al. 2016).

4.8 Other Laboratory Findings

Anemia (54–91%), raised lactate dehydrolase (25–65%), hypoalbuminemia (76–88%), and peripheral eosinophilia can be seen in patients with EATLs (Hovenga et al. 2003; Di Sabatino et al. 2012).

4.9 Differential Diagnoses

These include precursor lesions like refractory celiac disease (RCD) mainly type 2 and other types of intestinal lymphomas. Figure 2 shows simple diagnostic approach to EATL.

RCD Some EATL cases are preceded by a low-grade precursor lesion that clinically resembles RCD. RCD is characterised by persistence of histological lesions such as raised IELs, crypt hyperplasia and villous atrophy even after one year of strict GFD (van de Water et al. 2010; Wierdsma et al. 2016). RCD is commonly divided into two subtypes:

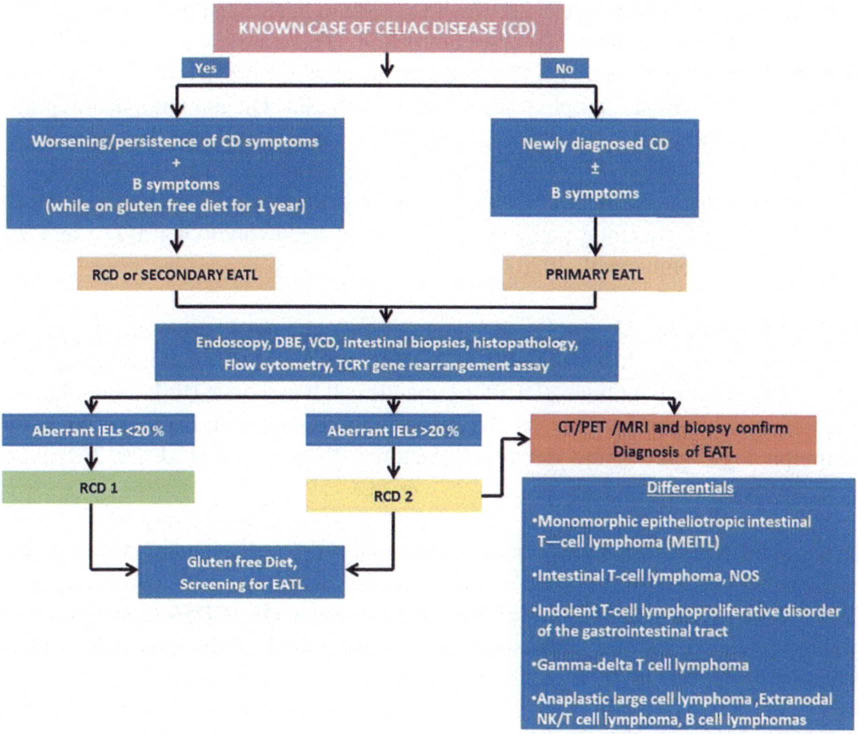

Fig. 2 Diagnostic algorithm of EATL

Type 1 RCD The lesions in RCD1 show normal intraepithelial lymphocytes, with normal surface T-cell receptor (TCR), CD3, and CD8 expression. T-cell receptor gene rearrangement assay shows normal polyclonal pattern (Wierdsma et al. 2016).

Type 2 RCD RCD2 shows no atypia in the intraepithelial lymphocytes, but there is loss of surface T-cell receptor, CD3, or CD8 expression. T-cell receptor gene rearrangement assay reveals monoclonal pattern. Type 2 RCD lesions are considered as precursor to EATL, whereas type 1 RCD is unlikely to progress to EATL. FDG-PET and video capsule endoscopy can be used to diagnose early lesions of RCD (van de Water et al. 2010; Wierdsma et al. 2016).

Monomorphic Epitheliotropic Intestinal T-Cell Lymphoma (MEITL) Monomorphic epitheliotropic intestinal T-cell lymphoma was classified as "type 2 EATL" in previous WHO classification (Swerdlow et al. 2016). There is no association between MEITL and CeD. MEITL has more incidence in Hispanics and Asian population as compared to EATL which is commonly seen in geographical regions with a high incidence of CeD (Northern Europe) (Di Sabatino et al. 2012; Tomita et al. 2020).

On histology, the tumor cells are small to medium in size and monomorphic and frequently infiltrate the intestinal crypt epithelium. There are no inflammatory cells in the background (as seen in EATL), and the surrounding normal mucosa may show increased IELs, villous atrophy, and crypt hyperplasia. On immunophenotyping, tumor cells demonstrate CD3 (95%), CD8 (63%), and CD56 (73%) and are usually CD4 negative. Gamma-delta T-cell receptors are expressed in up to 80% of the patients (Di Sabatino et al. 2012). Genetics analysis may show mutations in *SETD2* gene, JAK/STAT signaling pathway, and gains of 8q24 (involving *MYC*) and 9q may be found (Tomita et al. 2020).

B-Cell Lymphomas Mature B cell lymphomas are the most common subtypes of lymphoma of the small intestine. They include marginal zone lymphoma (MZL), diffuse large B-cell lymphoma (DLBCL), mantle cell lymphoma, Burkitt lymphoma, and follicular lymphoma. They can be easily distinguished from EATL on the basis of morphology and immunophenotypic features (Di Sabatino et al. 2012; Sandal et al. 2018).

Extranodal NK/T-Cell Lymphoma Extranodal NK/T-cell lymphoma commonly presents as facial mass and rarely involves the gastrointestinal tract. The tumor cells demonstrate NK/T-cell markers (CD2, cytoplasmic CD3, CD56, and cytotoxic granule proteins) and are universally positive for Epstein-Barr virus (EBV) (van Vliet and Spagnolo 2020).

Anaplastic Large Cell Lymphoma (ALCL) A few cases of EATL express CD30 and may mimic ALCL; however, both can be differentiated on the basis of cellular morphology and homogenous CD30 positivity in ALCL. It clinically presents as painless lymphadenopathy and is not associated with celiac disease as in EATL (Ondrejka and Jagadeesh 2016; van Vliet and Spagnolo 2020).

Gamma-Delta T-Cell Lymphoma It commonly involves the liver and spleen (hepatosplenic gamma-delta T-cell lymphoma), but rarely it involves the gastrointestinal tract or the skin (i.e., non-hepatosplenic or mucocutaneous gamma-delta T-cell lymphomas). Similar to EATL, tumor cells are positive CD2, surface CD3, CD7, CD56, and negative CD4, CD5, CD8, or B-cell markers. EATL has clonal T-cell receptor beta rearrangements and typically expresses CD103, as compared to gamma-delta T-cell lymphoma which exhibits T-cell receptor delta rearrangements (Foppoli and Ferreri 2015).

Intestinal T-Cell Lymphoma, NOS It is a nonspecific disease entity, which comprises of T-cell lymphomas arising from the intestines that are not classifiable into EATL or monomorphic epitheliotropic intestinal T-cell lymphoma or the cases where definite diagnosis is not achieved due to inadequate biopsy (Arber et al. 2016).

Indolent T-Cell Lymphoproliferative Disease This is a clonal T-cell lymphoproliferative disorder which commonly involves the small intestine and colon but can occur at any site in the gastrointestinal tract. The tumor cells infiltrate the lamina propria only with no invasion of the intestinal epithelium. The tumor cells are positive for CD2, CD3, CD4, CD5, variable CD7, and CD8. They demonstrate alpha-beta T-cell receptor arrangement with very low Ki-67 proliferation index (usually <10%). Most cases are indolent, but some subsets may have aggressive course with poor response to chemotherapy (Perry et al. 2013).

4.10 Management

4.10.1 Pretreatment Evaluation

After establishing the diagnosis of EATL by histopathological and immunophenotyping, the next step is to see the patient's performance status (i.e., ECOG) and associated comorbidities, by careful history and physical examination (Tse et al. 2012).

Routine investigations before treatment initiation including complete blood count (CBC), liver and renal function tests, serum electrolytes, lactate dehydrogenase (LDH), uric acid, and viral markers (HBs Ag, anti-HCV, HIV) should be done at the baseline. To see the extent or bulk of the disease, CECT scan of the chest, abdomen, and pelvis or whole body FDG- PET/CT is performed. Unilateral bone marrow aspiration and biopsy is performed to see lymphomatous infiltration (Cheson et al. 2014). The 2D echocardiography or MUGA scan is performed to assess the baseline cardiac function. The patients of childbearing age should be counseled about post-chemotherapy infertility and the fertility-preserving methods (Di Sabatino et al. 2012).

5 Treatment

Definite treatment protocol for peripheral T-cell lymphomas (PTCL) including EATL is lacking due to rarity of the disease and lack of data of randomized clinical trials. Most of the treatment protocols are based on retrospective studies and expert opinions. From historical point of view, treatment included surgery, chemotherapy, or combination (Al Somali et al. 2021). Most of the CeD patients have low BMI due to underlying disease, which negatively impacts treatment outcome. Various studies have reported a 5-year survival of 8–60% (Sieniawski et al. 2010, Malamut et al. 2013; Al Somali et al. 2021).

Surgery Surgery is commonly performed for the complications like perforation, intestinal obstruction, hemorrhage, and stenosis. The incidence of emergency versus elective laparotomy varies from study to study (Sieniawski et al. 2010). Postoperative complications like leakage and infection are more common in emergency procedures with an overall rate of 35%. The reported 5-year survival with elective and urgent procedures is 34% and 18%, respectively (van de Water et al. 2015).

There was no intestinal perforation in postoperative cases during induction therapy, so resection of lesions is performed by many physicians before definite therapy (van de Water et al. 2015).

5.1 Frontline Induction Therapy

Ideally, EATL patients should be encouraged to participate in well-planned clinical trials due to lack of definite treatment protocol. This is not always possible in the "real world"; hence, in era of targeted therapy, we recommended a commonly proposed treatment algorithm.

CD30 Negative Patients In patients with absence of CD30 expression (less than 10% of tumor cells), anthracycline-based regimen CHOP (cyclophosphamide, doxorubicin, vincristine, and prednisone) is the standard of treatment. CHOEP (CHOP plus etoposide) have been tried in medically fit and younger patients (less than 60 years) without any improvement in survival (Al Somali et al. 2021). More intensive protocols like infusional EPOCH (etoposide, prednisolone, vincristine, cyclophosphamide, and doxorubicin) and hyper-CVAD (cyclophosphamide, vincristine, doxorubicin, and dexamethasone) provide similar clinical results at the cost of greater toxicity (Wöhrer et al. 2004; Al Somali et al. 2021).

Both CHOP and CHOEP are given as 21-day cycle. Etoposide (100 mg/m^2) is given on days 1 through day 3 in CHOEP along with standard CHOP regimen. Common adverse effects are grade 3 or 4 neutropenia followed by infusion-related reaction, nausea, and vomiting with treatment related mortality of 1% with CHOP (Sitzia et al. 1997; Sandal et al. 2019).

CD30 Positive Patients For patients who are CD30 positive (i.e., ≥10%), BV plus CHP (i.e., brentuximab vedotin plus cyclophosphamide, doxorubicin, prednisone) regimen is preferred over standard CHOP. BV plus CHP is scheduled every 3 weeks for a total of six cycles similar to CHOP with brentuximab at dose of 1.8 mg/kg intravenously on day 1 of each cycle (Horwitz et al. 2019; Al Somali et al. 2021). This regimen is based on largest multicenter, double-blind, ECHELON-2 trial that has shown superior survival in BV plus CHP arm compared to CHOP in PTCL, although it included only three CD30 positive EATL patients. The 3-year progression-free survival (PFS) and overall survival (OS) are 52.9% and 73%, respectively, with overall response rate (ORR) of 79% in CD30 positive patients (Horwitz et al. 2019; Al Somali et al. 2021). Overall the treatment related toxicities were similar between two regimes with similar adverse event related deaths (3 and 4%, respectively). Grade ≥3 diarrhea (6%), neutropenia (35%), peripheral sensory neuropathy (4%), and nausea (2%) are more common in patients treated with BV plus CHP than CHOP (Fabbri et al. 2017; Horwitz et al. 2019).

5.2 Consolidation Therapy and Prognosis

Despite up-front therapy, i.e., surgery or induction chemotherapy or combinations, most of patients of EATL have poor survival (5-year OS 9–20%) (Daum et al. 2003; Al Somali et al. 2021). The outcomes are improved with consolidation therapy in the form of high dose chemotherapy followed by autologous stem cell therapy (HDCT/ASCT). But most of the patients are unable to undergo transplant due to poor performance status and early relapse. BEAM therapy (BCNU, etoposide, cytarabine, melphalan) is the most commonly used pretransplant regimen with tolerable toxicity. ASCT have given 4-year progression-free survival and overall survival of 54% and 59%, respectively (Jantunen et al. 2013). Data about allogeneic stem cell transplant is very limited and considered on a case-by-case only or in clinical trials (Schmitz et al. 2018). Figure 3 showed current treatment approach to EATL.

5.3 Predictive Factors

The International Prognostic Index (IPI), which is most commonly used in non-Hodgkin lymphoma (NHL), is not much helpful in case of EATL (Arber et al. 2016). Surgical removal of tumor along with chemotherapy gives best overall survival. Malnutrition and low albumin due to malabsorption because of prior celiac

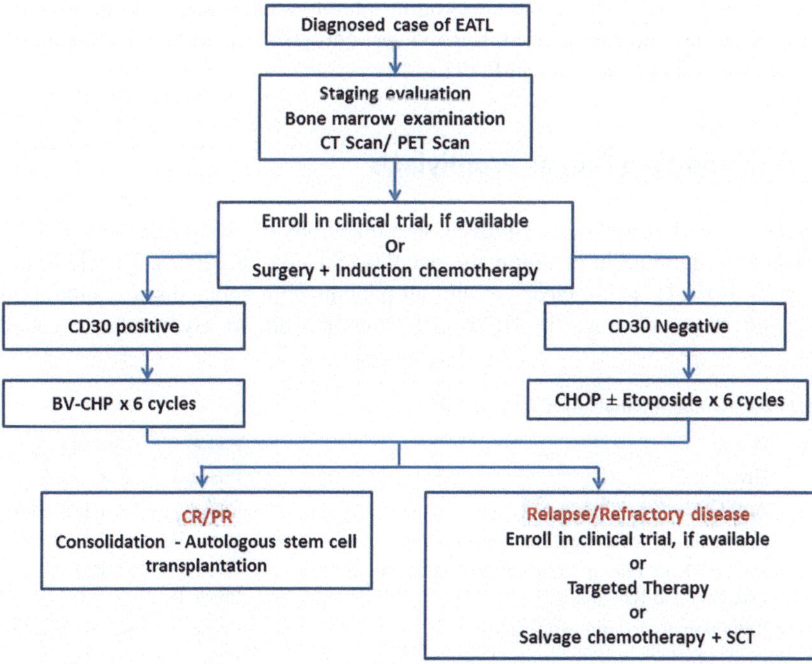

Fig. 3 Current treatment approach to EATL

disease leads to short survival (Cellier et al. 2000). The Prognostic Index for T-cell lymphoma (PIT) and an EATL prognostic index (EPI) are other prognostic indices which predict better survival than IPI (Delabie et al. 2011; de Baaij et al. 2015).

5.4 Relapse/Refractory Disease

Treatment of relapse/refractory EATL is not clear due to rarity of disease. Most clinicians prefer to treat it as an aggressive lymphoma. The best way is to recruit patients in a clinical trial, if available. Otherwise, salvage chemotherapy, preferably platinum based (e.g., GDP, DHAP) (full forms), may be given, followed by autologous stem cell transplant if not done earlier (Park et al. 2015).

6 Future Perspective

Whereas most of patients are frail, have poor performance status at the time of diagnosis, and are not candidate for intensive chemotherapy, this subgroup can be candidate for target therapy. Target antibody against IL15 has shown good response against RCD2 in many clinical trials (Vicari et al. 2017; Asri et al. 2020). Romidepsin and belinostat (histone deacetylase inhibitors) have been tried as in PTCL-NOS in small studies with poor outcome (Campbell and Thomas 2017; Maruyama et al. 2017). PI3K and checkpoint inhibitors have been studied in other T-cell lymphomas and can be tried in EATL patients (Horwitz et al. 2018). Table 1 shows the various clinical trials in EATL.

6.1 Gluten-Free Diet as Prophylaxis

Adherence to gluten-free diet to prevent complications of celiac disease was well established in literature in symptomatic patients (Silano et al. 2008). Its role in the prevention of EATL is not clear yet, but all patients with celiac disease should be advised for strict adherence to GFD on follow-up visit. In asymptomatic celiac

Table 1 Future ongoing trials for EATL and RCD

S. no.	Drugs	Phase	Clinical trial
1	Nivolumab	Phase II	NCT03075553
2	Anti-CD30-CAR T cells	Phase I	NCT03049449
3	Brentuximab vedotin and pembrolizumab	Phase II	NCT04795869
4	CAR-pNK cell immunotherapy in CD7 positive leukemia and lymphoma	Phase I Phase II	NCT02742727
5	Brentuximab vedotin single agent	Phase II	NCT02588651
6	Doxorubicin, azacitidine, romidepsin, and duvelisib	Phase I	NCT04639843
7	AMG 714 for RCD2	Phase II	NCT02633020

disease, the role of gluten-free diet in the prevention of complications is not clear (Smedby et al. 2005; Silano et al. 2008).

7 Conclusion

Enteropathy-associated T-cell Lymphoma is a rare T-cell lymphoma which has associations with celiac disease. Treatment is surgery and chemotherapy followed by autologous stem cell therapy. It has aggressive course and very poor outcome with all the currently available therapies for peripheral T-cell lymphoma (PTCL). Target therapies or recruiting more patients in clinical trials may give some better OS in the future. Early detection of celiac disease by implementation of common surveillance may improve outcome.

References

Al Somali Z, Hamadani M, Kharfan-Dabaja M, Sureda A, El Fakih R, Aljurf M (2021) Enteropathy-associated T cell lymphoma. Curr Hematol Malig Rep 16:140–147

Amiot A, Allez M, Treton X, Fieschi C, Galicier L, Joly F, Gornet JM, Oksenhendler E, Lémann M, Bouhnik Y (2012) High frequency of fatal haemophagocytic lymphohistiocytosis syndrome in enteropathy-associated T cell lymphoma. Dig Liver Dis 44:343–349

Arber DA, Orazi A, Hasserjian R, Thiele J, Borowitz MJ, Le Beau MM, Bloomfield CD, Cazzola M, Vardiman JW (2016) The 2016 revision to the World Health Organization classification of myeloid neoplasms and acute leukemia. Blood 127:2391–2405

Asri N, Rostami-Nejad M, Rezaei-Tavirani M, Razzaghi M, Asadzadeh-Aghdaei H, Zali MR (2020) Novel therapeutic strategies for celiac disease. Middle East J Dig Dis 12:229–237

Berman EL, Zauber NP, Rickert RR, Diss TC, Isaacson PG (1998) Enteropathy-associated T cell lymphoma with brain involvement. J Clin Gastroenterol 26:337–341

Biagi F, Corazza GR (2010) Mortality in celiac disease. Nat Rev Gastroenterol Hepatol 7:158–162

Campbell P, Thomas CM (2017) Belinostat for the treatment of relapsed or refractory peripheral T-cell lymphoma. J Oncol Pharm Pract 23:143–147

Cellier C, Delabesse E, Helmer C, Patey N, Matuchansky C, Jabri B, Macintyre E, Cerf-Bensussan-N, Brousse N (2000) Refractory sprue, coeliac disease, and enteropathy-associated T-cell lymphoma. French Coeliac Disease Study Group. Lancet 356:203–208

Chander U, Leeman-Neill RJ, Bhagat G (2018) Pathogenesis of enteropathy-associated T cell lymphoma. Curr Hematol Malig Rep 13:308–317

Cheson BD, Fisher RI, Barrington SF, Cavalli F, Schwartz LH, Zucca E, Lister TA (2014) Recommendations for initial evaluation, staging, and response assessment of Hodgkin and non-Hodgkin lymphoma: the Lugano classification. J Clin Oncol 32:3059–3068

Chott A, Dragosics B, Radaszkiewicz T (1992) Peripheral T-cell lymphomas of the intestine. Am J Pathol 141:1361–1371

Corrao G, Corazza GR, Bagnardi V, Brusco G, Ciacci C, Cottone M, Sategna Guidetti C, Usai P, Cesari P, Pelli MA, Loperfido S, Volta U, Calabró A, Certo M (2001) Mortality in patients with coeliac disease and their relatives: a cohort study. Lancet 358:356–361

Daum S, Ullrich R, Heise W, Dederke B, Foss HD, Stein H, Thiel E, Zeitz M, Riecken EO (2003) Intestinal non-Hodgkin's lymphoma: a multicenter prospective clinical study from the German Study Group on Intestinal non-Hodgkin's Lymphoma. J Clin Oncol 21:2740–2746 de Baaij LR, Berkhof J, van de Water JM, Sieniawski MK, Radersma M, Verbeek WH, Visser OJ, Oudejans JJ, Meijer CJ, Mulder CJ, Lennard AL, Cillessen SA (2015) A new and validated clinical prognostic model (EPI) for enteropathy-associated T-cell lymphoma. Clin Cancer Res 21:3013–3019

Delabie J, Holte H, Vose JM, Ullrich F, Jaffe ES, Savage KJ, Connors JM, Rimsza L, Harris NL, Müller-Hermelink K, Rüdiger T, Coiffier B, Gascoyne RD, Berger F, Tobinai K, Au WY, Liang R, Montserrat E, Hochberg EP, Pileri S, Federico M, Nathwani B, Armitage JO, Weisenburger DD (2011) Enteropathy-associated T-cell lymphoma: clinical and histological findings from the international peripheral T-cell lymphoma project. Blood 118:148–155

Di Sabatino A, Biagi F, Gobbi PG, Corazza GR (2012) How I treat enteropathy-associated T-cell lymphoma. Blood 119:2458–2468

Fabbri A, Cencini E, Gozzetti A, Schiattone L, Bocchia M (2017) Therapeutic use of brentuximab vedotin in CD30+ hematologic malignancies. Anticancer Agents Med Chem 17:886–895

Ferretti F, Branchi F, Orlando S, Roncoroni L, Barigelletti G, Fabiano S, Vecchi M, Penagini R, Doneda L, Elli L (2020) Effectiveness of capsule endoscopy and double-balloon enteroscopy in suspected complicated celiac disease. Clin Gastroenterol Hepatol

Foppoli M, Ferreri AJ (2015) Gamma-delta t-cell lymphomas. Eur J Haematol 94:206–218

Gale J, Simmonds PD, Mead GM, Sweetenham JW, Wright DH (2000) Enteropathy-type intestinal T-cell lymphoma: clinical features and treatment of 31 patients in a single center. J Clin Oncol 18:795–803

Gough KR, Read AE, Naish JM (1962) Intestinal reticulosis as a complication of idiopathic steatorrhoea. Gut 3:232–239

Green PH, Cellier C (2007) Celiac disease. N Engl J Med 357:1731–1743

Hadithi M, Mallant M, Oudejans J, van Waesberghe JH, Mulder CJ, Comans EF (2006) 18F-FDG PET versus CT for the detection of enteropathy-associated T-cell lymphoma in refractory celiac disease. J Nucl Med 47:1622–1627

Horwitz SM, Koch R, Porcu P, Oki Y, Moskowitz A, Perez M, Myskowski P, Officer A, Jaffe JD, Morrow SN, Allen K, Douglas M, Stern H, Sweeney J, Kelly P, Kelly V, Aster JC, Weaver D, Foss FM, Weinstock DM (2018) Activity of the PI3K-δ,γ inhibitor duvelisib in a phase 1 trial and preclinical models of T-cell lymphoma. Blood 131:888–898

Horwitz S, O'Connor OA, Pro B, Illidge T, Fanale M, Advani R, Bartlett NL, Christensen JH, Morschhauser F, Domingo-Domenech E, Rossi G, Kim WS, Feldman T, Lennard A, Belada D, Illés Á, Tobinai K, Tsukasaki K, Yeh SP, Shustov A, Hüttmann A, Savage KJ, Yuen S, Iyer S, Zinzani PL, Hua Z, Little M, Rao S, Woolery J, Manley T, Trümper L (2019) Brentuximab vedotin with chemotherapy for CD30-positive peripheral T-cell lymphoma (ECHELON-2): a global, double-blind, randomised, phase 3 trial. Lancet 393:229–240

Hovenga S, de Graaf H, Joosten P, van den Berg GA, Storm H, Langerak AW, Kluin PM, Kibbelaar RE (2003) Enteropathy-associated T-cell lymphoma presenting with eosinophilia. Neth J Med 61:25–27

Isaacson P, Wright DH (1978) Intestinal lymphoma associated with malabsorption. Lancet 1:67–70

Jantunen E, Boumendil A, Finel H, Luan JJ, Johnson P, Rambaldi A, Haynes A, Duchosal MA, Bethge W, Biron P, Carlson K, Craddock C, Rudin C, Finke J, Salles G, Kroschinsky F, Sureda A, Dreger P (2013) Autologous stem cell transplantation for enteropathy-associated T-cell lymphoma: a retrospective study by the EBMT. Blood 121:2529–2532

Kim DH, Lee D, Kim JW, Huh J, Park SH, Ha HK, Suh C, Yoon SM, Kim KJ, Choi KD, Ye BD, Byeon JS, Song HJ, Jung HY, Yang SK, Kim JH, Myung SJ (2014) Endoscopic and clinical analysis of primary T-cell lymphoma of the gastrointestinal tract according to pathological subtype. J Gastroenterol Hepatol 29:934–943

Lindfors K, Ciacci C, Kurppa K, Lundin KEA, Makharia GK, Mearin ML, Murray JA, Verdu EF, Kaukinen K (2019) Coeliac disease. Nat Rev Dis Primers 5:3

Malamut G, Cellier C (2015) Refractory celiac disease: epidemiology and clinical manifestations. Dig Dis 33:221–226

Malamut G, Chandesris O, Verkarre V, Meresse B, Callens C, Macintyre E, Bouhnik Y, Gornet JM, Allez M, Jian R, Berger A, Châtellier G, Brousse N, Hermine O, Cerf-Bensussan N, Cellier C (2013) Enteropathy associated T cell lymphoma in celiac disease: a large retrospective study. Dig Liver Dis 45:377–384

Mallant M, Hadithi M, Al-Toma AB, Kater M, Jacobs M, Manoliu R, Mulder C, van Waesberghe JH (2007) Abdominal computed tomography in refractory coeliac disease and enteropathy associated T-cell lymphoma. World J Gastroenterol 13:1696–1700

Maruyama D, Tobinai K, Ogura M, Uchida T, Hatake K, Taniwaki M, Ando K, Tsukasaki K, Ishida T, Kobayashi N, Ishizawa K, Tatsumi Y, Kato K, Kiguchi T, Ikezoe T, Laille E, Ro T, Tamakoshi H, Sakurai S, Ohtsu T (2017) Romidepsin in Japanese patients with relapsed or refractory peripheral T-cell lymphoma: a phase I/II and pharmacokinetics study. Int J Hematol 106:655–665

Matsumoto T, Nakamura S, Esaki M, Yada S, Moriyama T, Yanai S, Hirahashi M, Yao T, Iida M (2010) Double-balloon endoscopy depicts diminutive small bowel lesions in gastrointestinal lymphoma. Dig Dis Sci 55:158–165

McBride OM, Skipworth RJ, Leitch D, Yalamarthi S (2010) Cavitating mesenteric lymph node syndrome in association with coeliac disease and enteropathy associated T-cell lymphoma: a case report and review of the literature. Case Rep Med 2010:478269

Murray A, Cuevas EC, Jones DB, Wright DH (1995) Study of the immunohistochemistry and T cell clonality of enteropathy-associated T cell lymphoma. Am J Pathol 146:509–519

Obermann EC, Diss TC, Hamoudi RA, Munson P, Wilkins BS, Camozzi ML, Isaacson PG, Du MQ, Dogan A (2004) Loss of heterozygosity at chromosome 9p21 is a frequent finding in enteropathy-type T-cell lymphoma. J Pathol 202:252–262

Ondrejka S, Jagadeesh D (2016) Enteropathy-associated T-cell lymphoma. Curr Hematol Malig Rep 11:504–513

Park BB, Kim WS, Suh C, Shin DY, Kim JA, Kim HG, Lee WS (2015) Salvage chemotherapy of gemcitabine, dexamethasone, and cisplatin (GDP) for patients with relapsed or refractory peripheral T-cell lymphomas: a consortium for improving survival of lymphoma (CISL) trial. Ann Hematol 94:1845–1851

Perfetti V, Brunetti L, Biagi F, Ciccocioppo R, Bianchi PI, Corazza GR (2012) TCRβ clonality improves diagnostic yield of TCRγ clonality in refractory celiac disease. J Clin Gastroenterol 46:675–679

Perry AM, Warnke RA, Hu Q, Gaulard P, Copie-Bergman C, Alkan S, Wang HY, Cheng JX, Bacon CM, Delabie J, Ranheim E, Kucuk C, Hu X, Weisenburger DD, Jaffe ES, Chan WC (2013) Indolent T-cell lymphoproliferative disease of the gastrointestinal tract. Blood 122: 3599–3606

Roberti A, Dobay MP, Bisig B, Vallois D, Boéchat C, Lanitis E, Bouchindhomme B, Parrens MC, Bossard C, Quintanilla-Martinez L, Missiaglia E, Gaulard P, de Leval L (2016) Type II enteropathy-associated T-cell lymphoma features a unique genomic profile with highly recurrent SETD2 alterations. Nat Commun 7:12602

Rohatiner A, d'Amore F, Coiffier B, Crowther D, Gospodarowicz M, Isaacson P, Lister TA, Norton A, Salem P, Shipp M et al (1994) Report on a workshop convened to discuss the pathological and staging classifications of gastrointestinal tract lymphoma. Ann Oncol 5:397–400

Sandal R, Mishra K, Jandial A, Lad D (2018) Goitre, lymphoma and the doughnut sign. BMJ Case Rep 2018

Sandal R, Porchezhian P, Mishra K, Jandial A, Khadwal A, Malhotra P (2019) Mees lines. Qjm 112:633

Schmitt-Gräff A, Daum S, Hummel M, Zemlin M, Stein H, Riecken EO (1996) Presence of clonal T-cell receptor gene rearrangements provides evidence of widespread intramucosal intestinal T-cell lymphoma. Z Gastroenterol 34:680–685

Schmitz N, Lenz G, Stelljes M (2018) Allogeneic hematopoietic stem cell transplantation for T-cell lymphomas. Blood 132:245–253

Sieniawski M, Angamuthu N, Boyd K, Chasty R, Davies J, Forsyth P, Jack F, Lyons S, Mounter P, Revell P, Proctor SJ, Lennard AL (2010) Evaluation of enteropathy-associated T-cell lymphoma comparing standard therapies with a novel regimen including autologous stem cell transplantation. Blood 115:3664–3670

Silano M, Volta U, Vincenzi AD, Dessì M, Vincenzi MD (2008) Effect of a gluten-free diet on the risk of enteropathy-associated T-cell lymphoma in celiac disease. Dig Dis Sci 53:972–976

Sitzia J, North C, Stanley J, Winterberg N (1997) Side effects of CHOP in the treatment of non-hodgkin's lymphoma. Cancer Nurs 20:430–439

Smedby KE, Akerman M, Hildebrand H, Glimelius B, Ekbom A, Askling J (2005) Malignant lymphomas in coeliac disease: evidence of increased risks for lymphoma types other than enteropathy-type T cell lymphoma. Gut 54:54–59

Swerdlow SH, Campo E, Pileri SA, Harris NL, Stein H, Siebert R, Advani R, Ghielmini M, Salles GA, Zelenetz AD, Jaffe ES (2016) The 2016 revision of the World Health Organization classification of lymphoid neoplasms. Blood 127:2375–2390

Tomita S, Kikuti YY, Carreras J, Sakai R, Takata K, Yoshino T, Bea S, Campo E, Missiaglia E, Bouilly J, Letourneau A, de Leval L, Nakamura N (2020) Monomorphic epitheliotropic intestinal T-cell lymphoma in Asia frequently shows SETD2 alterations. Cancers (Basel) 12

Tse E, Gill H, Loong F, Kim SJ, Ng SB, Tang T, Ko YH, Chng WJ, Lim ST, Kim WS, Kwong YL (2012) Type II enteropathy-associated T-cell lymphoma: a multicenter analysis from the Asia Lymphoma Study Group. Am J Hematol 87:663–668

van de Water JM, Cillessen SA, Visser OJ, Verbeek WH, Meijer CJ, Mulder CJ (2010) Enteropathy associated T-cell lymphoma and its precursor lesions. Best Pract Res Clin Gastroenterol 24:43–56

van de Water JM, Nijeboer P, de Baaij LR, Zegers J, Bouma G, Visser OJ, van der Peet DL, Mulder CJ, Meijerink WJ (2015) Surgery in (pre)malignant celiac disease. World J Gastroenterol 21: 12403–12409

van Vliet C, Spagnolo DV (2020) T- and NK-cell lymphoproliferative disorders of the gastrointestinal tract: review and update. Pathology 52:128–141

Van Weyenberg SJ, Meijerink MR, Jacobs MA, van Kuijk C, Mulder CJ, van Waesberghe JH (2011) MR enteroclysis in refractory celiac disease: proposal and validation of a severity scoring system. Radiology 259:151–161

Varghese D, Haseer Koya H, Cherian SV, Mead K, Sharma A, Sharma N, Knohl SJ, Benjamin S (2013) Hemophagocytic lymphohistiocytosis: an uncommon presentation of enteropathy-associated T-cell lymphoma. J Clin Oncol 31:e226–e230

Vicari AP, Schoepfer AM, Meresse B, Goffin L, Léger O, Josserand S, Guégan N, Yousefi S, Straumann A, Cerf-Bensussan N, Simon HU, Chvatchko Y (2017) Discovery and characterization of a novel humanized anti-IL-15 antibody and its relevance for the treatment of refractory celiac disease and eosinophilic esophagitis. MAbs 9:927–944

Wierdsma NJ, Nijeboer P, De Van Der Schueren MA, Berkenpas M, Van Bodegraven AA, Mulder CJ (2016) Refractory celiac disease and EATL patients show severe malnutrition and malabsorption at diagnosis. Clin Nutr 35:685–691

Wöhrer S, Chott A, Drach J, Püspök A, Hejna M, Hoffmann M, Raderer M (2004) Chemotherapy with cyclophosphamide, doxorubicin, etoposide, vincristine and prednisone (CHOEP) is not effective in patients with enteropathy-type intestinal T-cell lymphoma. Ann Oncol 15:1680 1683

Molecular Pathogenesis of Hodgkin's Lymphoma: Advances Through the Key Player LMP1 and 3D Nanotechnology

Bassem Sawan, Tina Petrogiannis-Haliotis, and Hans Knecht

Abstract

Background

Our knowledge of the main molecular facets interacting in the complex pathogenesis of Hodgkin's lymphoma (cHL) has significantly increased over the past three decades. Here, we review these findings and trace the way of a germinal center B-cell from the first hit up to the mononuclear Hodgkin (H) and end-stage multinuclear Reed-Sternberg (RS) cell of EBV-negative and EBV-associated LMP1 expressing cHL. We present an evidence-based 3D telomere/shelterin model of the multistep transformation of the B-lymphocyte to the mononuclear H- and finally multinuclear RS-cell end-stage tumor cell.

Methods

Since our laboratory has been active in Hodgkin's lymphoma research for over 30 years there was no need for special databases research. However, over 95% of the literature cited is easily identifiable in PubMed under the subheadings of this chapter.

B. Sawan
Department of Pathology, Saint George Hospital University Medical Center, Achrafieh Beirut, Lebanon

T. Petrogiannis-Haliotis
Department of Pathology, Jewish General Hospital, McGill University, Montreal, QC, Canada

H. Knecht (✉)
Division of Hematology, Department of Medicine, Jewish General Hospital, McGill University, Montreal, QC, Canada
e-mail: hans.knecht@mcgill.ca

© The Author(s), under exclusive license to Springer Nature Switzerland AG 2022
Interdisciplinary Cancer Research, https://doi.org/10.1007/16833_2022_57
Published online: 5 October 2022

Results

Under "Established Contributors in EBV-independent Pathogenesis" we shortly review the well-known signaling pathways involved in the way from the germinal center B-lymphocyte to the RS cell, and the defense mechanisms of the latter to escape cell death. Under "LMP1: The Key player in EBV-associated Pathogenesis" we summarize recent 3D nanotechnology-based evidence that the generation of H- and RS-cells is related to an increasingly disturbing telomere-shelterin interaction. In the case of EBV-associated cHL, this interaction is orchestrated by LMP1 through down-regulation of TRF2, facilitating BBF (Breakage-Bridge-Fusion) cycles during mitosis. Under "3D Nanotechnology: Breakthrough in molecular Pathogenesis" we present our evidence-based model of LMP1 as the golden key to the understanding of the transition of the germinal center B-lymphocyte to the multinucleated end-stage RS-cell. Finally, the overexpression of lamin A/C in H- and RS-cells is intriguing since lamin A/C is essential for maintaining the chromosomal nuclear territories, and this occurs partially through direct interaction with TRF2 bound to t-loops at interstitial telomeric sequences.

Conclusions

The presented data suggest that the telomere/shelterin complex is the turntable of the molecular pathogenesis of cHL and the sculptor of the unique 3D nuclear architecture of the H- and RS-cells. These findings support the notion that cHL is the first shelterin-telomere (DNA)-interaction-related malignant lymphoma where LMP1 and TRF2 act as key players.

Keywords

3D Q-FISH · Lamin A/C · LMP1 oncogene · Nanotechnology · Reed-Sternberg cell · Telomere · TRF2

1 Introduction

Lymphomas represent a group of malignant neoplasms arising from immune-reactive cells. Historically, mature lymphomas are divided into two main groups: Hodgkin lymphoma (HL) and non-Hodgkin lymphoma (NHL). NHL is further divided into B-NHL and T- & NK-cell lymphomas (Hodgkin 1832; Reed 1902; Lennert 1978). HL represents about 30% of all lymphomas. They can be distinguished from NHL clinically and histologically and are characterized by distinct molecular signatures and pathogenetic mechanisms (reviewed by Kadin 1994; Piris et al. 2020; Bienz et al. 2020).

The WHO Classification of Tumours of Haematopoietic and Lymphoid Tissues (Swerdlow et al. 2016) divides HL into nodular lymphocyte predominant Hodgkin lymphoma (NLPHL) and classic Hodgkin lymphoma (cHL). The focus of this chapter is cHL. Histologically, most cHL are characterized by a paucity of scattered large mononuclear tumor cells called Hodgkin (H) cells and bi- or multinuclear Reed-Sternberg (RS) cells, summarized as HRS-cells, in a mixed background of nonneoplastic inflammatory and accessory cells. In this setting, the tumor cells are often surrounded by T-cells in rosette-like formations (Piris et al. 2020; Bienz et al. 2020). Classic Hodgkin lymphoma (cHL) is subdivided into four histological subtypes: Nodular sclerosis (NS), mixed cellularity (MC), lymphocyte-rich (LR), and lymphocyte-depleted (LD). Common to the pathognomonic HRS-cells (regardless of subtype) is their immunoprofile that is typically CD30+, CD15+, CD45 LCA-, CD20-, PAX5+, MUM-1+, OCT-2-, and BOB.1- (Figs. 1 and 2). In EBV-associated cHL, H- and RS-cells are typically EBER+ and latent membrane protein 1 (LMP-1)+ (Fig. 3). In EBV-associated cHL the postulated normal counterpart is a pre-apoptotic germinal center (GC) B-cell, rescued by EBV-encoded latent genes (Swerdlow et al. 2016; Mancao et al. 2005; Thorley-Lawson 2015).

Polymerase chain reaction (PCR) techniques combined with single-cell microdissection allowed to molecularly analyze the mononuclear H-cells and bi- or multinuclear RS-cells from diagnostic lymph node biopsies. H- and RS-cells were revealed to be neoplastic clonal B-cells of germinal center (GC) origin in >98% of patients, having largely lost the typical B-cell gene expression program (Kuppers and Rajewsky 1998; Marafioti et al. 2000). True T-cell derived HL does exist but is extremely rare (<1%) (Tzankov et al. 2005a).

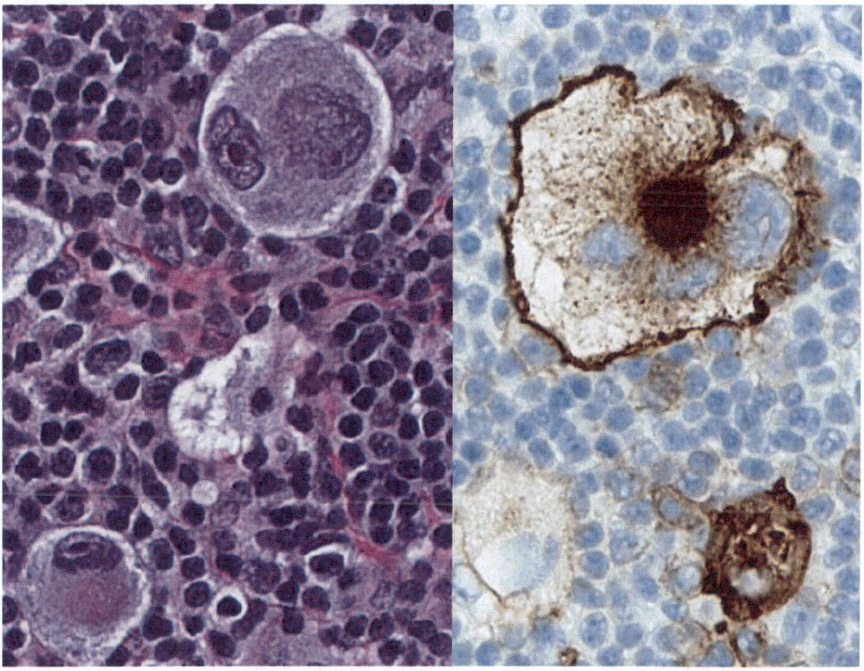

Fig. 1 EBV- cHL. Left: RS- and H-cells, H&E stain; Right: CD30+ (membranous, Golgi, and cytoplasmic granular)

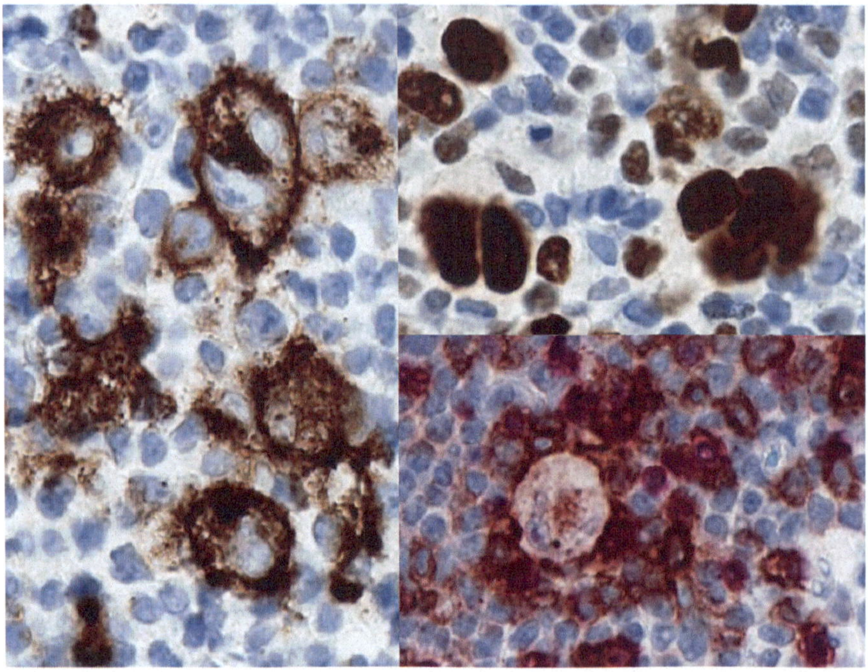

Fig. 2 EBV- cHL. (same patient as in Fig. 1). Left: CD15+ (cytoplasmic granular, Golgi, and membranous); Right, upper: MUM-1 (nuclear); Right, lower: CD4 (brown)/Ki-67 (magenta) dual stain illustrates CD4+ T-cell rosette around RS-cell

Through numerous epidemiological and molecular studies, our understanding of these specific features of cHL has considerably advanced in the past two decades (reviewed by Piris et al. 2020; Bienz et al. 2020; Cuceu et al. 2018). In particular, the risk to get cHL within 4 years after infectious mononucleosis (IM) is about 3.5–4 times elevated (Hjalgrim et al. 2003), the cytotoxic T-cell response appears to be HLA-A allele dependent (Hjalgrim et al. 2010), and EBV and HIV infection change the micro-environmental niche for H- and RS-cells (Carbone et al. 2017). Regular intake of Aspirin, a suppressor of NF-KB activation, reduces the risk to get cHL (Chang et al. 2004), whereas, through activation of major players of the NF-KB and JAK-STAT pathways (Schwarzer et al. 2012; Zahn et al. 2021), and constitutive activation of numerous further signaling pathways, H- and RS-cells actively orchestrate their complex microenvironment to create an immuno-suppressive environment (Cader et al. 2018).

However, only recently developed 3D nanotechnology methods, targeting the telomeres (Chuang et al. 2004; Louis et al. 2005), allowed to shed light on the transformation of mononuclear H- into bi- or multinucleated RS-cells and to uncover a pivotal role of the telomere/shelterin interaction in the molecular morphogenesis of this long time enigmatic, diagnostic tumor cell (Knecht et al. 2009, 2012; Guffei

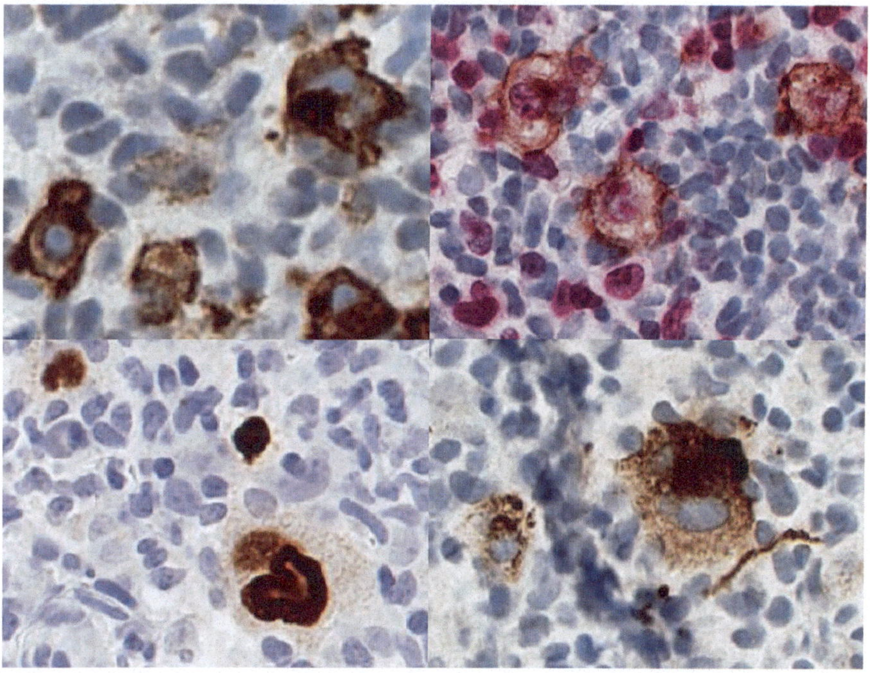

Fig. 3 EBV+ cHL. Left, upper: CD30 (cytoplasmic, Golgi, and membranous); Right, upper: CD30 (brown)/Ki-67 (magenta) dual stain shows CD30+ & Ki-67+ H- and RS-cells and numerous dividing inflammatory T-cells; Left, lower: EBER highlights nuclei, H- and RS-cells and small B-cells, Right, lower: LMP-1 immunostain (cytoplasmic granular and membranous). LMP1 positivity is associated with H- and RS-cells but not with EBER+ lymphocytes

et al. 2010). Moreover, 3D nanotechnology revealed distinct structural patterns of lamin A/C expression in H- and RS-ells (Contu et al. 2018).

As malignant tumors must acquire several hallmarks of cancer, they result from a stepwise accumulation of multiple genetic and epigenetic abnormalities. This process involves proto-oncogenes, tumor suppressor genes, evasion from apoptosis, evasion of host immune response, and genomic instability, to mention the most evident. Herein we focus on the most peculiar 3D epigenetic abnormalities related to the pathogenesis of cHL with emphasis on a major event, i.e., the dynamics of the telomere/shelterin complex, which is essential in the molecular pathogenesis of cHL.

2 Molecular Pathogenesis

There are likely two major multistep pathways resulting in classic Hodgkin lymphoma: one EBV-independent and one EBV-associated (Piris et al. 2020). Both pathways appear to be overlapping in that way that the activation of the EBV latency type II genes LMP1 and LMP2A mimics/replaces several important steps occurring in EBV-independent pathogenesis (Shannon-Lowe et al. 2017). In particular, it appears that in EBV-associated cHL most of the events leading from the GC B-cell to the H- and finally to the diagnostic bi- or multinucleated end-stage RS-cell are orchestrated by the multifunctional oncoprotein LMP1 (Bienz et al. 2020; Knecht et al. 2001) whereas in EBV-negative cHL a complex combination and cascade of multiple small genetic events terminates in the same results (Weniger and Küppers 2021). In this chapter, we briefly resume the main signaling pathways but focus on the molecular pathogenesis of EBV-associated cHL including the 3D nanotechnology discoveries allowing us to understand the morphogenesis of the diagnostic RS-cells.

3 Established Contributors of EBV-independent Pathogenesis

3.1 NF-κB Pathway

The involvement of the NF-kB pathway in tumorigenesis is well documented (Bargou et al. 1997; Emmerich et al. 2003). The NF-kB family, composed of Rel-A, Rel-B, cRel, p50, and p52, is a group of cytoplasmic proteins that translocate to the nucleus and act as transcription factors for proteins involved in inflammatory and anti-apoptotic signaling cascades (Gamboa-Cedeño et al. 2019). The activation of NF-kB is restricted by NF-kB inhibitors, which undergo ubiquitination and degradation upon activation of specific kinases. These kinases are regulated by the canonical and the non-canonical pathway. The canonical pathway is induced by inflammatory cytokines (namely TNF and IL-1) (Taniguchi and Karin 2018). The non-canonical pathway is activated by ligands such as CD40L and RANKL and stimulates the kinase NIK (MAP3K14) (Taniguchi and Karin 2018). Many mechanisms have been demonstrated to lead to NF-kB constitutional activation in Hodgkin lymphoma (Jost and Ruland 2007; Rothenberger et al. 1997). Detailed reviews on this subject have recently been published by three independent groups including ours (Cuceu et al. 2018; Bienz et al. 2020; Weniger and Küppers 2021).

3.2 JAK/STAT Pathway

The JAK-STAT signaling pathway requires activation of the JAK tyrosine kinases, which include JAK1, JAK2, JAK3, and TYK2 on the inner side of the cytoplasmic membrane (Hu et al. 2021). Each kinase is constitutively bound to a specific receptor such as the erythropoietin receptor, granulocyte colony-stimulating factor receptor, and interleukin receptors. When activated by the receptor's specific ligand, the associated JAK kinase is activated (Scott and Gandhi 2015). This leads to phosphorylation of the cytoplasmic STAT transcription factors and translocation of STAT to the nucleus (Rawlings et al. 2004). Target genes include namely those involved in proliferation, survival, and angiogenesis (Alvarez and Frank 2004). Many anomalies within the JAK-STAT pathway, leading to its constitutive activation in cHL, have been described (Vainchenker and Constantinescu 2013; Matsuki and Younes 2015) and recently reviewed (Cuceu et al. 2018, Bienz et al. 2020, Weniger and Küppers 2021).

3.3 MAPK/ERK, p53, FOX, ATM, and ATR Pathways

Involvement of these pathways in the pathogenesis of some cases of cHL has been proven but overall appears less important than the dysregulation of the NF-kB and JAK/STAT pathways. Relevant findings concerning these pathways have been dealt with recently (Cuceu et al. 2018, Bienz et al. 2020, Weniger and Küppers 2021).

3.4 Cyclin A

Cyclin A is a major regulator of cell cycle progression and occurs in two types; in the embryonic-specific form cyclin A1, and in the somatic form, cyclin A2. Cyclin A1 is prevalently expressed during meiosis and early in embryogenesis whereas cyclin A2 is expressed in dividing somatic cells (Yam et al. 2002). REF52 rat fibroblasts constitutively expressing an endoplasmic reticulum-associated cyclin A2 deletion variant show centrosome over-duplication, increased ploidy and transformation (Faivre et al. 2002). In a large series of 253 HL cases 250 cases strongly expressed cyclin A in 98% of H- and 99% RS-cells (Tzankov et al. 2005b). In a further, detailed study of 34 cases, composed of LMP1 positive (13) and LMP1 negative (21) samples, the mononuclear H-cells expressed cyclin A predominantly in the nuclei, whereas the multi-nucleated RS-cells expressed cyclin A aberrantly in the cytoplasm ($P < 0.001$) (Chang et al. 2009). Moreover, in the H-cell line L-428, LMP1 increased cytoplasmic cyclin A expression and multi-nucleation (Chang et al. 2009) confirming our earlier observation, that LMP1 is an inducer of multi-nucleation in EBV-negative H-cell lines (Knecht et al. 1996). A synopsis of these results was thoroughly resumed by Piris, Medeiros, and Chan (2020) and the morphogenesis of RS-cells was depicted by a nice scheme.

More than a decade later the molecular answer to these observations comes probably from a recent discovery. Cyclin A2, besides his well-known CDK activation, selectively and directly binds to the 3′ untranslated region (UTR) of Mre11 transcripts (Kanakkanthara et al. 2016). This interaction promotes Mre11 translation independent of CDK activity (Kanakkanthara et al. 2016; Baumann 2016). Hence, our earlier observation of Mre11 accumulation in RS- but not H-cells (Knecht et al. 2010a, b) and the cyclin A expression observed by Chang et al. (2009) mainly in the cytoplasm of multi-nucleated RS-cells, independent of LMP1 expression, make sense. In EBV-positive cases double-strand DNA breaks and unprotected single-strand overhangs, favored by LMP1-induced TRF2 down-regulation (Lajoie et al. 2015), lead to excessive Mre11 repair protein up-regulation through direct binding of cytoplasmic cyclin 2A to Mre11 mRNA transcripts, thereby promoting Mre11 translation.

3.5 The Tumoral Microenvironment

H- and RS-cells represent about 1 to maximal 5% of cells of a diagnostic lymph node. The bulk of residing cells is organized and forms the tumoral microenvironment. This reactive tumoral ecosystem is composed of lymphocytes, mainly CD4+ T-cells, B-cells, some plasma cells, myeloid cells (eosinophils, macrophages, mast cells), and stromal cells (fibroblasts, mesenchymal stromal cells, endothelial cells) (Mottok 2020; Aldinucci et al. 2019). The microenvironment is not reactive to cancer but rather the result of a complex cytokine-mediated recruitment of healthy cells. These are later "indoctrinated" by H- and RS-cells to work to the tumor's advantage to promote cancer cell survival, angiogenesis, and immune system evasion. The mechanisms by which H- and RS-cells attract such a diversity of accomplices are mediated through cytokine and growth factor release (Mottok 2020, Aldinucci et al. 2019). T-cells are the most prevalent cells of the cHL microenvironment. The CD40L-positive CD4+ T-cells are often encountered around H- and RS-cells, hence their nickname of rosetting T-cells (Carbone et al. 1995). H- and RS-cells can promote monocyte differentiation into M2 macrophages through secretion of M-CSF. M2 macrophages have a phenotype that favors tumor survival through angiogenesis and tumor growth stimulation, and anti-inflammatory properties (Aldinucci et al. 2010; Mantovani et al. 2017). Immune targeting of the microenvironment, especially T-lymphocytes and macrophages represents a major success in the treatment of HL (Mantovani et al. 2017; Carreau and Diefenbach 2019).

3.6 Epigenetic Abnormalities

Epigenetics embrace cellular changes related to modification of gene expression, which are not related to changes in the DNA sequence. Targeting distinct epigenetic processes as an alternative approach in cancer therapy without the risk to provoke further chromosomal changes represents a new, powerful therapeutic avenue (Lister et al. 2009). Most important mechanisms in the epigenetics of HL are DNA methylation and histone modification allowing changing the accessibility of chromosomal DNA (Ehlers et al. 2008). In particular, hyper-methylation of the KLF4 (Krüppel-like factor 4) promoter in Hodgkin cell lines results in KLF4 silencing associated with loss of cell cycle control and evasion from apoptosis, a characteristic of the H- and RS-cells (Guan et al. 2010). These and further epigenetic abnormalities in HL have recently been reviewed (Liu et al. 2020) and new successful epigenetic therapies have allowed to circumvent these important defense barriers of the H- and RS-cells (Meti et al. 2018).

3.7 Host-Immune System Evasion

Immune tolerance is an important mechanism by which H- and RS-cells promote their own survival. One mechanism of immune escape is the immunosuppressive effect of the microenvironment (Bair et al. 2018). In fact, CD8+ T-cells require β2-microglobulin (B2M) and major histocompatibility complex (MHC) class I antigens as targets to fulfill their role in the cytotoxic response of our immune system. However, in H- and RS- cells the expression of these molecules is down-regulated in 90% of cases (Reichel et al. 2015). Meanwhile, CD4+ helper T-cells require MHC class II to function but its expression is reduced or aberrant in 40% of cHL and lack of membranous HLA class II expression appears as adverse prognostic factor in cHL (Diepstra et al. 2007).Despite these attempts to camouflage, cHL antigen presentation does occur. In these cases, through their immune checkpoint receptors, H- and RS-cells are able to remain undetected. One prominent example is the programmed death 1 (PD-1) T-cell surface receptor which, when engaged by the ligand PD-L1 of the antigen-presenting cell (i.e., H- and RS-cells), induces transcription of genes aimed at suppressing its effector or cytotoxic functions (Meti et al. 2018). Several mechanisms can lead to high expression of PDL-1 in cHL such as amplification of the 9p24.1 chromosomal section, seen in a third of cHL cases (Green et al. 2010). This chromosomal region expresses the PD-L1 gene locus and effective anti-PD-1 therapy correlates with a diverse baseline TCR repertoire (Cader et al. 2020). It also promotes PD-L2 and JAK2 expressions (neighbor genes in the 9p24.1 region), both up-regulating PD-L1 expression on the cell surface, partly through JAK/STAT signaling (Gravelle et al. 2017). In EBV-positive cHL, EBV is also involved in host-immune system increasing PD-L1 and PD-L2 expression independently of the 9p24.1 copy number (Meti et al. 2018; Gravelle et al. 2017; Green et al. 2012).

3.8 Telomere and DDR

Telomeres are the nucleoprotein complexes at the ends of chromosomes. Telomeric DNA consists of multiple double-stranded TTAGGG repeats and ends in a single-stranded overhang of the G-rich strand (LeBel and Wellinger 2005). Furthermore, a number of specific proteins, either binding telomeric DNA directly or being associated with telomeric chromatin, called Shelterin complex, are found on telomeres (De Lange 2005). The shelterin protein TRF2 (telomere repeat binding factor 2) has emerged as the key-player in protecting the chromosomal ends against a full-blown DNA damage response (DDR) during the replicative cell cycle (Feuerhahn et al. 2015). We earlier identified major anomalies in TRF2 expression in H- and RS-cells (Knecht et al. 2009), as well accumulation of γ-H2AX and Mre11 in RS-cells, indicative of DNA double-strand breaks and repair activity, respectively (Knecht et al. 2010a, b). Hence, we hypothesized that the interaction between the telomere/shelterin complex with DDR mechanisms was a crucial turning plate of the molecular pathogenesis of cHL and the formation of the diagnostic RS-cells. (Research findings will be dealt with under LMP1 and 3D Nanotechnology).

4 LMP1: The Key Player in EBV-associated Pathogenesis

4.1 Transforming Power of LMP1 in Cell Morphology

The LMP1 multifunctional oncoprotein was first molecularly characterized in the IB4 cell line, established through infection of human cord blood lymphocytes by the EBV strain B95-8 by Fennewald et al. (1984) and was found to correspond to the hitherto immunologically identified LYDMA (lymphocyte-determined membrane antigen). In the following three decades significant progress in the characterization of the EBV-encoded oncogenes, in particular the clinically most relevant LMP1 has been achieved (Shannon-Lowe et al. 2017; Knecht et al. 2001; Dolcetti and Masucci 2003; Chang et al. 2017).

In infectious mononucleosis (IM) patients, lymph node and tonsillar biopsies contain few multi-nuclear large, LMP1-positive, CD30-positive RS-like cells, cytologically corresponding to textbook RS-cells (Reynolds et al. 1995; Knecht et al. 1993b). However, in an immune-competent host, or in B-cells having no further abnormality than episomal EBV genomes, these cells will disappear rapidly. Therefore, it appears logical to ask what will the expression of the LMP1 oncoprotein elicit in a RS-cell producing system like the EBV genome-negative Hodgkin cell lines? Transient transfection of the H-cell lines L-428, HD-MyZ (now classified as myelomonocytic) and the embryonic kidney cell line 293 results in a significantly high percentage of multinucleated RS- and RS-cell-like (293 cell line) cells (Knecht et al. 1996, 2001; Chang et al. 2009) proving that LMP1 is an inducer of multi-nuclear RS- and RS-like cells in lymphoid, myeloid and precursor kidney cells. Moreover, in an experimental mice system, the EBNA2-deleted Epstein-Barr virus (EBV) isolate, P3HR1, causes LMP1-expressing Hodgkin-like lymphomas (Li et al. 2020).

4.2 LMP1 Targets the 3D Telomere/Shelterin Complex

Telomerase activity in HL was first described by Brousset et al. (1997) and shorter telomere as well as less telomere signals in RS-cells of the HDLM-2 cell line and HL patient biopsy imprints by 2D FISH by Balta-Yildrim (2006). However, only 3D FISH analysis of the spatial telomere distribution and size within mononuclear H- and multi-nuclear RS-cells of the HD cell lines HDLM-2, L-428 L-1236, and lymph node biopsies allowed a mechanistic understanding of the transition from H- to RS-cells (Knecht et al. 2009). RS-cells show significantly shorter and significantly fewer telomeres in relation to the total nuclear volume when compared with H-cells, and telomere-poor "ghost" nuclei are often adjacent to one or two nuclei displaying huge telomeric aggregates. When analyzing in the same study concomitantly the shelterin proteins TRF1 and TRF2 in the HDLM-2 cell line we observed a mainly abnormal cytoplasmic localization of both shelterins, compatible with telomere uncapping (Knecht et al. 2009). The high expression of the DNA double-strand break sensor γ-H2AX and the repair proteins Mre11 and Rad51c identified in RS- but not H-cells, suggests that the telomere-shelterin complex is the turntable of the transition from H- to RS-cells. Chromosome painting revealed increasingly unequal chromosomal distribution and "Zebra" chromosomes, indicating multiple (BBF) breakage-bridge-fusion cycles during the transition from H- to RS-cells (Guffei et al. 2010). These findings were extended through spectral karyotyping (SKY) and super-resolution imaging (3D-SIM), where intra-nuclear DNA bridges were observed in RS-cells (Guffei et al. 2010). 3D-SIM showed a steadily increasing amount of submicron DNA structures and DNA-free spaces "holes" from lymphocytes up to H-cells, bi-nuclear and finally multi-nuclear RS-cells (Righolt et al. 2014). These changes do occur in H- and RS-cells irrespective of their EBV positivity or negativity. This progressive increase of very short telomeres, telomere loss, aggregate formation, chromosomal complexity, multi-nucleation, and generation of "ghost" nuclei (Fig. 4) results in end-stage tumor cells in which the extent to divide further is inversely proportional to their degree of nuclear damage (Knecht et al. 2010a, b; MacLeod et al. 2000).

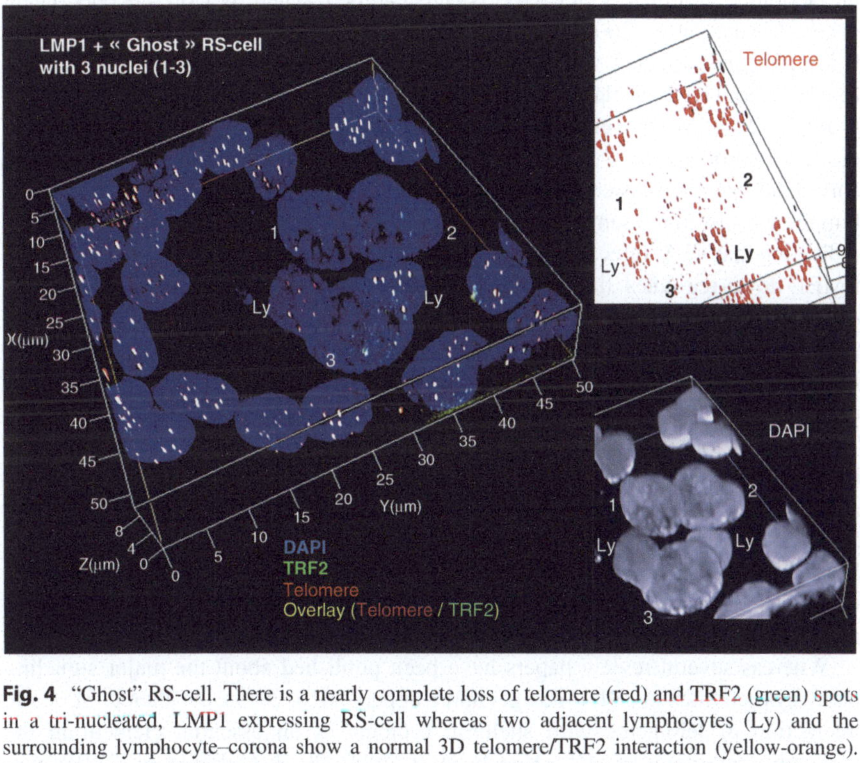

Fig. 4 "Ghost" RS-cell. There is a nearly complete loss of telomere (red) and TRF2 (green) spots in a tri-nucleated, LMP1 expressing RS-cell whereas two adjacent lymphocytes (Ly) and the surrounding lymphocyte–corona show a normal 3D telomere/TRF2 interaction (yellow-orange). Lower right is a DAPI staining in transparency mode to optimize morphological aspects. The two adjacent lymphocytes (Ly) are above (left) and below (right) the RS-cell nucleus 3. Upper right demonstrates that the surrounding and adjacent lymphocytes (Ly) contain mainly mid-sized telomere (red). The preserved 1:1 3D interaction with TRF2 is evident from the yellow-orange color seen in the main figure as a result of red-green overlaying. The left panel (transparency mode) reveals a true ghost RS-cell with very few short telomere and without any TRF2 signals

We recently showed that LMP1, stably transfected with an inducible tet-Off system in the EBV-negative post GC B-cell line BJAB (Floettmann et al. 1996), induced multi-nuclearity and increasing chromosomal complexity mainly by down-regulation of TRF2 (Lajoie et al. 2015). This TRF2 down-regulation was reversible upon LMP1 suppression. Analogous effects of LMP1 induction/suppression were also observed for TRF1 and POT1 (Lajoie et al. 2015).

5 3D Nanotechnology: Breakthrough in Molecular Pathogenesis

5.1 Progressive Telomere/Shelterin Disruption: A Common Denominator of Both EBV-associated cHL and EBV-independent cHL

Pre-apoptotic germinal center (GC) B-cells can be rescued by EBV-encoded latent genes (Mancao et al. 2005; Thorley-Lawson 2015) and EBV-negative circulating B-lymphocytes infected with the B95-8 strain show after 4 weeks γ-H2Ax and Mre11 foci, TRF2 displacement from telomeres rather than telomere shortening (Lacoste et al. 2010). Telomere dysfunction was documented by identification of dicentric chromosomes with interstitial telomeric signals, unique telomere-free chromatid ends and sister chromatid fusions. On the contrary, mitogen-stimulated lymphocytes from the same donors didn't show these changes (Lacoste et al. 2010), indicating that EBV infection targets the telomere/shelterin complex. In GC B-cells LMP1 down-regulates the expression of B-cell-specific genes and induces the expression of ID2, a negative regulator of B-cell differentiation (Vockerodt et al. 2008). However, constitutive NF-kB activation is observed in Hodgin cell lines and H- and RS-cells (Bargou et al. 1996, 1997), including cases expressing naturally occurring LMP1 deletion variants (Rothenberger et al. 1997; Knecht et al. 1993a). Telomere shortening and progressive telomere/shelterin complex disruption are identified in both H- and RS-cells of EBV-associated, LMP1-positive cHL, as well as in H- and RS-cells of EBV-negative cHL and H-cell lines (Knecht et al. 2009; 2017, Balta-Yildrim 2006). A common denominator of EBV-independent and EBV-associated cHL has thus included strong NF-kB activation combined with a mechanism generating complex chromosomal abnormalities and multi-nuclearity (Fig. 5) as recently proposed by our group (Bienz et al. 2020).

Whereas several review papers have been published about the major signaling pathways activated in cHL (Weniger and Küppers 2016, 2021), the altered, 3D steric interaction of telomeres with shelterin proteins as an essential element in the molecular pathology of cHL (Knecht et al. 2009, 2017; Guffei et al. 2010), was only recently recognized (Piris et al. 2020; Li et al. 2020).

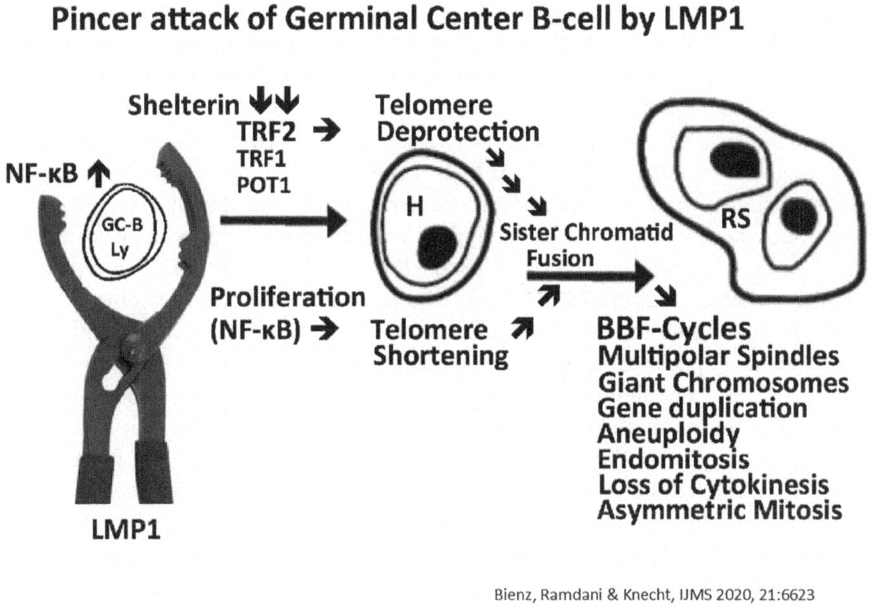

Fig. 5 Pathogenetic Model of EBV-associated Hodgkin Lymphoma. Continuous transformation from a GC B-lymphocyte into an H-cell and terminal RS-cell. A "crippled" GC B-lymphocyte (Mancao et al. 2005) or an EBV+ memory B-lymphocyte (Thorley-Lawson 2015) re-entering GC reaction undergoes LMP1 oncoprotein-induced transformation, targeting the telomere/shelterin complex through a "pincer attack". NF-kB-driven mitosis (Bargou et al. 1997) leads to telomere shortening and TRF2 down-regulation (Knecht et al. 2009; Lajoie et al. 2015), resulting in telomere deprotection, both in concert generate short deprotected telomeres, prone to undergo sister chromatid fusion as the origin of repeated BBF-cycles, resulting first in H-cells, and, after several subsequent mitotic cycles, in the end-stage multi-faceted, often telomere poor RS-cells (Knecht et al. 2009, 2010a, b; Guffei et al. 2010; Balta-Yildrim 2006). The constitutive NF-kB activation (Bargou et al. 1997; Rothenberger et al. 1997), the dysfunctioning telomere-shelterin complex, in particular TRF2 (Lajoie et al. 2015; Knecht et al. 2017), appears to emerge as a logical candidate for a common pathogenic denominator not only for EBV-positive (Knecht and Mai 2017a, b) but also for EBV-negative cHL (MacLeod et al. 2000)

5.2 Hodgkin and Reed-Sternberg Cells Show Lamin A/C Overexpression and Disruption of the Lamin A/C-TRF2-DNA Interaction

Lamins are type V intermediate filament proteins, associate physically with the inner nuclear membrane, tether heterochromatin in the periphery, and exert important scaffolding functions within the nucleus (Moir et al. 2000; Naetar et al. 2017). In human cells two types of lamin have been identified: B-type lamins, encoded by the *LMNB1* gene for lamin B1 (Lin and Worman 1995) and the *LMNB2* gene for lamin B2 (Höger et al. 1990), and A-type lamins, encoded by the *LMNA* gene, the alternative splicing of which produces lamin A and lamin C (Fisher et al. 1986). Recently, three-dimensional structured illumination microscopy (3-SIM) allowed to characterize a distinct fiber meshwork at the supra-molecular level for lamin A, C, B1, and B2 (Shimi et al. 2015), and cryo-electron microscopic tomography identified a coil-coiled, 51 nm long and 3.5 nm thick rod-like dimer with two lateral globular domains as the building stone of all lamins (Turgay et al. 2017; Makarov et al. 2019). Lamin dimers form head-to-tail polymers which laterally interact to form protofilaments (Makarov et al. 2019; de Leeuw et al. 2018). Stochastic optical reconstruction microscopy (STORM) further revealed that lamin B1 and lamin A/C form two concentric and overlapping nutshell-like networks adjacent to the inner nuclear membrane (INM) (Nmezi et al. 2019). Lamin B1 preferentially localizes directly adjacent to the INM, whereas lamin A/C localizes closer to the nucleoplasm (Nmezi et al. 2019). Lamin B1 and lamin B2 are constitutively expressed and appear to be necessary for cell survival (Yang et al. 2019), whereas lamin A/C, also distributed throughout the nuclear interior (Naetar et al. 2017), is involved in transcription (Naetar et al. 2017; Hutchinson and Worman 2004) and plays a crucial role in the regulation of mitotic spindle assembly and positioning (Qi et al. 2015). Distinct mutations in the coding sequence of the latter result in severe congenital disorders, known as laminopathies (Shimi et al. 2010), the most impressive being Hutchinson-Gilford progeria syndrome (HGPS) (De Sandre-Giovannoli et al. 2003).

Using immunohistochemistry techniques in lymphoid cells, lamin A/C expression is absent in centrocytes, centroblasts and mantle zone lymphocytes of reactive lymph node follicles (Jansen et al. 1997), aggressive diffuse large B-cell lymphoma (DLBCL) (Agrelo et al. 2005), malignant B- and T-lymphoblasts (Stadelmann et al. 1990), but identified in stimulated blood lymphocytes (Contu et al. 2018; Stadelmann et al. 1990; Rocha-Perugini and Gonzalez-Granado 2014; Contu 2020) as well as in H- and RS-cells (Contu et al. 2018; Jansen et al. 1997).

Telomeres are the nucleoprotein complexes at the ends of chromosomes where a number of specific proteins, -either binding telomere proteins directly or a protein complex, termed shelterin-, are directly associated with telomeric DNA (de Lange 2005, 2018). There is increasing evidence of tight interactions of lamins with telomeres and the shelterin complex. Human fibroblasts with a homozygous non-sense mutation of the lamin A/C gene and thus completely devoid of lamin A/C expression, show increased nuclear plasticity and telomere mobility (De Vos et al. 2010). Chromatin immuno-precipitation (ChIP) assays performed with lamin A/C antibody revealed binding of lamin A/C to telomeres and expression of a dominant negative telomere repeat binding factor 2 (TRF2) mutant leads to loss of telomere integrity and to the formation of DNA damage foci at telomeres (Gonzales-Suarez et al. 2009). Indeed, TRF2 is not only essential for formation and maintenance of the configuration of functional telomeres at chromosomal ends (Doksani et al. 2013), but also binds to interstitial telomeric sequences (ITS) (Simonet et al. 2011). On the other hand, lamin A/C is essential for maintaining the chromosomal nuclear territories and this occurs partially through direct interaction with TRF2 bound to t-loops at ITS (Wood et al. 2014), so-called interstitial t-loops (ITL). Impairment or loss of this 3D lamin A/C-TRF2 interaction at ITLs results in increased chromatin

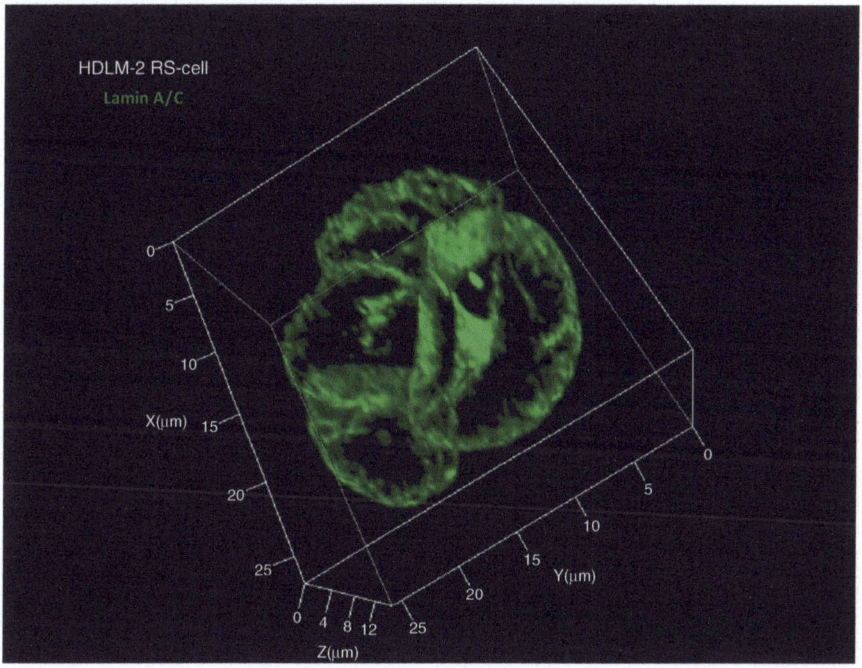

Fig. 6 LaminA/C overexpression in Hodgkin and Reed-Sternberg cells. Tetra-nucleated HDLM-2 Reed-Sternberg cell (3D transparency mode) shows abundant lamin A/C expression with formation of intra-nuclear septa and cleft formation

dynamics (Bronshtein et al. 2015), changes in gene expression, impaired chromosome stability and genomic integrity (Smith et al. 2018; Wood et al. 2015).

Recently, we have shown that the regular, spherically shaped lamin A/C organization, identified in activated lymphocytes, was disturbed in mononuclear H-cells and underwent further structural changes during the transformation of H- to binuclear and finally multinuclear RS-cells (Contu et al. 2018). This dynamic process of steadily increasing lamin A/C associated nuclear compartmentalization of individual nuclei was identified in both, H-cell lines (Fig. 6) and diagnostic H- and RS-cells of diagnostic lymph node biopsies (Contu et al. 2018; Contu 2020).

Thus, we hypothesized that the direct 3D interaction of lamin A/C of the INM with TRF2, localized at t-loops of ITS and telomeres undergoes substantial changes during the formation of H-cells and further transformation to RS-cells. To this end, we applied our combined quantitative 3D TRF2-telomere immuno Q-FISH (3D TRF2/Telo-Q-FISH) protocol (Knecht and Mai 2017a, b) and extended it to lamin A/C as a further target (3D Lamin A/C-TRF2/Telo-Q-FISH) (Contu 2020). Our findings reveal that the lamin A/C-TRF2-ITL and telomere interaction is completely disrupted in H- and RS-cells (Fig. 7) and contrasts with BJ-5ta control fibroblasts.

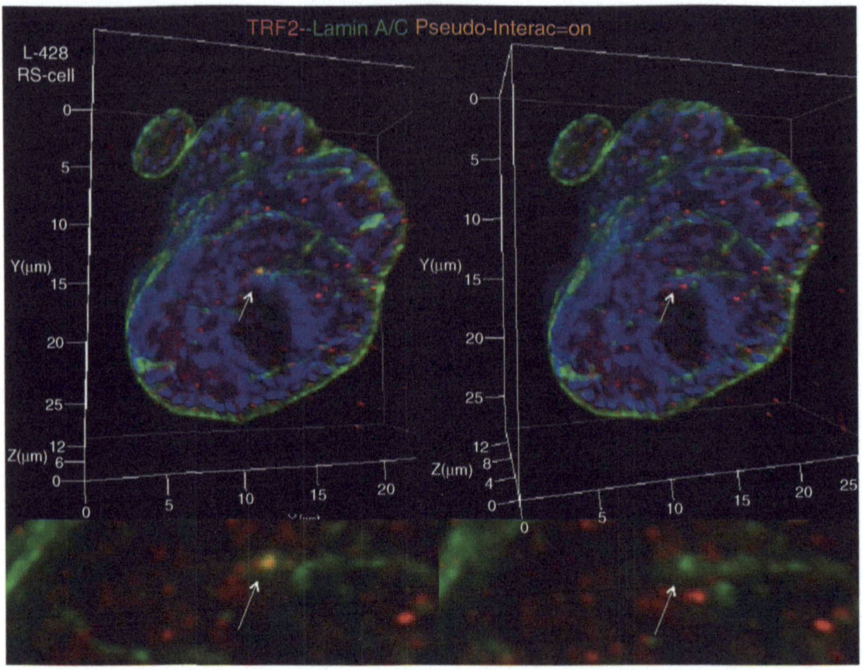

Fig. 7 TRF2-Lamin A/C pseudointeraction in a tri-nuclear L-428 Reed-Sternberg cell. Left panel shows a 3D TRF2-lamin overlay (orange) in transparency mode. Right panel, showing the same RS-cell in 3D after skipping (about 15° in the *X*-axis and 10° in the *Y*-axis), identifies the TRF2-lamin A/C interaction as pseudo-interaction due to superposition of the green (lamin A/C) and red signal (TRF2) in a given angle, no more present under a different angle. Nuclear DNA (Dapi) is blue. Arrows indicate the pseudo-interaction. At the bottom corresponding insets at higher magnification

Moreover there is a significant increase of lamin A/C accumulation at RNA level and even more impressively, at the protein level, from activated lymphocytes to H- and finally RS-cells (Contu 2020). In summary, our findings reveal that accumulation of Lamin A/C increasingly destroys the nuclear architecture in Hodgkin lymphoma tumor cells during the transition from the mononuclear H- to the bi- and finally multi-nuclear RS-cell.

6 Conclusion

Data and facts presented in this article underscore the unique and complex molecular pathogenesis of Hodgkin's lymphoma. Better understanding of the involvement of several signaling pathways, in particular NF-kB and JAK/STAT pathways, of epigenetic modifiers, and defense mechanisms adopted by the H- and RS-cells to circumvent anti-tumor immunity, shed light on many hidden facets of the true nature of H- and RS-cells and were at the origin of recent therapeutic successes.

However, one essential aspect of Hodgkin's lymphoma, its large-scale clinical presentation, the late relapses, sometimes after many years, the aggressiveness of some stage II cases, the excellent response (cure) of stage IV disease by chemotherapy only, were and are indicators that one or few essential elements in the molecular pathogenesis were still missing.

One such element is the transforming power of the EBV-encoded LMP1 multifunctional oncoprotein. Herein we present our and others' advancements made in basic and translational research, identifying LMP1 as a "golden key" to the pathogenesis of EBV-associated Hodgkin's lymphoma. The impact of LMP1 on the telomere/shelterin complex, in particular the down-regulation of TRF2, appears to represent a long time not recognized essential key-player in the molecular pathogenesis of EBV-associated Hodgkin's lymphoma. Analogous mechanistic 3D TRF2-telomere-interaction changes revealed by nanotechnology also in non-EBV associated cHL including the perturbation of the lamin A/C-TRF2-DNA interaction suggest that the telomere/shelterin complex is the turntable of the molecular pathogenesis of cHL and the sculptor of the unique 3D nuclear architecture of the H- and RS-cells. These findings support the notion that cHL is the first shelterin-telomere (DNA)-interaction-related malignant lymphoma where LMP1 and TRF2 act as key players.

Acknowledgments None.

Compliance with Ethical Standards The authors declare that there is no conflict of interest.

References

Agrelo R, Setien F, Espada J, Artiga MJ, Rodriguez M, Pérez-Rosado A, Sanchez-Aguilera A, Fraga MF, Piris MA, Esteller M (2005) Inactivation of the lamin A/C gene by CpG island promoter hypermethylation in hematologic malignancies, and its association with poor survival in nodal diffuse large B-cell lymphoma. J Clin Oncol 23:3940–3947. https://doi.org/10.1200/JCO.2005.11.650. Epub 2005 May 2

Aldinucci D, Gloghini A, Pinto A, De Filippi R, Carbone A (2010) The classical Hodgkin's lymphoma microenvironment and its role in promoting tumour growth and immune escape. J Pathol 221:248–263

Aldinucci D, Borghese C, Casagrande N (2019) Formation of the immunosuppressive microenvironment of classic Hodgkin lymphoma and therapeutic approaches to counter it. Int J Mol Sci 20:2416

Alvarez JV, Frank DA (2004) Genome-wide analysis of STAT target genes: elucidating the mechanism of STAT-mediated oncogenesis. Cancer Biol Ther 3:1045–1050. [CrossRef]

Bair SM, Mato A, Svoboda J (2018) Immunotherapy for the treatment of Hodgkin lymphoma: an evolving paradigm. Clin Lymphoma Myeloma Leuk 18(6):380–391. https://doi.org/10.1016/j.clml.2018.03.012

Balta-Yildrim Z (2006) Untersuchungen zu Telomerlängen bei Hodgkin- und Reed-Sternberg-Zellen. Doctoral thesis, RWTH Aachen

Bargou RC, Leng C, Krappmann D, Emmerich F, Mapara MY, Bommert K, Royer HD, Scheidereit C, Dörken B (1996) High-level nuclear NF-kappa B and Oct-2 is a common feature of cultured Hodgkin/Reed-Sternberg cells. Blood 87(10):4340–4347. PMID: 8639794

Bargou RC, Emmerich F, Krappmann D, Bommert K, Mapara MY, Arnold W, Royer HD, Grinstein E, Greiner A, Scheidereit C, Dörken B (1997) Constitutive nuclear factor-kappaB-RelA activation is required for proliferation and survival of Hodgkin's disease tumor cells. J Clin Invest 100:2961–2969

Baumann K (2016) Genome stability: Cyclin' on mRNA. Nat Rev Mol Cell Biol 17(11):676–677. https://doi.org/10.1038/nrm.2016.142. PMID: 27760992

Bienz M, Ramdani S, Knecht H (2020) Hodgkin's lymphoma: past, present, future. Int J Mol Sci 21:6623. https://doi.org/10.3390/ijms21186623

Bronshtein I, Kepten E, Kanter I et al (2015) Loss of lamin A function increases chromatin dynamics in the nuclear interior. Nat Commun 6:8044. https://doi.org/10.1038/ncomms9044

Brousset P, al Saati T, Chaouche N, Zenou RC, Schlaifer D, Chittal S, Delsol G (1997) Telomerase activity in reactiveand neoplastic lymphoid tissues: infrequent detection of activity in Hodgkin's disease. Blood 89:26–31

Cader FZ, Schackmann RCJ, Hu X, Wienand K, Redd R, Chapuy B, Ouyang J, Paul N, Gjini E, Lipschitz M et al (2018) Mass cytometry of Hodgkin lymphoma reveals a CD4 + regulatory T-cell-rich and exhausted T-effector microenvironment. Blood 132:825–836

Cader FZ, Hu X, Goh WL, Wienand K, Ouyang J, Mandato E, Redd R, Lawton LN, Chen PH, Weirather JL et al (2020) A peripheral immune signature of responsiveness to PD-1 blockade in patients with classical Hodgkin lymphoma. Nat Med 26(9):1468–1479. https://doi.org/10.1038/s41591-020-1006-1. Epub 2020 Aug 10

Carbone A, Gloghini A, Gattei V, Aldinucci D, Degan M, De Paoli P, Zagonel V, Pinto A (1995) Expression of functional CD40 antigen on Reed-Sternberg cells and Hodgkin's disease cell lines. Blood 85:780–789

Carbone A, Gloghini A, Caruso A, De Paoli P, Dolcetti R (2017) The impact of EBV and HIV infection on the microenvironmental niche underlying Hodgkin lymphoma pathogenesis. Int J Cancer 140:1233–1245

Carreau NA, Diefenbach CS (2019) Immune targeting of the microenvironment in classical Hodgkin's lymphoma: insights for the hematologist. Ther Adv Hematol 10:2040620719846451. https://doi.org/10.1177/2040620719846451. PMID: 31105921; PMCID: PMC6501496

Chang ET, Zheng T, Weir EG, Borowitz M, Mann RB, Spiegelman D, Mueller NE (2004) Aspirin and the risk of Hodgkin's lymphoma in a population-based case-control study. J Natl Cancer Inst 96:305–315

Chang KC, Chang Y, Jones D, Su IJ (2009) Aberrant expression of cyclin a correlates with morphogenesis of Reed-Sternberg cells in Hodgkin lymphoma. Am J Clin Pathol 132(1): 50–59. https://doi.org/10.1309/AJCPBDFR5L5UOAUZ. PMID: 19864233

Chang KC, Chen PC, Chang Y, Wu YH, Chen YP, Lai CH, Medeiros LJ, Su IJ, Wang HW (2017) Epstein-Barr virus latent membrane protein-1 up-regulates cytokines and correlates with older age and poorer prognosis in Hodgkin lymphoma. Histopathology 70(3):442–455. https://doi.org/10.1111/his.13085. Epub 2016 Nov 8. Erratum in: Histopathology. 2017 May;70(6): 1015. PMID: 27632954

Chuang TC, Moshir S, Garini Y, Chuang AY, Young IT, Vermolen B, van den Doel R, Mougey V, Perrin M, Braun M et al (2004) The three-dimensional organization of telomeres in the nucleus of mammalian cells. BMC Biol 2:12. https://doi.org/10.1186/1741-7007-2-12

Contu F (2020) Spatial organization of lamin A/C in Hodgkin's lymphoma and multiple myeloma. Doctoral thesis, University of Cagliari

Contu F, Rangel-Pozzo A, Trokajlo P, Wark L, Klewes L, Johnson NA, Petrogiannis-Haliotis T, Gartner JG, Garini Y, Vanni R et al (2018) Distinct 3D structural patterns of Lamin A/C expression in Hodgkin and Reed-Sternberg cells. Cancers (Basel) 10. https://doi.org/10.3390/cancers10090286

Cuceu C, Hempel WM, Sabatier L, Bosq J, Carde P, M'Kacher R (2018) Chromosomal instability in Hodgkin lymphoma: an in-depth review and perspectives. Cancers (Basel) 10. https://doi.org/10.3390/cancers10040091

de Lange T (2005) Shelterin: the protein complex that shapes and safeguards human telomeres. Genes Dev 19:2100–2110

de Lange T (2018) Shelterin-mediated telomere protection. Annu Rev Genet 52:223–247. https://doi.org/10.1146/annurev-genet-032918-021921. Epub 2018 Sep 12. PMID: 30208292

de Leeuw R, Gruenbaum Y, Medalia O (2018) Nuclear lamins: thin filaments with major functions. Trends Cell 28(1):34–45. https://doi.org/10.1016/j.tcb.2017.08.004

De Sandre-Giovannoli A, Bernard R, Cau P et al (2003) Lamin a truncation in Hutchinson-Gilford progeria. Science 300(5628):2055. https://doi.org/10.1126/science.1084125. Epub 2003 Apr 17

De Vos WH, Houben F, Hoebe RA, Hoebe RA, Hennekam R, van Engelen B, Manders EM, Ramaekers FC, Broers JL, Van Oostveldt P (2010) Increased plasticity of the nuclear envelope and hypermobility of telomeres due to the loss of A–type lamins. Biochim Biophys Acta 1800: 448–458. https://doi.org/10.1016/j.bbagen.2010.01.002. Epub 2010 Jan 14

Diepstra A, van Imhoff GW, Karim-Kos HE, van den Berg A, te Meerman GJ, Niens M, Nolte IM, Bastiaannet E, Schaapveld M, Vellenga E et al (2007) HLA class II expression by Hodgkin Reed-Sternberg cells is an independent prognostic factor in classical Hodgkin's lymphoma. J Clin Oncol 25:3101–3108. [CrossRef] [PubMed]

Doksani Y, Wu JY, de Lange T, Zhuang X (2013) Superresolution fluorescence imaging of telomeres reveals TRF2-dependent T-loop formation. Cell 155(2):345–356. https://doi.org/10.1016/j.cell.2013.09.048

Dolcetti R, Masucci MG (2003) Epstein-Barr virus: induction and control of cell transformation. J Cell Physiol 196(2):207–218. https://doi.org/10.1002/jcp.10263. PMID: 12811813

Ehlers A, Oker E, Bentink S, Lenze D, Stein H, Hummel M (2008) Histone acetylation and DNA demethylation of B cells result in a Hodgkin-like phenotype. Leukemia 22(4):835–841. https://doi.org/10.1038/leu.2008.12. Epub 2008 Feb 7. PMID: 18256685

Emmerich F, Theurich S, Hummel M, Haeffker A, Vry MS, Döhner K, Bommert K, Stein H, Dörken B (2003) Inactivating I kappa B epsilon mutations in Hodgkin/Reed-Sternberg cells. J Pathol 201:413–420

Faivre J, Frank-Vaillant M, Poulhe R, Mouly H, Jessus C, Bréchot C, Sobczak-Thépot J (2002) Centrosome overduplication, increased ploidy and transformation in cells expressing endoplasmic reticulum-associated cyclin A2. Oncogene 21(10):1493–1500. https://doi.org/10.1038/sj.onc.1205215. PMID: 11896577

Fennewald S, van Santen V, Kieff E (1984) Nucleotide sequence of an mRNA transcribed in l atent growth- transforming virus infection indicates that it may encode a membrane protein. J Virol 51(2):411–419. https://doi.org/10.1128/JVI.51.2.411-419.1984. PMID: 6086953; PMCID: PMC254453

Feuerhahn S, Chen LY, Luke B et al (2015) No DDRama at chromosome ends: TRF2 takes centre stage. Trends Biochem Sci 40:275–285

Fisher DZ, Chaudhary N, Blobel G (1986) cDNA sequencing of nuclear lamins A and C reveals primary and secondary structural homology to intermediate filament proteins. Proc Natl Acad Sci U S A 83:6450–6454. https://doi.org/10.1073/pnas.83.17.6450

Floettmann JE, Ward K, Rickinson AB, Rowe M (1996) Cytostatic effect of Epstein-Barr virus latent membrane protein-1 analyzed using tetracycline-regulated expression in B cell lines. Virology 223(1):29–40. https://doi.org/10.1006/viro.1996.0452. PMID: 8806537

Gamboa Cedeño AM, Castillo M, Xiao W, Waldmann TA, Ranuncolo SM (2019) Alternative and canonical NF-kB pathways DNA-binding hierarchies networks define Hodgkin lymphoma and Non-Hodgkin diffuse large B Cell lymphoma respectively. J Cancer Res Clin Oncol 145.1437 1448. https://doi.org/10.1007/s00432-019-02909-z

Gonzales-Suarez I, Redwood AB, Perkins SM et al (2009) Novel roles for A-type lamins in telomere biology and the DNA damage response pathway. EMBO J 28(16):2414–2427. https://doi.org/10.1038/emboj.2009.196. Epub 2009 Jul 23

Gravelle P, Burroni B, Péricart S, Rossi C, Bezombes C, Tosolini M, Damotte D, Brousset P, Fournié JJ, Laurent C (2017) Mechanisms of PD-1/PD-L1 expression and prognostic relevance in non-Hodgkinlymphoma: a summary of immunohistochemical studies. Oncotarget 8:44960–44975. [CrossRef]

Green MR, Monti S, Rodig SJ, Juszczynski P, Currie T, O'Donnell E, Chapuy B, Takeyama K, Neuberg D, Golub TR et al (2010) Integrative analysis reveals selective 9p24.1 amplification, increased PD-1ligand expression, and further induction via JAK2 in nodular sclerosing Hodgkin lymphoma and primary mediastinal large B-cell lymphoma. Blood 116:3268–3277. [CrossRef]

Green MR, Rodig S, Juszczynski P, Ouyang J, Sinha P, O'Donnell E, Neuberg D, Shipp MA (2012) Constitutive AP-1 activity and EBV infection induce PD-L1 in Hodgkin lymphomas and posttransplant lymphoproliferative disorders: implications for targeted therapy. Clin Cancer Res 18:1611–1618. [CrossRef]

Guan H, Xie L, Leithäuser F, Flossbach L, Möller P, Wirth T, Ushmorov A (2010) KLF4 is a tumor suppressor in B-cell non-Hodgkin lymphoma and in classic Hodgkin lymphoma. Blood 116(9): 1469–1478. https://doi.org/10.1182/blood-2009-12-256446. Epub 2010 Jun 2. PMID: 20519630

Guffei A, Sarkar R, Klewes L, Righolt C, Knecht H, Mai S (2010) Dynamic chromosomal rearrangements in Hodgkin's lymphoma are due to ongoing three-dimensional nuclear remodeling and breakage-bridge-fusion cycles. Haematologica 95:2038–2046. https://doi.org/10.3324/haematol.2010.030171

Hjalgrim H, Askling J, Rostgaard K, Hamilton-Dutoit S, Frisch M, Zhang JS, Madsen M, Rosdahl N, Konradsen HB, Storm HH, Melbye M (2003) Characteristics of Hodgkin's lymphoma after infectious mononucleosis. N Engl J Med 349:1324–1332

Hjalgrim H, Rostgaard K, Johnson PC, Lake A, Shield L, Little AM, Ekstrom-Smedby K, Adami HO, Glimelius B, Hamilton-Dutoit S et al (2010) HLA-A alleles and infectious mononucleosis suggest a critical role for cytotoxic T-cell response in EBV-related Hodgkin lymphoma. Proc Natl Acad Sci U S A 107:6400–6405. https://doi.org/10.1073/pnas.0915054107

Hodgkin T (1832) On some morbid appearances of the absorbent glands and spleen. Med Chir Trans 17:68–114. https://doi.org/10.1177/095952873201700106

Höger TH, Zatloukal K, Waizenegger I, Krohne G (1990) Characterization of a second highly conserved B-type lamin present in cells previously thought to contain only a single B-type lamin. Chromosoma 99(6):379–390. https://doi.org/10.1007/BF01726689

Hu X, Li J, Fu M, Zhao X, Wang W (2021) The JAK/STAT signaling pathway: from bench to clinic. Signal Transduct Target Ther 6(1):402. https://doi.org/10.1038/s41392-021-00791

Hutchinson CJ, Worman HJ (2004) A-type lamins: guardians of the soma? Nat Cell Biol 6:1062–1067

Jansen MP, Machiels BM, Hopman AH, Broers JL, Bot FJ, Arends JW, Ramaekers FC, Schouten HC (1997) Comparison of A and B-type lamin expression in reactive lymph nodes and nodular sclerosing Hodgkin's disease. Histopathology 31:304–312. https://doi.org/10.1046/j.1365-2559.1997.2820881.x

Jost PJ, Ruland J (2007) Aberrant NF-kappaB signaling in lymphoma: mechanisms, consequences, and therapeutic implications. Blood 109(7):2700–2707. https://doi.org/10.1182/blood-2006-07-025809

Kadin ME (1994) Pathology of Hodgkin's disease. Curr Opin Oncol 6:456–463

Kanakkanthara A, Jeganathan KB, Limzerwala JF, Baker DJ, Hamada M, Nam HJ, van Deursen WH, Hamada N, Naylor RM, Becker NA et al (2016) Cyclin A2 is an RNA binding protein that controls Mre11 mRNA translation. Science 353(6307):1549–1552. https://doi.org/10.1126/science.aaf7463. PMID: 27708105; PMCID: PMC5109925

Knecht H, Mai S (2017a) LMP1 and dynamic progressive telomere dysfunction: a major culprit in EBV-associated Hodgkin's lymphoma. Viruses 9(7):164. https://doi.org/10.3390/v9070164. PMID: 28654015; PMCID: PMC5537656

Knecht H, Mai S (2017b) The use of 3D telomere FISH for the characterization of the nuclear architecture in EBV-positive Hodgkin's lymphoma. Methods Mol Biol 1532:93–104. https://doi.org/10.1007/978-1-4939-6655-4_6

Knecht H, Bachmann E, Brousset P, Sandvej K, Nadal D, Bachmann F, Odermatt BF, Delsol G, Pallesen G (1993a) Deletions within the LMP1 oncogene of Epstein-Barr virus are clustered in

Hodgkin's disease and identical to those observed in nasopharyngeal carcinoma. Blood 82(10): 2937–2942. PMID: 8219183

Knecht H, Brousset P, Bachmann E, Sandvej K, Odermatt BF (1993b) Latent membrane protein 1: a key oncogene in EBV-related carcinogenesis? Acta Haematol 90(4):167–171. https://doi.org/10.1159/000204451. PMID: 8140854

Knecht H, McQuain C, Martin J, Rothenberger S, Drexler HG, Berger C, Bachmann E, Kittler EL, Odermatt BF, Quesenberry PJ (1996) Expression of the LMP1 oncoprotein in the EBV negative Hodgkin's disease cell line L-428 is associated with Reed-Sternberg cell morphology. Oncogene 13(5):947–953. PMID: 8806684

Knecht H, Berger C, Rothenberger S, Brousset P (2001) The role of Epstein-Barr virus in neoplastic transformation. Oncology 60:289–302

Knecht H, Sawan B, Lichtensztejn D, Lemieux B, Wellinger RJ, Mai S (2009) The 3D nuclear organization of telomeres marks the transition from Hodgkin to Reed-Sternberg cells. Leukemia 23:565–573. https://doi.org/10.1038/leu.2008.314

Knecht H, Brüderlein S, Mai S, Möller P, Sawan B (2010a) 3D structural and functional characterization of the transition from Hodgkin to Reed-Sternberg cells. Ann Anat 192:302308. Epub 2010 Aug 6

Knecht H, Sawan B, Lichtensztejn Z, Lichtensztejn D, Mai S (2010b) 3D telomere FISH defines LMP1-expressing Reed-Sternberg cells as end-stage cells with telomere-poor 'ghost' nuclei and very short telomeres. Lab Investig 90(4):611–619. https://doi.org/10.1038/labinvest.2010.2. Epub 2010 Feb 8. PMID: 20142802

Knecht H, Kongruttanachok N, Sawan B, Brossard J, Prévost S, Turcotte E, Lichtensztejn Z, Lichtensztejn D, Mai S (2012) Three-dimensional telomere signatures of Hodgkin- and Reed-Sternberg cells at diagnosis identify patients with poor response to conventional chemotherapy. Transl Oncol 5(4):269–277. https://doi.org/10.1593/tlo.12142. Epub 2012 Aug 1. PMID: 22937179; PMCID: PMC3431037

Knecht H, Johnson NA, Haliotis T, Lichtensztejn D, Mai S (2017) Disruption of direct 3D telomere-TRF2 interaction through two molecularly disparate mechanisms is a hallmark of primary Hodgkin and Reed-Sternberg cells. Lab Investig 97(7):772–781. https://doi.org/10.1038/labinvest.2017.33. Epub 2017 Apr 24. PMID: 28436953

Kuppers R, Rajewsky K (1998) The origin of Hodgkin and Reed/Sternberg cells in Hodgkin's disease. Annu Rev Immunol 16:471–493. https://doi.org/10.1146/annurev.immunol.16.1.471

Lacoste S, Wiechec E, Dos Santos Silva AG, Guffei A, Williams G, Lowbeer M, Benedek K, Henriksson M, Klein G, Mai S (2010) Chromosomal rearrangements after ex vivo Epstein-Barr virus (EBV) infection of human B cells. Oncogene 29(4):503–515. https://doi.org/10.1038/onc.2009.359. Epub 2009 Nov 2. PMID: 19881539

Lajoie V, Lemieux B, Sawan B, Lichtensztejn D, Lichtensztejn Z, Wellinger R, Mai S, Knecht H (2015) LMP1 mediates multinuclearity through downregulation of shelterin proteins and formation of telomeric aggregates. Blood 125(13):2101–2110. https://doi.org/10.1182/blood-2014-08-594176. Epub 2015 Jan 7. PMID: 25568351; PMCID: PMC4424269

LeBel C, Wellinger RJ (2005) Telomeres: what's new at your end? J Cell Sci 118:2785–2788

Lennert K (1978) Malignant lymphomas other than Hodgkin's disease. In: Handbuch der Speziellen Pathologischen Anatomie und Histologie. Springer Verlag, Heidelberg. ISBN:978-3-642-81094-7

Li C, Romero-Masters JC, Huebner S, Ohashi M, Hayes M, Bristol JA, Nelson SE, Eichelberg MR, Van Sciver N, Ranheim EA, Scott RS, Johannsen EC, Kenney SC (2020) EBNA2-deleted Epstein-Barr virus (EBV) isolate, P3HR1, causes Hodgkin-like lymphomas and diffuse large B cell lymphomas with type II and Wp-restricted latency types in humanized mice. PLoS Pathog 16(6):e1008590. https://doi.org/10.1371/journal.ppat.1008590. PMID: 32542010; PMCID: PMC7316346

Lin F, Worman HJ (1995) Structural organization of the human gene (LMNB1) encoding nuclear lamin B1. Genomics 27(2):230–236. https://doi.org/10.1006/geno.1995.1036

Lister R, Pelizzola M, Dowen RH, Hawkins RD, Hon G, Tonti-Filippini J, Nery JR, Lee L, Ye Z, Ngo QM et al (2009) Human DNA methylomes at base resolution show widespread epigenomic differences. Nature 462(7271):315–322. https://doi.org/10.1038/nature08514. Epub 2009 Oct 14. PMID: 19829295; PMCID: PMC2857523

Liu Y, Huang R, Liu L, Meng Y, Liu X (2020) Epigenetic abnormalities of classical Hodgkin lymphoma and its effect on immune escape. Cell Biochem Funct 38(3):242–248. https://doi.org/10.1002/cbf.3463. Epub 2019 Nov 11. PMID: 31709594

Louis SF, Vermolen BJ, Garini Y, Young IT, Guffei A, Lichtensztejn Z, Kuttler F, Chuang TC, Moshir S, Mougey V et al (2005) c-Myc induces chromosomal rearrangements through telomere and chromosome remodeling in the interphase nucleus. Proc Natl Acad Sci U S A 102:9613–9618. https://doi.org/10.1073/pnas.0407512102

MacLeod RA, Spitzer D, Bar-Am I, Sylvester JE, Kaufmann M, Wernich A, Drexler HG (2000) Karyotypic dissection of Hodgkin's disease cell lines reveals ectopic subtelomeres and ribosomal DNA at sites of multiple jumping translocations and genomic amplification. Leukemia 14(10):1803–1814. https://doi.org/10.1038/sj.leu.2401894. PMID: 11021756

Makarov AA, Zou J, Houston DR, Spanos C, Solovyova AS, Cardenal-Peralta C, Rappsilber J, Schirmer EC (2019) Lamin A molecular compression and sliding as mechanisms behind nucleoskeleton elasticity. Nat Commun 10(1):3056. https://doi.org/10.1038/s41467-019-11063-6

Mancao C, Altmann M, Jungnickel B, Hammerschmidt W (2005) Rescue of "crippled" germinal center B cells from apoptosis by Epstein-Barr virus. Blood 106:4339–4344. https://doi.org/10.1182/blood-2005-06-2341

Mantovani A, Marchesi F, Malesci A, Laghi L, Allavena P (2017) Tumour-associated macrophages as treatment targets in oncology. Nat Rev Clin Oncol 14:399–416

Marafioti T, Hummel M, Foss HD, Laumen H, Korbjuhn P, Anagnostopoulos I, Lammert H, Demel G, Theil J, Wirth T et al (2000) Hodgkin and reed-sternberg cells represent an expansion of a single clone originating from a germinal center B-cell with functional immunoglobulin gene rearrangements but defective immunoglobulin transcription. Blood 95:1443–1450

Matsuki E, Younes A (2015) Lymphomagenesis in Hodgkin lymphoma. Semin Cancer Biol 34:14–21. https://doi.org/10.1016/j.semcancer.2015.02.002. Epub 2015 Feb 25. PMID: 25725205

Meti N, Esfahani K, Johnson NA (2018) The role of immune checkpoint inhibitors in classical Hodgkin lymphoma. Cancers (Basel) 10(6):204. https://doi.org/10.3390/cancers10060204. PMID: 29914088; PMCID: PMC6025119

Moir RD, Yoon M, Khuon S, Goldman RD (2000) Nuclear lamins A and B1: different pathways of assembly during nuclear envelope formation in living cells. J Cell Biol 151(6):1155–1168. https://doi.org/10.1083/jcb.151.6.1155

Mottok A (2020) Microenvironment in classical Hodgkin lymphoma. Pathologe 41(3):254–260. https://doi.org/10.1007/s00292-020-00774-z

Naetar N, Ferraioli S, Foisner R (2017) Lamins in the nuclear interior – life outside the lamina. J Cell Sci 130(13):2087–2096. https://doi.org/10.1242/jcs.203430

Nmezi B, Xu J, Fu R et al (2019) Concentric organization of A- and B-type lamins predicts their distinct roles in the spatial organization and stability of the nuclear lamina. Proc Natl Acad Sci U S A 116:4307–4315. https://doi.org/10.1073/pnas.1810070116

Piris MA, Medeiros LJ, Chang KC (2020) Hodgkin lymphoma: a review of pathological features and recent advances in pathogenesis. Pathology 52:154–165. https://doi.org/10.1016/j.pathol.2019.09.005

Qi R, Xu N, Wang G et al (2015) The lamin-A/C-LAP2α-BAF1 protein complex regulates mitotic spindle assembly and positioning. J Cell Sci 128:2830–2841. https://doi.org/10.1242/jcs.164566. Epub 2015 Jun 19

Rawlings JS, Rosler KM, Harrison DA (2004) The JAK/STAT signaling pathway. J Cell Sci 117:1281–1283. [CrossRef] [PubMed]

Reed DM (1902) On the pathological changes in Hodgkin's disease, with especial reference to its relation to tuberculosis. Johns Hopkins Hosp Rep 10:133–196

Reichel J, Chadburn A, Rubinstein PG, Giulino-Roth L, Tam W, Liu Y, Gaiolla R, Eng K, Brody J, Inghirami G et al (2015) Flow sorting and exome sequencing reveal the oncogenome of primary Hodgkin and Reed-Sternberg cells. Blood 125:1061–1072. [CrossRef]

Reynolds DJ, Banks PM, Gulley ML (1995) New characterization of infectious mononucleosis and a phenotypic comparison with Hodgkin's disease. Am J Pathol 146(2):379–388. PMID: 7531953; PMCID: PMC1869866

Righolt CH, Guffei A, Knecht H, Young IT, Stallinga S, van Vliet LJ, Mai S (2014) Differences in nuclear DNA organization between lymphocytes, Hodgkin and Reed-Sternberg cells revealed by structured illumination microscopy. J Cell Biochem 115(8):1441–1448. https://doi.org/10.1002/jcb.24800. PMID: 24590512; PMCID: PMC4231252

Rocha-Perugini V, Gonzalez-Granado JM (2014) Nuclear envelope lamin-A as a coordinator of T cell activation. Nucleus 5:396–401

Rothenberger S, Bachmann E, Berger C, McQuain C, Odermatt BF, Knecht H (1997) Natural 30 bp and 69 bp deletion variants of the LMP1 oncogene do stimulate NF-κB mediated transcription. Oncogene 14:2123–2126. https://doi.org/10.1038/sj.onc.1201032

Schwarzer R, Dörken B, Jundt F (2012) Notch is an essential upstream regulator of NF-κB and is relevant for survival of Hodgkin and Reed-Sternberg cells. Leukemia 26:806–813

Scott LM, Gandhi MK (2015) Deregulated JAK/STAT signalling in lymphomagenesis, and its implications for the development of new targeted therapies. Blood Rev 29:405–415. [CrossRef]

Shannon-Lowe C, Rickinson AB, Bell AI (2017) Epstein-Barr virus-associated lymphomas. Philos Trans R Soc Lond Ser B Biol Sci 372. https://doi.org/10.1098/rstb.2016.0271

Shimi T, Butin-Israeli V, Adam SA et al (2010) Nuclear lamins in cell regulation and disease. Cold Spring Harb Symp Quant Biol 75:525–531

Shimi T, Kittisopikul M, Tran J, Goldman AE, Adam SA, Zheng Y, Jaqaman K, Goldman RD (2015) Structural organization of nuclear lamins A, C, B1, and B2 revealed by superresolution microscopy. Mol Biol Cell 26(22):4075–4086. https://doi.org/10.1091/mbc.E15-07-0461. Epub 2015 Aug 26

Simonet T, Zaragosi LE, Philippe C et al (2011) The human TTAGGG repeat factors 1 and 2 bind to a subset of interstitial telomeric sequences and satellite repeats. Cell Res 21:1028–1038. https://doi.org/10.1038/cr.2011.40. Epub 2011 Mar 22

Smith ED, Garza-Gongora AG, MacQuarrie KL, Kosak ST (2018) Interstitial telomeric loops and implications of the interaction between TRF2 and lamin A/C. Differentiation 102:19–26. https://doi.org/10.1016/j.diff.2018.06.001

Stadelmann B, Khandjian E, Hirt A, Lüthy A, Weil R, Wagner HP (1990) Repression of nuclear lamin A and C gene expression in human acute lymphoblastic leukemia and non-Hodgkin's lymphoma cells. Leuk Res 14:815–821. https://doi.org/10.1016/0145-2126(90)90076-l

Swerdlow SH, Campo E, Pileri SA, Harris NL, Stein H, Siebert R, Advani R, Ghielmini M, Salles GA, Zelenetz AD et al (2016) The 2016 revision of the World Health Organization classification of lymphoid neoplasms. Blood 127:2375–2390. https://doi.org/10.1182/blood-2016-01-643569

Taniguchi K, Karin M (2018) NF-κB, inflammation, immunity and cancer: coming of age. Nat Rev Immunol 18:309–324. https://doi.org/10.1038/nri.2017.142

Thorley-Lawson DA (2015) EBV persistence – introducing the virus. Curr Top Microbiol Immunol 390:151–209. https://doi.org/10.1007/978-3-319-22822-8_8

Turgay Y, Eibauer M, Goldman AE, Shimi T, Khayat M, Ben-Harush K, Dubrovsky-Gaupp A, Sapra KT, Goldman RD, Medalia O (2017) The molecular architecture of lamins in somatic cells. Nature 543(7644):261–264. https://doi.org/10.1038/nature21382. Epub 2017 Mar 1

Tzankov A, Bourgau C, Kaiser A, Zimpfer A, Maurer R, Pileri SA, Went P, Dirnhofer S (2005a) Rare expression of T-cell markers in classical Hodgkin's lymphoma. Mod Pathol 18:1542–1549. https://doi.org/10.1038/modpathol.3800473

Tzankov A, Zimpfer A, Went P, Maurer R, Pileri SA, Geley S, Dirnhofer S (2005b) Aberrant expression of cell cycle regulators in Hodgkin and Reed-Sternberg cells of classical Hodgkin's lymphoma. Mod Pathol 18(1):90–96. https://doi.org/10.1038/modpathol.3800276. PMID: 15389259

Vainchenker W, Constantinescu SN (2013) JAK/STAT signaling in hematological malignancies. Oncogene 32:2601–2613. https://doi.org/10.1038/onc.2012.347. Epub 2012 Aug 6. [CrossRef] [PubMed]

Vockerodt M, Morgan SL, Kuo M, Wei W, Chukwuma MB, Arrand JR, Kube D, Gordon J, Young LS, Woodman CB, Murray PG (2008) The Epstein-Barr virus oncoprotein, latent membrane protein-1, reprograms germinal centre B cells towards a Hodgkin's Reed- Sternberg-like phenotype. J Pathol 216(1):83–92. https://doi.org/10.1002/path.2384

Weniger MA, Küppers R (2016) NF-κB deregulation in Hodgkin lymphoma. Semin Cancer Biol 39:32–39. https://doi.org/10.1016/j.semcancer.2016.05.001. Epub 2016 May 21. PMID: 27221964

Weniger MA, Küppers R (2021) Molecular biology of Hodgkin lymphoma. Leukemia 35:968–981

Wood AM, Rendtlew Danielsen JM, Lucas CA, Rice EL, Scalzo D, Shimi T, Goldman RD, Smith ED, Le Beau MM, Kosak ST (2014) TRF2 and lamin A/C interact to facilitate the functional organization of chromosome ends. Nat Commun 5:5467. https://doi.org/10.1038/ncomms6467

Wood AM, Laster K, Rice EL, Kosak ST (2015) A beginning of the end: new insights into the functional organization of telomeres. Nucleus 6:172–178. https://doi.org/10.1080/19491034.2015.1048407

Yam CH, Fung TK, Poon RY (2002) Cyclin A in cell cycle control and cancer. Cell Mol Life Sci 59(8):1317–1326. https://doi.org/10.1007/s00018-002-8510-y. PMID: 12363035

Yang SH, Jung HJ, Coffinier C et al (2019) Are B-type lamins essential in all mammalian cells? Nucleus 2(6):562–569. https://doi.org/10.4161/nucl.2.6.18085. Epub 2011 Nov 1

Zahn M, Kaluszniak B, Möller P, Marienfeld R (2021) The PTP1B mutant PTP1BΔ2-4 is a positive regulator of the JAK/STAT signalling pathway in Hodgkin lymphom. Carcinogenesis 42(4): 517–527. https://doi.org/10.1093/carcin/bgaa144

Updates on Molecular Pathogenesis of Non-Hodgkin's Lymphoma

Hassan Abolhassani

Abstract

Recent advances in molecular characterization of different types of cancers improved refining the diagnostic criteria, prediction of prognosis, and identification of correct targets for drug development. Considering the board functional heterogenicity and property to transform in lymphoid malignancies, varied pathogenic mechanisms have been proposed underlying B cell, T cell, or lineage negative lymphomas. These current pieces of evidence show that although intrinsic genetic variations, epigenetic alterations, oncogenic viruses, and microenvironmental interactions are implicated in all subtypes of lymphomas, unique subtypes present accumulation for particular recurrent aberrations targeting specific pathways.

Keywords

Antigen receptor · B cell lymphomagenesis · Co-stimulatory signals · Cytokines · DNA repair · Epigenetic modifications · Immune evasion · Non-B cell lymphoma · Non-Hodgkin's lymphoma

---

H. Abolhassani (✉)
Division of Clinical Immunology, Department of Biosciences and Nutrition, Karolinska Institute, Stockholm, Sweden

Research Center for Immunodeficiencies, Children's Medical Center, Tehran University of Medical Sciences, Tehran, Iran
e-mail: Hassan.abolhassani@ki.se

© The Author(s), under exclusive license to Springer Nature Switzerland AG 2023
Interdisciplinary Cancer Research, https://doi.org/10.1007/16833_2023_146
Published online: 17 February 2023

1 Introduction

As mentioned before in volume 4, chapter 2, entitled "Immune Dysregulation, Leukemia, and Lymphoma," non-Hodgkin's lymphoma (NHL) is a less frequent type of hematologic malignancy that develops in the secondary lymphoid organs and is typically recorded in patients over >60 years and slightly more often in males (Board 2002; Kos et al. 2021). NHL constitutes about 80% of all lymphomas and can be classified according to the origin of malignant cells (based on morphology, genetics, and immunophenotype of the neoplastic cells) where mainly B cells are known to have the majority of the involved cases (80–90%), followed by T cells (<10%) and lineage negative lymphomas (<1%) (Holdsworth et al. 2021; Rangoonwala and Cascella 2021; Thida and Gohari 2021; Wang et al. 2021). All these three main classes of lymphomas can be subdivided based on the error in the various stages of development (Popovic et al. 2021; Sorge et al. 2021). To better characterize the molecular heterogeneity among morphologically similar lymphoma phenotypes, advanced immunohistochemistry, high-throughput genomic and transcriptomic profiling, and single-cell assays were implemented recently, and now >50 distinct subtypes of lymphoma have been described (Ganapathi et al. 2021; Koh 2021; Kos et al. 2021).

Main NHL entities of B cell lymphoma (BL) according on the level of developmental arrests can be identified as pre-germinal center cells [mantel BL], germinal center cells [primary mediastinal BL, follicular BL, germinal center diffuse large BL (DLBCL, most common subtype ~25–30%), Burkitt BL], post-germinal center memory cells [mutated chronic leukemia, hairy cell leukemia, splenic marginal zone BL, mucosa-associated lymphoid tissue BL], and post-germinal center plasma cells [lymphoplasmacytic lymphoma (Waldenstrom macroglobulinemia), primary effusion lymphoma, activated B cell DLBCL, multiple myeloma] (Xie et al. 2015; Ganapathi et al. 2021; Koh 2021). Similarly, T cell lymphomas (TL) can be subdivided based on the affected stage of T cells to leukemic (adult TL, T cell prolymphocytic leukemia, T cell large granular leukemia), nodal (anaplastic large cell TL, angioimmunoblastic TL, follicular TL, nodal TFH-like TL, other peripheral TL), and extra nodal (enteropathy-associated TL, monomorphic epitheliotropic intestinal TL, indolent T cell lymphoproliferative disorder, hepatosplenic TL, systemic EBV-positive TL of childhood, cutaneous TL [including Sezary syndrome, mycosis fungoides-cutaneous TL, anaplastic large cell TL, primary cutaneous CD4 small/medium TL, primary cutaneous acral CD8 TL, primary cutaneous gamma-delta TL]) entities (Rangoonwala and Cascella 2021; Wang et al. 2021). Lineage negative lymphomas are extremely rare and have been reported in a few case reports as natural killer (NK)/TL and innate lymphocyte cells type 2 (ILC2) (Thida and Gohari 2021; Wang et al. 2021).

In the following sections, updated evidence about the underlying molecular pathogenesis of these subclasses of NHL are provided based on the known intrinsic genetic alterations (antigen receptor, downstream second messengers, co-stimulatory signals, cytokine signaling), epigenetic dysregulation, oncogenic viruses, and micro-environmental interactions (Fig. 1).

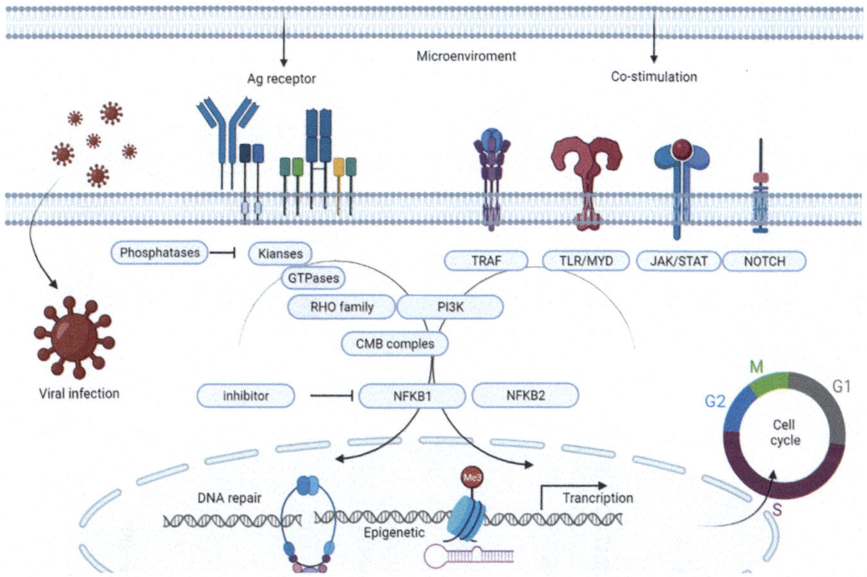

Fig. 1 Several pathways are dysregulated in NHL leading to signals driving proliferation and survival

2 Antigen Receptor and Downstream Signaling Dysregulation

The development of antigen receptor on both B and T cells requires somatic recombination and proper DNA repair machinery for VDJ rearrangement. Moreover, terminal B cell development is dependent on further class switching recombination of immunoglobulin heavy chain and somatic hypermutations in the VDJ region which also require double and single DNA strand lesions, respectively (Fowler and Davis 2013; Burger and Wiestner 2018). Therefore, tight regulation of these processes within both lymphocytes is very crucial to direct the mutagenesis targeted and prevent pathogenic chromosomal translocations (Kenkre and Kahl 2012; Young et al. 2015). The main surveillance security check is epigenetic elements that open the frame of targeted regions and DNA repair molecules which keep the lesion to their adjacent proximity. In line with this notion, random genetic alteration in both epigenetic modifiers and DNA repair elements has been linked with lymphomagenesis with B cell (Table 1) or non-B cell (Table 2) origin (Warner et al. 2013; Young et al. 2015; Burger and Wiestner 2018). Moreover, the off-target activity of mutagenesis factors, particularly AID protein (affecting regions with AA**G**TT/CA**G**CT = **RG**YW motif), has a significant role in this process.

Table 1 Most frequent molecular mechanisms of pathogenesis in different types of B cell origin NHL based on the developmental stage arrest

NHL subtype	Ag receptor and downstream signaling	Co-stimulation, cytokine, survival signals, and cell cycles	DNA repair	Epigenetic	Viral oncogenesis	Microenvironment
B cell NHL (BL)						
MCL		TNFAIP3[a] CCND1 CDK4 TP53[a]	t(11:14) CDKN2A[a] ATM[a]	P16/INK4A SOX9, HOXA9 AHR, NR2F2 ROBO1 miR-15a/miR17-92/miR-155/ miR-127-3p/miR-615-3p/miR-29/ miR-18b/miR-150		
FL	TNFAIP3[a] BTK FOXO1	JAK1/JAK3 STAT3/STAT6 NOTCH TNFRSF14[a] FAS MLL2[a] SOCS1 BCL2	t(14:18)	EZH2 KLF4 MLL2 KMT2D MEF2B/C[a] CREBBP/EP300[a] miR17-92/miR-21	HCV	HVEM[a] EPHA7[a]
GCB-DLBCL	CD79B PRKCB LYN PTPN6 PTEN[a] CARD11 GRB2 TNFAIP3 TBL1XR1 KLHL6 NFKBIE	IL6 STAT3/STAT6 NOTCH2 DTX1 SPEN TLR2 MYD88 MYC IRF4 FAS[a] BCL2	t(14:18) t(3:x) MDM2 UBE2A ZNF423 PIM1 BTG1 CCND3	EZH2 P16/INK4A KLF4 MLL2[a] KMT2D MGMT MEF2B/C[a] CREBBP/EP300[a] SMAD1 HIST1H1E/HIST1H1C/ HIST1H1B/HIST1H2BK/	HCV	HLAB HLADMA HLAA HLAC B2M CD70 CD58 CD83 CIITA[a] PDL1 GNA13

	ZC3H12A NFKBIA BRAF KRAS	BCL6 REL TP53[a]		HIST1H1D/HIST1H2BC HIST1H2AC/HIST1H2AM/ HIST2H2BE/HIST1H3B miR17-92/miR-127-3p/miR-222/ miR-181a/miR-150		RHOA CXCR4
Burkitt		MYC TCF3 CCND3 ID3[a] TP53[a]	t(8:14) CDKN2A[a]	KLF4 miR17-92/miR-155/miR-296-3p	HIV EBV	
MCLL				miR-155/miR-29 miR-150/miR-15a/16-1/miR-34a		
MZBL		CCND1	t(11:14)		HCV	
MALT	API2-MALT1	NALT1 FOXP3	t(8:14) t(14:X) t(11;18)	miR-34a		*H. pylori* infection
HCL	NFKB1A/E[a]	JAK2 CD274 TNFAIP3[a] TP53[a] FAS[a] SOCS1[a]	KLF4	JMJD2C miR-155		CIITA[a]
LL		IGH-PAX5	t(9:14)			
ABC-DLBCL	CD79B CD79A PRKCB PTEN[a] LYN PTPN6 CARD11 GRB2	IL6 STAT3/STAT6 TLR2 MYD88 MYC IRF4 CCND3 PRDM1	t(3:x) CNV (9p24) UBE2A ZNF423 PIM1 BTG1 CCND3 CDKN2A[a] SPIB	P16/INK4A KLF4 MLL2[a] KMT2D MGMT CREBBP/EP300[a] SMAD1 HIST1H1E/HIST1H1C/ HIST1H1B/HIST1H2BK/		HLAB HLADMA HLAA HLAC B2M CD70 CD58 CD83

(continued)

Table 1 (continued)

NHL subtype	Ag receptor and downstream signaling	Co-stimulation, cytokine, survival signals, and cell cycles	DNA repair	Epigenetic	Viral oncogenesis	Microenvironment
	TNFAIP3 TBL1XR1 KLHL6 NFKBIE ZC3H12A NFKBIA BRAF KRAS	MEF2B IRF8 EBF1 ETS1 POU2F2 ZFP36L1 MEF2B/C[a] pFAS[a] BCL2 BCL6 BCL11A TP53[a]		HIST1H1D/HIST1H2BC HIST1H2AC/HIST1H2AM/ HIST2H2BE/HIST1H3B miR-155/miR-127-3p/miR-222/ miR-181a/miR-150		CIITA[a] PDL1 GNA13 RHOA CXCR4
PMBL		STAT6 JAK2 CD274 IRF4 TNFAIP3[a] REL TP53[a] SOCS1[a]		JMJD2C miR-155		CIITA[a]
MM		CCND1 FCGR3 MAF	t(11:14) t(1:14) t(14:)	miR16-1		

MCL Mantle cell BL, *MZBL* marginal zone BL, *SBL* splenic BL, *MALT* mucosa-associated lymphoid tissue BL, *PMBL* primary mediastinal BL, *FL* follicular BL, *GCB-DLBCL* germinal center diffuse large BL, *MCLL* mutated chronic lymphocytic leukemia, *HCL* hairy cell leukemia, *LL* lymphoplasmacytic lymphoma, *ABC-DLBCL* activated B cell diffuse large BL, *MM* multiple myeloma

[a]Loss of function

Table 2 Most frequent molecular mechanisms of pathogenesis in different types of non-B cell origin NHL based on the developmental stage arrest

NHL subtype	Ag receptor and downstream signaling	Co-stimulation, cytokine, survival signals, and cell cycles	DNA repair	Epigenetic	Viral oncogenesis	Microenvironment
T cell NHL (TL)						
ATLL	CARD11 VAV1 FYN	ZEB1[a] ICOS-CD28 CTLA4-CD28		TET1[a] TET2[a] SETD2		PDL1 CCR4
SS	PLCG1 CARD11	IL13/IL15/IL21 ZEB1[a]		TET1[a] SETD2 KMT2B KMT2C BCOR	CMV HTLV1	
PTCL	VAV1 DUSP22[a] RHOA[a] CARD11 KRAS ITK SYK FER FYN	JAK1 JAK3 STAT3 STAT5b CTLA4-CD28 GATA3-TBX21	t(19:x) t(5:9)	DNMT3A[a] TET1[a] TET2[a] SETD2 BCORL1		PDL1
ALCL	VAV1 PTPN6[a] DUSP22[a]	IL9/IL22/IL6/ NPM1 ALK (positive or negative) DUSP22 IRF4 JAK2-PCM1 TYK2-PCM1	t(2;5) t(6:x)	DNMT3A[a] TET2[a] miR17-92/miR-512-3p/miR-886-5p/3p/ miR-708/miR-135b/miR-146a/miR-155	HTLV2	PDL1
AITL	VAV1 RHOA[a] CARD11 KRAS	CD28 ICOS-CD28	t(19:x)	IDH2 DNMT3A[a] TET2[a]	EBV	PDL1

(continued)

Table 2 (continued)

NHL subtype	Ag receptor and downstream signaling	Co-stimulation, cytokine, survival signals, and cell cycles	DNA repair	Epigenetic	Viral oncogenesis	Microenvironment
EATL	KRAS NRAS			TET1[a] TET2[a] SETD2		
HSTCL		JAK1 JAK3 STAT3 STAT5b		TET2[a] TET3[a] SETD2		
MF	VAV1 PLCG1 CARD11 MAP2K1 NFKB1 FYN	IL13/IL15/IL21 RTLPR ZEB1[a]		TET1[a]	EBV CMV HHV-6 HHV-7 HHV8 HTLV1 HTLV2	
Lineage negative NHL						
NKTL	PTPN6[a]	PTRPK[a] CTLA4-CD28	t(x,x)	EZH2 BCOR MLL2 miR-21/miR-150		HLA-DPB1 FAM160A1 PDL1
ILC	CD79a	CD127 IL1R FOXP3				PDL1 PDL2

ATLL Adult TL, *SS* Sezary syndrome, *PTCL* peripheral TL, *ALCT* anaplastic large cell TL, *AITL* angioimmunoblastic TL, *EATL* enteropathy-associated TL, *HSTCL* hepatosplenic TL, *MF* mycosis fungoides-cutaneous TL and primary cutaneous anaplastic large cell TL, *NKTL* natural killer T cell lymphoma, *ILC* innate lymphocyte cell lymphoma

[a]Loss of function

Although the immunoglobulin region is the most frequent region involved in this translocation, other non-antigen receptor locations that are in high transcription load during the early B or T cell development can be affected including BCL6, API, MALT1, CTLA4, CD28, ICOS, TYK2, JAK2, PCM1, ITK, SYK, FER, and VAV1 (Tables 1 and 2) (Warner et al. 2013; Xu et al. 2021).

Translocation between active open chromosomal regions is not the only reason for lymphomagenesis due to defects in DNA damage-sensing elements, but also copy number variation in the antigen receptor downstream signaling proteins has been reported as well due to underlying DNA repair defect (Kenkre and Kahl 2012; Fowler and Davis 2013). Overactivation of associated signaling factors mainly kinases have been also documented due to somatic mutation (leading to gain of function) or transcriptional deregulation (Xu et al. 2021). Moreover, alteration in genomic levels due to point mutations or large deletion particularly in the inhibitory elements of this signaling mainly phosphatases has been identified in different types of NHL (shown with [a] on Tables 1 and 2) (Carey et al. 2000; Lucas et al. 2004). Altogether these mutations can resemble constitutional overactivation of CARD-BCL10-MALT1 and canonical NFKB signaling similar to recurrent antigen stimulation which leads to continuous survival and cycle cell checkpoint eversion (Lucas et al. 2004; Kenkre and Kahl 2012; Fowler and Davis 2013; Young et al. 2015).

3 Co-stimulation, Cytokine, and Survival Signal Dysregulation

According to the known physiologic function of naive B and T cells, activation is dependent on the sequential integration of extra immune synapses apart from cognate antigens (Lulla and Heslop 2016). This additional stimulation is required for initial cell division, induction of class switching and somatic hypermutation, as well as promoting survival and differentiation (Ke et al. 2008; Komohara and Harada 2019). It is also well known that antigen signaling cannot drive the differentiation of activated cells into specific lymphocyte subtypes without cooperation with signaling pathways induced by co-stimulatory molecules which harmonize chemokines, integrins, and metabolites (Lulla and Heslop 2016; Wilcox 2016; Waldmann and Chen 2017). Four major classes of these required signals are provided by co-stimulatory membrane proteins, cytokine receptors, toll-like receptors (TLR), and NOTCH signaling (Fig. 1) (Jacobsen and Weinstock 2018; Strati et al. 2018). The regulation of these signals will be disturbed by both intrinsic (mutations resulting in overactivation of these four pathways) and extrinsic factors (microenvironment of the lymphoma) (Tables 1 and 2) (Zappasodi et al. 2015; Schwarzbich and Witzens-Harig 2017).

4 Epigenetic and Transcriptional Dysregulation

As mentioned above, the epigenetic regulators have a significant impact on the transcriptional state of developmental genes within the lymphocytes and chromatin modulation (Ribeiro et al. 2019; Ahmed and Feldman 2020; Isshiki and Melnick 2021; Lai and Wang 2021). The generation of functional mature lymphocyte is the outcome of different cell lineage selection, unique commitment, and cell differentiation (Bakhshi and Georgel 2020; Duy et al. 2020; Yang and Green 2020). These gradual transitions happen in a stepwise process and should be well regulated at the transcriptomics level. In line with this notion, deleterious mutations in proteins with crucial function in DNA methylation (particularly DNMT3A and TET family), histone modification (mainly EZH2, SETD2, CREBBP, MLL2, KMT2 family, BCOR family), and direct mutation within miRNA that have been linked to HIST1 regions have been reported in several subtypes of NHL (Tables 1 and 2) (Bakhshi and Georgel 2020; Yang and Green 2020).

5 Oncogenic Viruses

The major known environmental pathogen for lymphomagenesis is exemplified by disruption or inhibition of T cell activity due to HIV infection (mainly at AIDS stage) which provide a condition for Epstein-Barr virus (EBV)-driven uncontrolled lymphoproliferation and malignant transformation (Chen and Chuang 2020; Shindiapina et al. 2020; Ganapathi et al. 2021). EBV proteins in different latency phases can be involved in lymphomagenesis including EBNA1 (promoting the development and antiapoptotic signals by decreasing the protein tyrosine phosphatase receptor type K), LMP1 (activating constitutively CD40 receptor homolog and enhances transforming properties and immune eversion factors PDL1), and LMP2A (mimics the antigen receptor signal and suppresses expression of EBF1 and E2A) (Saha and Robertson 2019; Hue et al. 2020; Sukswai et al. 2020). Cytomegalovirus (CMV) has also been linked to lymphomagenesis due to its main function in disturbing human leukocyte antigen (HLA) expression enabling escaping innate immune cytotoxicity by NK cells, disturbing physiologic lymphocyte signaling, invading complement system activation, and enhancing immunosuppressive chemokines. Other known Herpes virus families can be involved in T cell origin NHL mainly mycosis fungoides and a common subtype of cutaneous T cell lymphoma (Table 2) (Bispo et al. 2020; Chen and Chuang 2020). Human T-lymphotropic virus type 1,2 (HTLV1-2) is also associated with the same type of lymphoma due to the function of its main TAX protein which can repress PTPN6 and also transactivate the gamma chain receptor for IL2 and IL15 (Chen and Chuang 2020; Consul et al. 2020). Finally, the impact of hepatitis C virus (HCV) is well known in the development of chronic liver cancers, but recently it has been also reported in association with B cell origin NHL, due to continuous and recurrent antigenic stimulation activating expansion of specific lymphocyte clones either monoclonal or oligoclonal statues (Table 1) (Bispo et al. 2020; Consul et al. 2020).

6 Microenvironmental Interaction Dysregulation and Immune Evasion

Genetic alteration within the HLA region and combination of certain HLA alleles and haplotypes are related to the risk of NHL development (Araujo-Ayala et al. 2021; Liu et al. 2021; Menter et al. 2021; Pileri et al. 2021a). The phenomenon may be resulted in the direct involvement of HLA molecule antigen presentation, or by indirect linkage to some other unknown non-HLA genes (Pileri et al. 2021b). Indeed, virus- and cancer-specific (neo)antigens can be presented by HLA stimulating mainly adaptive anti-tumor immunity (Kumar et al. 2021; Willard et al. 2021). However, HLA expression could be suppressed by the tumor cells in a selected group of patients and NHL entities, which facilitate immune evasion (Tables 1 and 2) (Tamma et al. 2020; Araujo-Ayala et al. 2021; Liu et al. 2021). The second immune evasion mechanism which has been described in NHL cases is overexpression of negative regulators like PDL1 (Choi 2021; Kumar et al. 2021; Pileri et al. 2021b). NHL is distinct from other major cancer cells regarding the dysregulation of PDL1 expression due to its specific physiologic presences within the normal lymphocyte origin (Brockelmann et al. 2020; Menter et al. 2021). Nonmalignant lymphocytes require activation to express constitutive PDL1 mainly via active JAK/STAT3 signaling. The absence of HLA expression has been linked with higher severity and diminished immune responses (Choi 2021; Willard et al. 2021). On the other hand, and independent from the status of HLA, the PDL1 overproduction can provide protective suppression against immune surveillance (Csizmar and Ansell 2021). Other immune and non-immune cells in the microenvironment of lymphoma also have a significant role in the induction and prolongation of malignancy (Csizmar and Ansell 2021; Ioannou et al. 2021). Moreover, these surrounding cells can be influenced by other environmental agents like *Helicobacter pylori* infection's impact on the merged expression of both Th1 and Th2 signals from follicular helper T cells exhibiting unregulated activation stimulus for autologous B cell proliferation and decreased cytotoxicity against malignant B cells (via perforin- and FAS-FAS ligand-mediated killing mechanism) (Table 1) (Liu et al. 2021).

7 Conclusion

Advances in understanding of the molecular basis of signaling pathways both at antigenic receptor and co-stimulatory elements as well as epigenetic markers integrated by microenvironmental boosters can direct scientists to personalized medicine against specific subtypes of lymphoma. Progression in this field can provide novel therapeutic modalities and immunosuppressive strategies governing a better quality of life mainly in patients with poor prognosis. Future studies can also address the identification of potential biomarkers which can distinguish a patient's refractory to treatment and suggest adjuvant targeted therapy. In conclusion, this approach will eventually drive mechanism-based treatment toward curing the majority of lymphoma patients.

Acknowledgments None.

Compliance with Ethical Standards The authors declare that there is no conflict of interest.

References

Ahmed N, Feldman AL (2020) Targeting epigenetic regulators in the treatment of T-cell lymphoma. Expert Rev Hematol 13:127–139

Araujo-Ayala F, Perez-Galan P, Campo E (2021) Vulnerabilities in the tumor and microenvironment in follicular lymphoma. Hematol Oncol 39(Suppl 1):83–87

Bakhshi TJ, Georgel PT (2020) Genetic and epigenetic determinants of diffuse large B-cell lymphoma. Blood Cancer J 10:123

Bispo JAB, Pinheiro PS, Kobetz EK (2020) Epidemiology and etiology of leukemia and lymphoma. Cold Spring Harb Perspect Med 10:a034819

Board PATE (2002) Adult non-Hodgkin lymphoma treatment (PDQ(R)): patient version. PDQ Cancer Information Summaries, Bethesda

Brockelmann PJ, de Jong MRW, Jachimowicz RD (2020) Targeting DNA repair, cell cycle, and tumor microenvironment in B cell lymphoma. Cell 9:2287

Burger JA, Wiestner A (2018) Targeting B cell receptor signalling in cancer: preclinical and clinical advances. Nat Rev Cancer 18:148–167

Carey GB, Donjerkovic D, Mueller CM, Liu S, Hinshaw JA, Tonnetti L, Davidson W, Scott DW (2000) B-cell receptor and Fas-mediated signals for life and death. Immunol Rev 176:105–115

Chen BJ, Chuang SS (2020) Lymphoid neoplasms with plasmablastic differentiation: a comprehensive review and diagnostic approaches. Adv Anat Pathol 27:61–74

Choi YS (2021) Immuno-oncology for B-cell lymphomas. Blood Res 56:S70–S74

Consul N, Menias CO, Lubner MG, Katabathina VS, Chahinian RA, Mansour J, Elsayes KM (2020) A review of viral-related malignancies and the associated imaging findings. AJR Am J Roentgenol 214:W1–W10

Csizmar CM, Ansell SM (2021) Engaging the innate and adaptive antitumor immune response in lymphoma. Int J Mol Sci 22:3302

Duy C, Beguelin W, Melnick A (2020) Epigenetic mechanisms in leukemias and lymphomas. Cold Spring Harb Perspect Med 10:a034959

Fowler N, Davis E (2013) Targeting B-cell receptor signaling: changing the paradigm. Hematology Am Soc Hematol Educ Program 2013:553–560

Ganapathi KA, Brown LE, Prakash S, Bhargava P (2021) New developments in non-Hodgkin lymphoid malignancies. Pathology 53:349–366

Holdsworth F, Worku D, Bretton AL, Vella C, Walker E (2021) A guide to Hodgkin and non-Hodgkin lymphomas: similarities and differences. Br J Nurs 30:S16–S22

Hue SS, Oon ML, Wang S, Tan SY, Ng SB (2020) Epstein-Barr virus-associated T- and NK-cell lymphoproliferative diseases: an update and diagnostic approach. Pathology 52:111–127

Ioannou N, Jain K, Ramsay AG (2021) Immunomodulatory drugs for the treatment of B cell malignancies. Int J Mol Sci 22:8572

Isshiki Y, Melnick A (2021) Epigenetic mechanisms of therapy resistance in diffuse large B cell lymphoma (DLBCL). Curr Cancer Drug Targets 21:274–282
Jacobsen ED, Weinstock DM (2018) Challenges and implications of genomics for T-cell lymphomas. Hematology Am Soc Hematol Educ Program 2018:63–68
Ke X, Wang J, Li L, Chen IH, Wang H, Yang XF (2008) Roles of CD4+CD25(high) FOXP3+ Tregs in lymphomas and tumors are complex. Front Biosci 13:3986–4001
Kenkre VP, Kahl BS (2012) The future of B-cell lymphoma therapy: the B-cell receptor and its downstream pathways. Curr Hematol Malig Rep 7:216–220
Koh Y (2021) Genomics of diffuse large B cell lymphoma. Blood Res 56:S75–S79
Komohara Y, Harada M (2019) Potential mechanisms of spontaneous regression in patients with B-cell lymphoma; the significance of co-stimulatory molecules in lymphoma cells. J Clin Exp Hematop 59:207–210
Kos IA, Thurner L, Bittenbring JT, Christofyllakis K, Kaddu-Mulindwa D (2021) Advances in lymphoma molecular diagnostics. Diagnostics (Basel) 11:2174
Kumar E, Pickard L, Okosun J (2021) Pathogenesis of follicular lymphoma: genetics to the microenvironment to clinical translation. Br J Haematol 194:810–821
Lai P, Wang Y (2021) Epigenetics of cutaneous T-cell lymphoma: biomarkers and therapeutic potentials. Cancer Biol Med 18:34–51
Liu Y, Zhou X, Wang X (2021) Targeting the tumor microenvironment in B-cell lymphoma: challenges and opportunities. J Hematol Oncol 14:125
Lucas PC, McAllister-Lucas LM, Nunez G (2004) NF-kappaB signaling in lymphocytes: a new cast of characters. J Cell Sci 117:31–39
Lulla P, Heslop HE (2016) Checkpoint inhibition and cellular immunotherapy in lymphoma. Hematology Am Soc Hematol Educ Program 2016:390–396
Menter T, Tzankov A, Dirnhofer S (2021) The tumor microenvironment of lymphomas: insights into the potential role and modes of actions of checkpoint inhibitors. Hematol Oncol 39:3–10
Pileri A, Guglielmo A, Grandi V, Violetti SA, Fanoni D, Fava P, Agostinelli C, Berti E, Quaglino P, Pimpinelli N (2021a) The microenvironment's role in mycosis Fungoides and Sezary syndrome: from progression to therapeutic implications. Cell 10:2780
Pileri SA, Tripodo C, Melle F, Motta G, Tabanelli V, Fiori S, Vegliante MC, Mazzara S, Ciavarella S, Derenzini E (2021b) Predictive and prognostic molecular factors in diffuse large B-cell lymphomas. Cell 10:675
Popovic LS, Matovina-Brko G, Popovic M, Popovic M, Cvetanovic A, Nikolic I, Kukic B, Petrovic D (2021) Immunotherapy in the treatment of lymphoma. World J Stem Cells 13:503–520
Rangoonwala HI, Cascella M (2021) Peripheral T-cell lymphoma. StatPearls, Treasure Island
Ribeiro ML, Reyes-Garau D, Armengol M, Fernandez-Serrano M, Roue G (2019) Recent advances in the targeting of epigenetic regulators in B-cell non-Hodgkin lymphoma. Front Genet 10:986
Saha A, Robertson ES (2019) Mechanisms of B-cell oncogenesis induced by Epstein-Barr virus. J Virol 93:e00238-19
Schwarzbich MA, Witzens-Harig M (2017) Cellular immunotherapy in B-cell malignancy. Oncol Res Treat 40:674–681
Shindiapina P, Ahmed EH, Mozhenkova A, Abebe T, Baiocchi RA (2020) Immunology of EBV-related lymphoproliferative disease in HIV-positive individuals. Front Oncol 10:1723
Sorge I, Georgi TW, Hirsch FW (2021) Lymphoma in children and adolescents. Radiologe 61:611–618
Strati P, Patel S, Nastoupil L, Fanale MA, Bollard CM, Lin AY, Gordon LI (2018) Beyond chemotherapy: checkpoint inhibition and cell-based therapy in non-Hodgkin lymphoma. Am Soc Clin Oncol Educ Book 38:592–603
Sukswai N, Lyapichev K, Khoury JD, Medeiros LJ (2020) Diffuse large B-cell lymphoma variants: an update. Pathology 52:53–67
Tamma R, Ranieri G, Ingravallo G, Annese T, Oranger A, Gaudio F, Musto P, Specchia G, Ribatti D (2020) Inflammatory cells in diffuse large B cell lymphoma. J Clin Med 9:2418
Thida AM, Gohari P (2021) Extranodal NK-cell lymphoma. StatPearls, Treasure Island

Waldmann TA, Chen J (2017) Disorders of the JAK/STAT pathway in T cell lymphoma pathogenesis: implications for immunotherapy. Annu Rev Immunol 35:533–550
Wang H, Fu BB, Gale RP, Liang Y (2021) NK−/T-cell lymphomas. Leukemia 35:2460–2468
Warner K, Weit N, Crispatzu G, Admirand J, Jones D, Herling M (2013) T-cell receptor signaling in peripheral T-cell lymphoma – a review of patterns of alterations in a central growth regulatory pathway. Curr Hematol Malig Rep 8:163–172
Wilcox RA (2016) A three-signal model of T-cell lymphoma pathogenesis. Am J Hematol 91:113–122
Willard P, McKay J, Yazbeck V (2021) Role of antibody-based therapy in indolent non-Hodgkin's lymphoma. Leuk Res Rep 16:100275
Xie Y, Pittaluga S, Jaffe ES (2015) The histological classification of diffuse large B-cell lymphomas. Semin Hematol 52:57–66
Xu W, Berning P, Lenz G (2021) Targeting B-cell receptor and PI3K signaling in diffuse large B-cell lymphoma. Blood 138:1110–1119
Yang H, Green MR (2020) Harnessing lymphoma epigenetics to improve therapies. Hematology Am Soc Hematol Educ Program 2020:95–100
Young RM, Shaffer AL 3rd, Phelan JD, Staudt LM (2015) B-cell receptor signaling in diffuse large B-cell lymphoma. Semin Hematol 52:77–85
Zappasodi R, de Braud F, Di Nicola M (2015) Lymphoma immunotherapy: current status. Front Immunol 6:448

Intraocular Lymphoma

Konstantina Riri, Dimitrios Kalogeropoulos, Andreas Mitsios, Ioanna Ploumi, and Sofia Androudi

Abstract

Intraocular lymphomas are rare, non-Hodgkin, large B-cell lymphomas affecting the vitreous, the retina, the retinal pigment epithelium (RPE) or the uvea tract. Vitreoretinal lymphomas (VRLs) are the most common subtype and are associated with central nervous system (CNS) involvement. Both the eye and the brain share mutual characteristics and are targeted by the spreading lymphoma cells.

Intraocular lymphoma is a rare invasion of intraocular malignancy. More common form is VRL involving the vitreous and the retinal pigment epithelium. The uvea tract is affected in metastatic lymphomas. VRL is firmly correlated with CNS involvement. Diagnosis of VRL is challenging as its initial appearance is like intermediate or posterior uveitis. Treatment strategies are debated when concurrent CNS involvement is not confirmed.

Intraocular lymphoma is a great mimic of chronic uveitis and taking into consideration the malignant nature of lymphoma disease, every ophthalmologist should be aware of it and a multidisciplinary approach is vital for the patient.

K. Riri · A. Mitsios · S. Androudi (✉)
Department of Ophthalmology, University of Thessaly, Larissa, Greece
e-mail: androudi@otenet.gr

D. Kalogeropoulos
Department of Ophthalmology, University of Ioannina, Ioannina, Greece

I. Ploumi
Northeastern University, College of Science, Boston, MA, USA

Massachusetts Eye and Ear, Fluorescein Department, Boston, MA, USA

© The Author(s), under exclusive license to Springer Nature Switzerland AG 2022
Interdisciplinary Cancer Research, https://doi.org/10.1007/16833_2022_101
Published online: 23 December 2022

Keywords

Chemotherapy · Intraocular lymphoma · Methotrexate · MYD88 · PCNSL uveitis · Primary central nervous lymphoma · Vitreoretinal lymphoma · Vitritis · VRL

1 Introduction

Intraocular lymphomas are rare non-Hodgkin lymphomas of large B-cells origin in the majority of cases, including the following types: vitreoretinal lymphomas (primary or secondary), primary uveal/choroidal lymphomas, and secondary intraocular lymphomas (SIOL).

Vitreoretinal lymphomas (VRL) are the most common form and constitute a subtype of malignant primary central nervous system lymphomas (PCNSL). They consist of highly aggressive diffuse large B lymphocytes. Primary uveal or choroidal lymphomas are small-cell, mildly aggressive varieties of mucosa-associated lymphoid tissue and lack the involvement of the central nervous system (CNS). On the other hand, SIOL is much less frequent and is associated with hematogenous dissemination from systemic lymphomas. SIOL usually affects the uvea and more specifically, the choroid. However, in rare cases, systemic metastatic lymphomas with the primary site in the nasopharynx, the testes, or the skin may involve the vitreous and the retina (systemic metastatic retinal lymphoma) (Cao et al. 2011; Tiu Teo et al. 2019; Salomão et al. 2013; Jahnke et al. 2006).

Infrequently, large B-cell systematic lymphomas may affect the choroid, the ciliary body, and the iris with aggressive behavior and a high chance of concomitant systematic involvement but without malignant participation of the CNS. However, diffuse large B-cell lymphomas (DLBCL) may be intravascular or angiotropic, in which case CNS and choroid involvement may occur (Fonkem et al. 2014).

The most common intraocular form is the primary vitreoretinal lymphoma (PVRL) and it affects the vitreous, the retina, and the retinal pigment epithelium (RPE) (Chan et al. 2011; Coupland et al. 2002). There is a strong correlation between PVRL and PCNSL, as the majority of patients with ophthalmic manifestations will present CNS lymphoma (Chan et al. 2011; Mochizuki and Singh 2009).

VRL and PCNSL are DLBCL, which is the most common non-Hodgkin lymphoma. Secondary VRL occurs after the onset of PCNSL, less frequently after systemic DLBCL, and more commonly after testicular lymphoma (Coupland et al. 2005a, b, c; Cimino et al. 2009).

Clinical suspicion of intraocular lymphoma is delayed due to non-specific intraocular inflammation as the first sign. Moreover, vitreoretinal lymphomas may be treated as posterior uveitis with corticosteroids and short-term improvement is feasible. Diagnosis is also challenging in terms of a proven biopsy. Considering it represents 0.01% of all ocular disorders, lymphoma constitutes a sporadic case of ophthalmic malignancy (Chan et al. 2011; Mochizuki and Singh 2009).

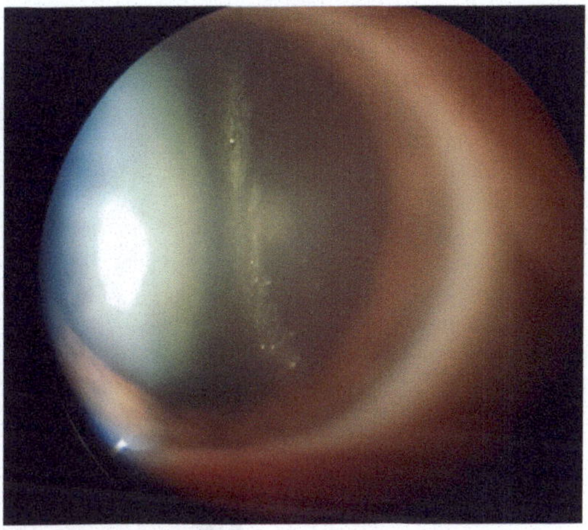

Fig. 1 Slit lamp photograph of the left eye of 68-year-old woman with floaters in both eyes. Anterior vitreous sheets and infiltration with lymphoma cells are disclosed. (Image courtesy of Dr. Riri)

Pinpointing the disease is a complex matter. VRL typically occurs as a posterior and/or intermediate uveitis. Harvesting and maintaining vitreous samples can be hard (Gonzales and Chan 2007) and general hospitals usually lack cytopathologists experienced with this entity. Consequently, several diagnostic tests have emerged to back up this gap, such as flow cytometry, immunohistochemistry molecular detection of clonal gene rearrangements, and cytokine profiling of intraocular fluid (Davis et al. 1992; Wang et al. 2011; Cassoux et al. 2007). Taking action without the presence of concurrent CNS disease is disputed. There is evidence to support the use of intraocular chemotherapy with methotrexate (Frenkel et al. 2008), or rituximab (Itty and Pulido 2009; Sen et al. 2008). Some recommend using systemic chemotherapy with courses of high-dose methotrexate along with radiation, or another adjuvant local ocular treatment (Berenbom et al. 2007; Stefanovic et al. 2010). Regarding cases of lymphoma with solely ophthalmic manifestations, combined systemic and radiotherapy regimens have been shown to prolong survival rates from the point of ocular symptoms presentation (Hormigo et al. 2004). The mere use of local therapy has no impact on the overall prognosis (Grimm et al. 2008). Palliative local treatment is a preferable alternative to prophylactic systemic therapy (Grimm et al. 2008).

Being highly alert and suspicious of lymphoma cases is a challenge every ophthalmologist should be aware of. The collection of ample samples as well as cooperating with a pathologist to verify the diagnosis and a hemato-oncologist to take part in the therapy selection process are equally important. Treatment options should deliberate over both ocular and non-ocular involvement (Figs. 1, 2, 3, 4 and 5).

2 Pathogenesis of VRL

The currently available data on the subject of the pathogenetic mechanisms of VRL suggest that malignant lymphocytes spread to the brain and the eye through systemic circulation (Coupland and Damato 2008). Supporting evidence of the

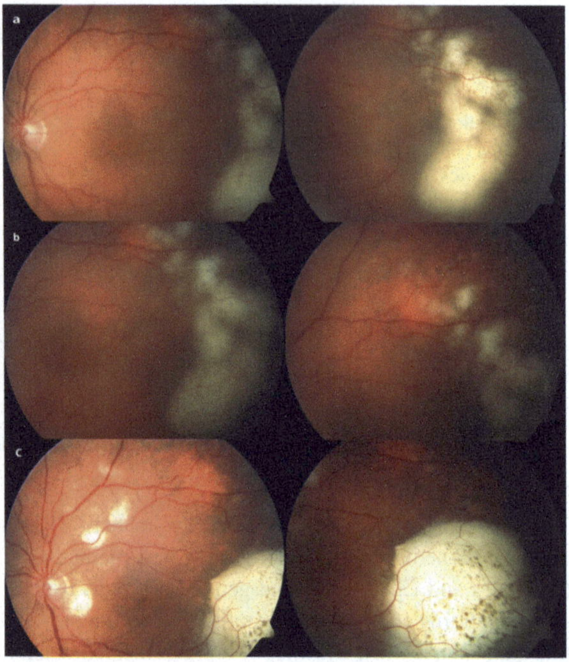

Fig. 2 (**a**) First presentation with whitish lesions in the fundoscopy in a patient 57-year-old with VRL (**b**) Multifocal creamy/white subretinal lesions (**c**) Color images were taken after pars plana vitrectomy (PPV). Brain MRI became positive a couple of months later. Patient had Sx, radiation, eventually died. (Image courtesy of Dr. Androudi)

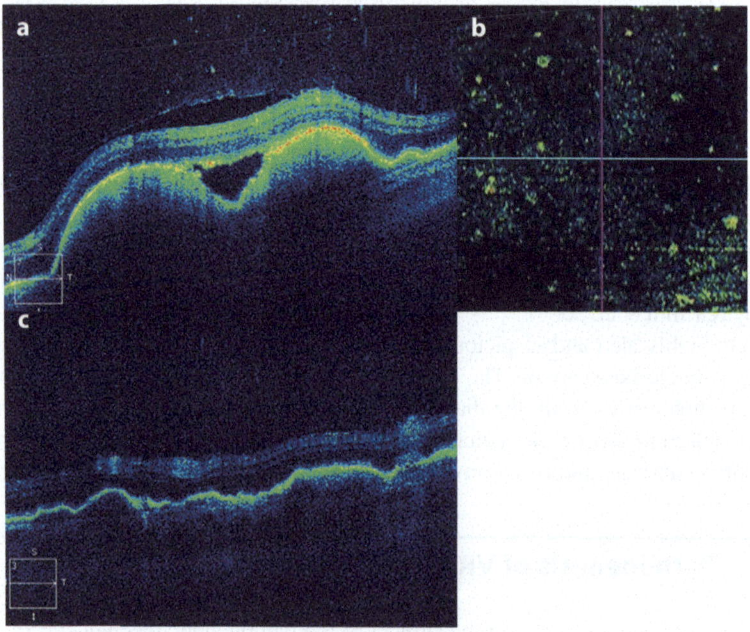

Fig. 3 OCT of the left eye of 57-year-old patient with VRL. (**a**) Subretinal nodules beneath RPE (**b**) Hyperreflective foci in the anterior vitreous (**c**) Spike-like lesions

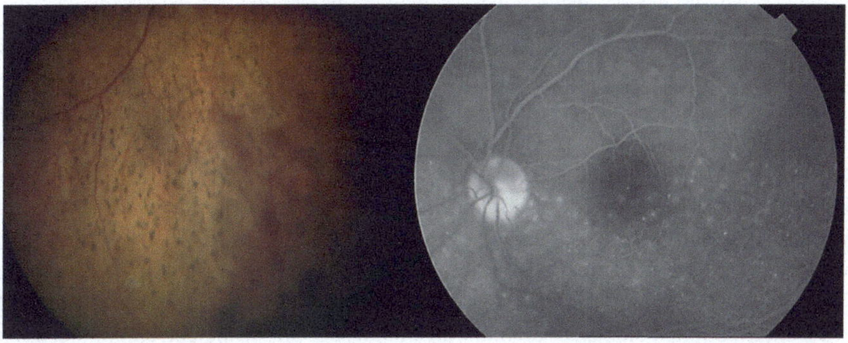

Fig. 4 Fundus image and FA with leopard spot pattern of a 13-year-old patient with secondary intraocular lymphoma (SIOL) from testicular lymphoma. Vitreous biopsy was negative in this case. Due to the patient's history of testicular lymphoma, the patient was given intravitreal methotrexate with dramatic improvement. Pediatric oncologist refused to give systemic treatment in the absence of a positive biopsy for intraocular lymphoma. The patient died a couple of months later due to disseminated disease. (Image courtesy of Dr. Androudi)

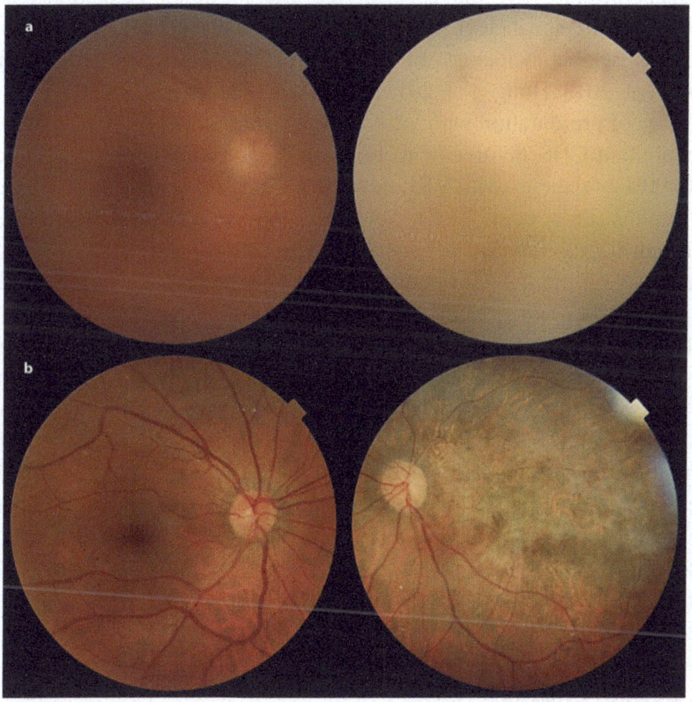

Fig. 5 Fundus images of a 39-year-old man with VRL (**a**) before and (**b**) after 1-year treatment with local methotrexate. (Image courtesy of Dr. Androudi)

aforementioned theory exists, as blood marrow and blood samples of PCNSL patients have shown cells with clonal transpositions similar to the ones in CNS tumors (Grimm et al. 2008). Additionally, poor immune barriers and permissive retinal endothelial receptors (Zamora et al. 2007) could potentially enable malignant cell retinal intrusion, and thus an ensuing ocular clonal proliferation. Vitreous and RPE diaspora is a possibility, but further dispersion is blocked by Bruch's membrane. It seems that choroidal vessels act as possible portals in metastatic systemic lymphoma, contrary to the mechanism involved in VRL, with Bruch's membrane serving once more as a blockade and containing the malignancy in the uveal tract. On sporadic occasions, iris metastases from the uvea have been observed (Coupland and Damato 2008; Velez et al. 2002a, b). Considering the strong correlation between PCNSL and VRL, it is necessary for molecular strategies to be expanded and elucidate the pathogenetic mechanisms.

3 Epidemiology

PVRL is a subtype of PCNSL and represents less than 1% of non-Hodgkin's lymphomas and 4–6% of brain malignancies (Sagoo et al. 2014; Meunier et al. 2004). Although it is a rare disease, the global incidence of PVRL and PCNSL has increased in the last year (Levasseur et al. 2013; Hong et al. 2011; Sjö 2009; Haldorsen et al. 2007).

At the time of presentation with vitreoretinal lymphoma, 16–34% of patients may have concomitant CNS involvement. From the literature, a significant percentage, that is, 50–90% of patients with VRL develop a CNS and or spinal cord disease after 16 to 24 months (Pe'er et al. 2009). According to an extensive retrospective survey by Kimura et al., the interval between ocular disease and CNS involvement was approximately 21.7 months. This study speculates the necessity of early detection of the disease and the pathogenetic correlation between VRL and PCNSL (Kimura et al. 2012). Regarding patients with PCNSL, 25% of them will have VRL simultaneously. The incidence of PCNSL is significantly higher in immunocompetent patients with AIDS or with transplantation history. The typical phenotype of the patient with PCNSL is someone in the fifth or sixth decade of his life and commonly with immunocompromised status (Brown et al. 1994; Hoffman et al. 2003). However, there are reports of VRL in younger patients, particularly with compromised immune status (Sobrin et al. 2005). There is no sex predilection, albeit women may be affected more commonly according to the literature (Sagoo et al. 2014; Cassoux et al. 2000; Fend et al. 2016; Llorenç et al. 2019).

4 Ocular Characteristics

Primary intra-ocular lymphoma (PIOL) is the most common neoplastic masquerade syndrome. The initial presentation of the disease looks like posterior uveitis. According to recently published consensus recommendations for the diagnosis of VRL, the diagnostic pillars are clinical history, anterior segment findings, vitreous and posterior segment findings, behavior treated with systemic corticosteroids, ocular fluid analysis, and ocular imaging (Carbonell et al. 2021). The main clinical manifestation is painless loss of vision with floaters in a patient 50 years or older, usually involving one eye and subsequently the fellow eye. Patients will have bilateral disease in more than 80% (Freeman et al. 1987).

Regarding anterior segment examination, findings may be unremarkable. Keratic precipitates in the corneal endothelium may be present in a minority of cases. Slit-lamp examination may reveal mild reaction with flare and cells. As far as posterior segment findings are concerned, the hallmark is vitritis with clumps or sheets of vitreous and large B-lymphoma cells (Chan and Gonzalez 2007).

Vitreous haze in different patterns is present in 92% of eyes with VRL and in 15% was the only sign of the disease in a study of 26 eyes of 13 patients with VRL. The "aurora-borealis" pattern was described in 46%, a non-specific pattern in 38%, and a "string of pearls" pattern was noticed in 23% of patients (Marchese et al. 2019).

Retinal lesions consist of multifocal white patches without clear borders, involving the outer retina in 50% at the onset of ocular disease (Kimura et al. 2012; Fardeau et al. 2009). The lymphoma lesions may be in clusters and RPE infiltration is possible. Then, a leopard spot appearance may occur (Chan et al. 2011). The lesions may increase in size or resolve automatically with subsequent RPE atrophy or fibrosis (Goto et al. 2006; Mantopoulos and Cebulla 2015; Kase et al. 2012). The noticeable difference between VRL and chronic posterior uveitis is a "quiet" eye with dense vitritis and a decrease in visual acuity not compatible with the grade of vitritis. Visual acuity is better than expected. In uveitis cases, the inflammation leads to an increase of cytokines like IL-6 (interleukins) and accumulation of inflammatory cells in the vitreous (Valentincic et al. 2011).

However, in cases of VRL, the vitreous behaves as a scaffold for the neoplastic cells. The more specific proliferation of lymphoma cells occurs in situ in the vitreous and it is not the result of the immune process. In contrast to uveitis cases, in VRL the predominant cytokines like IL-10 are anti-inflammatory and induce the growth of lymphocytes. Therefore, VRL concerns typically a "quiet" eye without redness and pain (Davis 2013). Moreover, vitreous infiltration in VRL cases is massive, without macular edema because the inflammation is not the generative cause of vitritis. Retinal hemorrhages, vasculitis, detachment, and necrosis are unlike to occur. Posterior synechiae or cystoid macula edema advocates for a different diagnosis than VRL (Gonzales and Chan 2007; Faia and 2009). VRL is an imitator of uveitis and when it is treated with systemic corticosteroids in the initial phase it seems to respond to treatment. This contributes to further delay of lymphoma diagnosis and the time until the right diagnosis is over (12 months) (Chan et al. 2011; Grimm et al. 2008).

5 Imaging (Optical Coherence Tomography—Fundus Autofluorescence—Fluorescein Angiography)

Multimodal imaging is the mainstay of diagnosis in VRL with a wide range of findings from none to extensive.

Spectral-domain optical coherence tomography (SD-OCT) offers an optical biopsy of the retina and thus is useful for specifying the precise location of accumulated lymphoma cells. OCT helps the ophthalmologist in diagnosing VRL and staging the disease (Lee and Goldstein 2020; Barry et al. 2018; Liu et al. 2012). OCT findings highlight infiltrations in the inner retina, deposits in the outer retina, sub-RPE material affecting the complex of RPE-Bruch's membrane (Barry et al. 2018; Liu et al. 2012), loss of ellipsoid zone, spike-like lesions in the inner retina (Fonkem et al., 2014), and vertical hyperreflective areas (Deák et al. 2019). OCT may be useful to detect early disease in the fellow eye without obvious clinical signs via evaluation of the vitreous. In some cases, OCT may indicate hyperreflective material in the vitreous. In VRL, the neoplastic material affects the RPE-Bruch's membrane complex in a superficial layer, in contrast to choroidal lymphoma cases where the affected region is deep (Shields et al. 2015).

Fundus Autofluorescence (FAF) pattern speculates the activity of the disease (Casady et al. 2014). More specifically, subretinal lesions beneath RPE tend to be hyperautofluorescent due to dysfunction of RPE and subsequent accumulation of lipofuscin. Retinal lesions above the RPE are hypoautofluorescent due to the blockage of endogenous fluorophores (Ishida et al. 2010). Dark areas may reflect either the deposits of lymphoma cells or the degenerative RPE. As a consequence, the character of autofluorescence is depending on the position of the lymphoma lesions. The hyperfluorescent profile of the spots tends to change to hypofluorescent after treatment with intravitreal methotrexate (Lee and Goldstein 2020). Nowadays, FAF imaging is a cornerstone in evaluating the response to treatment and detecting early relapsing disease.

Fluorescein angiography (FA) highlights focal lesions in the outer retina with early and late hypofluorescence (Fardeau et al. 2009). The hallmark common pattern is the "leopard spot" with multifocal round hyperfluorescent spots. These areas remain hypofluorescent in the late phases of FA and correlate to deep, creamy lesions in the fundoscopy. The "leopard spot" pattern is due to accumulated material between RPE and Bruch's membrane (Fardeau et al. 2009). In VRL, cystoid macular edema with petaloid leakage in FA is not expected, in contrast to the uveitis cases where the inflammation is the generative etiology. However, leakage may be noticed along the vessels and is more common along the veins and the vasculitis concerns the periphery (Cassoux et al. 2000; Fardeau et al. 2009; Velez et al. 2002a, b).

Thus, ultra-wide FA plays an important role in underlying peripheral vasculitis. According to current literature, with the addition of ultra-wide FA in the quiver of multimodal imaging, peripheral vasculitis is detected in more than 77% of VRL patients (Fardeau et al. 2009; Velez et al. 2002a, b).

Indocyanine angiography (ICGA) is preferred in uveal involvement cases with lesions affecting mostly the choroid. Therefore, the diagnostic value of ICGA is limited in the VRL, where the point of interest is the retina. Still, it is possible to show a few areas of hypofluorescence but they will definitely be less than FA imaging (Fardeau et al. 2009; Lavine et al. 2019).

Provided that the majority of patients with VRL will have concomitant involvement of CNS, neuroimaging is vital as well as detecting early symptoms. There is a

3-month delay between symptoms and diagnosis of CNS disease (Grimm et al. 2008). Misdiagnosed PCNSL drives worsened prognosis and overall survival.

6 Differential Diagnosis

Intraocular lymphoma may masquerade as a non-specific chronic intermediate or posterior uveitis. Infectious differential diagnosis includes chronic endophthalmitis, acute retinal necrosis, cytomegalovirus retinitis, ocular toxoplasmosis, syphilis retinitis, and pneumocystis choroiditis. Other causes of non-infectious uveitis are Birdshot choroidopathy, serpiginous choroiditis, Vogt-Koyanagi-Harada syndrome, acute posterior multifocal placoid pigment epitheliopathy, sarcoidosis, sympathetic ophthalmia, Whipple's disease.

Neoplastic diseases such as amelanotic melanoma, choroidal lymphoma, metastatic malignancy, and uveal lymphoma must also be excluded.

The cornerstone to the differential diagnosis algorithm in VRL is better visual acuity in contrast with the severity of clinical findings, absence of posterior synechiae and cystoid macula edema, lack of disc swelling and epiretinal membrane (Davis 2004), presence of migrating retinal lesions, especially in the absence of vitritis and spontaneous resolving of large subretinal lesions due to the function of CD8+ T cells and natural killer cells (Tan et al. 2018; Rajagopal and Harbour 2011; Peterson et al. 1993; Raparia et al. 2009; Sen et al. 2009).

7 Analysis of Intraocular Samples

7.1 Collections of Intraocular Samples

When there is clinical suspicion of VRL, collection and analysis of intraocular samples are important steps toward diagnosis confirmation and tumor characteristics identification. Different sites can be used to access ocular tissue based on tumor location including vitreous, retina, and anterior chamber (Parver and Font 1979; Lobo and Lightman 2003; Palexas et al. 1995). Vitreous samples are commonly used during initial diagnostic investigation and they can be collected through fine needle aspiration and pars plana vitrectomy techniques (Char et al. 2005). With the fine needle aspiration method, a needle of 21- to 25-gauge is inserted into the vitreous chamber through the pars plana and the vitreous is aspirated into the attached syringe (Caya et al. 1985). Although sample size can be limited, fine needle aspiration is a safe procedure that can be quickly performed in the outpatient setting with the additional benefit of limited tissue disruption. A 40% success rate in initial diagnosis has been published with this technique (Lobo and Lightman 2003). In addition, a small retrospective study of patients with a possible diagnosis of VRL demonstrated successful diagnostic confirmation solely by fine needle aspiration of the vitreous (Lobo and Lightman 2003; Choi et al. 2006).

The collection of vitreous samples with pars plana vitrectomy offers several advantages over fine needle aspiration. This surgical technique optimizes sample size and removes vitreous opacities and thus improves vision in symptomatic patients (Gonzales and Chan 2007; Sagoo et al. 2014; Coupland et al. 2004). To obtain the samples, a complete standard three-port pars plana vitrectomy is used (Sen et al. 2009; Choi et al. 2006). A 3-cc syringe is incorporated into the vitrector aspiration line to draw the sample (Sen et al. 2009; Bardenstein 1998). First, 1–2 cc of undiluted vitreous is aspirated with aspiration-only mode to prevent mechanical tissue disruption from the vitrector (Sen et al. 2009). The sample is then added to a cell culture medium for a total of 3–5 cc and immediately sent to the ocular pathology laboratory for further histopathological and molecular analysis (Sen et al. 2009; Whitcup et al. 1993). From a dilute sample collected with infusion both specimens, along with the vitreous wash in the cassette, are transported for microbiologic cultures (Sen et al. 2009; Dworkin et al. 2002; Gupta et al. 2007).

Chorioretinal biopsy represents an alternative option to collect intraocular samples, especially when subretinal lesions are present or if vitreous cells are detected in low counts. Several approaches have been described to obtain appropriate samples, including subretinal aspiration where a Chiba biopsy needle and a bent 27-gauge needle are used to create a retinotomy and facilitate tissue aspiration from the deeper layers of a lesion while also preserving anatomic boundaries of the tumor (Levy-Clarke et al. 2001; Pavan et al. 1995). Furthermore, the vitreous cutter can also be used to directly remove retinal segments with low cut rates, while a more invasive transscleral technique achieves excision of retinal and sub-RPE infiltrates through a partial thickness scleral flap (Choi et al. 2006; Kirmani et al. 1987; Johnston et al. 2004).

Anterior chamber paracentesis can also be useful to obtain aqueous humor samples for cytopathological and genomic analysis (Merle-Béral et al. 2004). Although this method is minimally invasive and easily performed in the outpatient setting, the associated smaller sample size and limited tumor cell counts decrease overall diagnostic efficiency. Most commonly, anterior chamber paracentesis is performed through a 27-or 30-gauge needle on a tuberculin syringe with specific attention paid to avoid iris or lens trauma. To decrease potential complications associated with the traditional approach, a study group in Japan has described a technique to safely aspirate 300 μL of aqueous humor through limbal paracentesis by using a 30-gauge needle that is attached to a specially designed disposable pipette (Kitazawa et al. 2017; Takase et al. 2022).

Despite the optimization of the abovementioned methods, it is anticipated that multiple samples might be needed for a definite diagnosis of VRL in some patients, which can further delay appropriate treatment (Takase et al. 2022). In addition, lymphoma cells are prone to rapid degradation, within 60 min of collection, and for this reason, gentle handling and rapid transportation of the specimens to the laboratory are needed (Merle-Béral et al. 2004; Whitcup et al. 2000). Corticosteroid treatment has also been shown to decrease the diagnostic yield of vitrectomy through cytolytic mechanisms and these medications should be discontinued, where possible, 2 weeks before the diagnostic workup (Whitcup et al. 2000). Finally, if the visual prognosis for the affected eye is poor or there are contraindications to treatment, then diagnostic enucleation might be more appropriate (Trudeau et al. 1988).

7.2 Cytology

Histopathological examination of intraocular tissues is essential in VRL diagnosis. As discussed previously, rapid preparation and gentle handling of specimens are very important steps to decrease false-negative results. Following collection, diluted and undiluted samples are centrifuged in order to concentrate cells before further processing (Davis 2004). Several stains are available to highlight cellular morphology including hematoxylin-eosin and Papanicolaou stains, however, Diff Quick staining and Giemsa are considered better for malignant B-lymphocytes identification (Gonzales and Chan 2007; Ljung et al. 1988). Light microscopy of vitreous samples reveals populations of large neoplastic lymphocytes interspersed with inflammatory cells and necrotic debris (Sagoo et al. 2014; Whitcup et al. 1993; Zaldivar et al. 2004; Akpek et al. 1999a, b). Malignant cells of B-cell origin demonstrate pleomorphism with scant basophilic cytoplasm, increased nucleus: cytoplasm ratio, round or oval hyperchromatic nuclei with irregular margins, coarse chromatin, conspicuous nucleoli, and poor intercellular cohesiveness (Sagoo et al. 2014; Ljung et al. 1988; Chan and Wallace 2004; Davis et al. 2005; Karma et al. 2007; Tuaillon and Chan 2001; Chan et al. 2002; Char et al. 1988; Farkas et al. 2004). Inflammatory infiltrates contain an aggregate of fibrin, macrophages, neutrophils, and small lymphocytes that can confound the differential diagnosis of neoplastic processes from reactive inflammation given the decreased cellularity of specimens (Whitcup et al. 1993). Moreover, lymphocytes undergoing cytolysis are also common because of the increased fragility of the neoplastic cells. Cytologic examination of chorioretinal tissue shows medium to large atypical lymphocytes with increased mitotic features, basophilic cytoplasm, oval-shaped nuclei and prominent nucleoli, while specimens from enucleated eyes reveal malignant infiltration in the subretinal space, between the retinal pigment epithelium and Bruch's membrane and within the neurosensory retina (Mastropasqua et al. 2015; Coupland et al. 2003). Atypical lymphoid cells are usually located in the sub-RPE space but vitreous, retina, and optic nerve are also involved (Sen et al. 2009; Coupland et al. 2003).

According to a study, the positive predictive value of cytologic examination for the diagnosis of VRL was 99–100% while the negative predictive value was 61–81%, but a large Japanese study of 217 patients with VRL reported that only 44.5% of patients were classified as positive (either strongly suggestive or conclusive for malignancy) based on cytologic findings (Coupland et al. 2002; Kimura et al. 2012). The most common morphologic characteristics noted, based on a publication reviewing 13 vitreous specimens positive or highly suspicious for malignant lymphoma, were the large cellular size (2–4 times larger diameter than a mature lymphocyte) and the prominent nucleoli (Sagoo et al. 2014).

7.3 Immunocytology and Flow Cytometry

Although cytologic examination remains the gold standard for diagnosing, additional techniques are necessary to increase diagnostic efficiency. Immunocytology and flow cytometry are powerful tools that can augment phenotyping in lymphomas by identifying specific antigens and measuring specific cellular characteristics. Monoclonal antibodies linked to an enzyme, or a fluorescent substance, are used in both methods to target specific markers of interest. Following antigen-antibody interaction, the labeling agent is activated allowing visualization of cellular markers either through microscopic examination of slide fixated specimens for immunocytological testing or via laser-mediated single-cell processing of suspensions passing through flow cytometer machines (Davis et al. 1992; Sagoo et al. 2014).

In most cases, PIOL is of monoclonal B-cell origin and for this reason, B-cell specific markers like CD19, CD20, CD22, CD79a, and PAX5 are commonly expressed (Davis et al. 1992; Davis 2013; Faia and Chan 2009; Davis et al. 1997; Intzedy et al. 2008). BCL6 and CD10 are additional markers that are specific to germinal centers of B-cell lymphomas (Rajagopal and Harbour 2011; Wilson et al. 1992; Sugita et al. 2009). Moreover, kappa (κ) and lambda (λ) light-chain immunoglobulins can also be used to identify neoplastic B-cell populations (Davis 2013; Lobo and Lightman 2003; Davis et al. 1997; Intzedy et al. 2008). Normally, equal amounts of both immunoglobulins are determined for heterogeneous B lymphocytes. On the contrary, monoclonal B-cell populations overexpress one light chain over the other and a κ:λ ratio of 3 or 0.6 is highly suggestive of lymphoma (Sagoo et al. 2014; Davis et al. 1997; Wilson et al. 1992). Primary intraocular T-cell lymphomas represent a minority of cases and useful markers include CD3, CD4, and CD8 9 (Coupland et al. 2005a, b, c; Goldey et al. 1989). However, immunocytological markers for T-cell lymphomas are limited and might also be expressed in inflammatory conditions complicating differential diagnosis (Hoffman et al. 2003; White et al. 1999).

When compared to immunocytology, flow cytometry has the advantage of simultaneous testing for different monoclonal antibodies in the same specimen and thus maximizing detection capacity over single antibody use in microscopic slides (Sagoo et al. 2014). Intraocular samples though, especially from vitreous, contain limited numbers of disease-specific cells and this can negatively affect the sensitivity of flow cytometry. In addition, there are cases of B-cell lymphoma that are not expressing surface immunoglobulin markers which can further decrease cellular recognition (Sagoo et al. 2014; Takase et al. 2022).

7.4 Immunoglobulin Heavy Chain Gene Rearrangement

Molecular techniques, like polymerase chain reaction (PCR), have also been successfully used as adjunctive tools to characterize intraocular lymphomas (Takase et al. 2022; Katai et al. 1997).

Monoclonality of B lymphocytes is strongly suggestive of lymphomas and PCR can demonstrate immunoglobulin heavy chain (IgH) gene rearrangements (Shen et al. 1998; Chan 2003). Several genetic loci are implicated but the most common sequence reported is at the third complementarity-determining region (CDR3) which is part of the variable region of the IgH (Sagoo et al. 2014; Choi et al. 2006; Chan 2003). Chan et al. reported 100% monoclonality at this specific site for 60 cases of PIOL that were tested. Additional genetic locations supporting the clonal expansion of B cells are FR2 and FR3 loci, while TCR gene rearrangements in the TCR γ region can be detected in T-cell lymphomas (Sen et al. 2009; Gorochov et al. 2003). PCR is also useful to identify chromosomal translocation, which is responsible for gene rearrangement and resulting overexpression of the IgH promoter region on chromosome 14 with the proto-oncogene gene BCL-2 on chromosome 18 (Wallace et al. 2006).

This type of molecular analysis can be applied to vitreous samples, paraffin-embedded fixated tissue biopsies, subretinal infiltrates as well as cerebral tissue (Coupland et al. 2005a, b, c). Through IgH-PCR testing the clonality of the B lymphocytes that were detected in vitreous aspirates and cerebral specimens from a patient with VRL/PCNSL was associated with the same progenitor cell, supporting the hypothesis of a single disease entity affecting two different systems (Coupland et al. 2005a, b, c; Higashide et al. 2004).

Although sensitivity and specificity of the method are determined to be 46–96% and 85–100% respectively, there were cases where PCR failed to demonstrate gene rearrangements possibly due to somatic mutations within the immunoglobulin gene (Kimura et al. 2012; Sugita et al. 2009; Baehring et al. 2005; Tanaka et al. 2021; Abu Samra et al. 2018). For this reason, identification of better suited primers and PCR optimization are needed to improve detection rates (van Dongen et al. 2003). Moreover, findings interpretation should also take into consideration results from other diagnostic methods in the context of a multidisciplinary approach rather than a single gold standard test.

7.5 Cytoline Analysis

Detection of several interleukins (IL) can be helpful in the diagnostic workup of lymphomas, especially to differentiate VRL from other conditions associated with uveitis. Both, aqueous and vitreous samples have been used in different studies to augment PIOL diagnosis and monitor treatment response (Cassoux et al. 2007; Merle-Béral et al. 2004). IL-10 is believed to be produced by malignant B lymphocytes to assist with cellular growth, differentiation, and suppression of cell-mediated immunity. (Davis 2004) Several studies have reported increased concentrations of IL-10 in cases of VRL, whereas IL-6 and IL-12 were associated more commonly with inflammatory conditions (Chan 2003; Benjamin et al. 1994; Wolf et al. 2003; Chan et al. 1995). There is also evidence that detection of IL-10 in undiluted samples of aqueous or vitreous greater than 150 pg/mL and greater than 50 pg/mL in diluted specimens is diagnostic of B-lymphocytic PIOL (Cassoux et al.

2007). The increased IL-10:IL-6 ratio has also been used as an index of PIOL, with some groups supporting sensitivity of 74–88% and specificity of 75–85% compared with uveitis cases (Chan et al. 1995; Whitcup et al. 1997). Moreover, a sensitivity of 0.89 and specificity of 0.93 were estimated for raised IL-10 levels in the aqueous, supporting its use as a potential screening marker (Cassoux et al. 2007). However, there is no consensus regarding the applicability of IL-10:IL-6 or IL-10:IL-12 ratios since increases in both have been noted in patients with uveitis that was not related to neoplastic processes (Akpek et al. 1999a, b). Furthermore, it is not documented if a similar increase in IL-10 levels is also associated with T-cell PIOL.

Additional cytokines and chemokines that were found to be elevated in the vitreous of PIOL cases include IL-8, Fas ligand, IP-10, MCP-1, Granzyme A and BCA-1 (Usui et al. 2012). Specifically, BCA-1 or B-cell attracting chemokine (BLC) is found in the lymph nodes, spleen, and Peyer's patches and it is selective for B-lymphocytes homing (Choi et al. 2006). The receptor for BCA-1 is CXCR5 (BLR-1) and it is also considered a B-cell homing chemokine (Choi et al. 2006). CXCR5 was identified in subretinal PIOL infiltrates while BCA-1 was expressed only in RPE cells (Choi et al. 2006). Takeda et al. suggested that soluble vascular endothelial growth factor receptors sVEGFR1 and sVEGFR2 could facilitate differential diagnosis of systemic metastatic retinal lymphoma, PCNSL and PIOL (Takeda et al. 2015). In terms of prognosis, levels of sIL-2Rα were associated with subretinal infiltration by PIOL cells while increased intravitreal IL-35 levels at the time of diagnosis indicated a poorer prognosis (Takeda et al. 2015; Balamurugan et al. 2020).

7.6 Genetics

Understanding genetics of cancer is vital to better appreciate cell origin, pathogenetic mechanisms, prognosis, and even response to treatment. Analysis of gene expression and genetic mutation profiling has been studied in intraocular lymphomas but evidence is generally limited due to the decreased numbers of tumor cells available in these cases.

Besides the IgH gene rearrangements that have been discussed previously, MYD88 L265P mutation is reported to be associated with VRL (Takase et al. 2022; Narasimhan et al. 2020; Bonzheim et al. 2015). Chromosome 3p22.2 is the genetic locus for myeloid differentiation primary response 88 (MYD88) gene which encodes for MYD88 protein (Narasimhan et al. 2020). This protein mediates intracellular signaling via Toll-like receptors resulting in increased activation of NF-κB pathway that promotes tumor growth (Wang et al. 2014). The causative mutation is a single base nucleotide substitution (guanine instead of adenine) at position 265 with the final amino acid sequence containing proline in place of lysine (Wang et al. 2014). B cells are continuously activated as a result and this pathway has been also described as contributing pathogenetic mechanism in several proliferative conditions, including lymphoplasmacytic lymphomas, Waldenstrom macroglobulinemia, chronic lymphocytic leukemia, diffuse large B-cell lymphoma and primary central nervous system (CNS) lymphoma (Treon et al. 2012; Swerdlow et al. 2016a, b; Fernández-Rodríguez et al. 2014; Puente et al. 2011). In PIOL, independent studies estimated increased prevalence of MYD88 L265P mutation (69–86%) (Bonzheim et al. 2015; Raja et al. 2016; Pulido et al. 2015).

CD79BY196 is another mutation that is commonly described in relevance to CNS lymphomas and was also detected in 35% of PIOL cases (Takase et al. 2022; Swerdlow et al. 2016a, b). The protein product of CD79B gene is the Ig-beta protein of the B-cell antigen receptor (BCR) and similarly to MYD88L265P, CD79BY196 is an upregulating mutation causing uncontrolled BCR activation and subsequent NF-κB production in the absence of appropriate antigen stimulation (Swerdlow et al. 2016a, b; Yonese et al. 2019).

BCL6 gene is found to be important for T-cell specific immunity as well as the normal development of germinal centers related to B-cell maturation. BCL6 mutations could lead to unregulated growth of B lymphocytes and thus promote neoplastic growth (Schneider et al. 2011). Moreover, analysis of diffuse large B-cell lymphomas has pointed out its role as a proto-oncogene, since translocations affecting this gene were found in this condition (Chen et al. 1998). Further studies correlated BCL6 overexpression with prognosis and in PIOL BCL6 positivity was related to poorer outcomes (Phan and Dalla-Favera 2004).

A novel approach in cancer diagnosis that is also promising for patient allocation to treatment and follow-up monitoring is cell-free DNA analysis (cfDNA). cfDNA is produced by cells and can be isolated in blood or other bodily fluids (Lv and Liu 2021). Neoplastic cells also secrete analogous DNA which is called circulating tumor DNA (ctDNA) and originates from the degradation of dead cells (Lv and Liu 2021). This DNA fraction is small in length, usually less than 200 nucleotide bases but it is considered a tumor-specific biomarker and a potential substrate for mutation analysis (Lv and Liu 2021). There is emerging evidence that ctDNA could be useful in the investigation of lymphomas and this concept of liquid biopsy was used to study non-Hodgkin lymphoma (NHL) (Lv and Liu 2021). Similarly, there is pre-published evidence of a research group in China that used cfDNA analysis to characterize mutations in patients with PIOL and uveitis. Their findings suggested a sensitivity of 100% in cfDNA mutation analysis of aqueous humor and vitreous for patients with PIOL (Chen et al. 2021). Higher frequencies of mutant variants were noted for MYD88, CD79B, PIM1, ETV6, and IRF4 genes (Chen et al. 2021).

8 Treatment

The treatment of PCNSL is substantially different from that of systemic non-Hodgkin lymphoma. Treatments that have been successful in extra-CNS non-Hodgkin lymphomas were ineffective for PCNSL, as the majority of chemotherapeutic agents could not cross the blood-brain barrier (Schultz et al. 1996; O'Neill et al. 1999; Mead et al. 2000). In contrast to other brain tumors, the role of resection in the PCNSL treatment is only for diagnostic purposes (Hunt et al. 2006). Initially, until the early 1990s, whole-brain radiation treatment was the gold standard of PCNSL. However, relapse of the disease will invariably occur, leading to a median overall survival of only 12 months (Nelson et al. 1992).

Due to the infrequency of intraocular lymphomas, defining a standard and optimal treatment is challenging and many aspects of their management remain controversial (Kalogeropoulos et al. 2019). Therapeutic modalities for intraocular lymphoma include intravitreal chemotherapy, systemic chemotherapy, and radiotherapy, which can be used alone or in the context of a combined therapeutic strategy. Choosing the appropriate treatment depends on the stage of the disease, the presence or absence of CNS involvement, and the patient's performance status (Kim et al. 2006). In general, the current trend for the treatment of intraocular lymphoma without CNS or systemic involvement is local treatment (i.e., intraocular methotrexate and/or ocular radiation) to minimize systemic toxicities (Grimm et al. 2008).

A multidisciplinary assessment of patients with VRL and/or PCNSL is mandatory (Hoang-Xuan et al. 2015; Nguyen et al. 2016). In 2011, the significance of a collaboration between ophthalmologists, pathologists, and oncologists (either neuro-oncologists or hemato-oncologists) was underlined leading to the following guidelines (Chan et al. 2011):

1. Without CNS or systemic involvement: topical treatment (i.e., intravitreal injection of methotrexate or rituximab). Alternatively, low-dose stereotactic external beam radiotherapy (30–35 Gy) to the involved eye and follow-up. Intravitreal treatment combined with systemic chemotherapy has been proposed for cases with bilateral involvement.
2. With CNS involvement: high dose of systemic methotrexate (+/− systemic rituximab) combined with topical treatment. If the patient does not respond to treatment and is not fit for a more aggressive treatment (e.g., autologous stem cell transplantation), ocular and brain radiotherapy should be considered.

On the contrary, the Neuro-Oncology Society (National Cancer Action Team Rare Tumor Guidelines, June 2011) proposed that high-dose systemic methotrexate followed by integral-globe radiotherapy for ocular only or concurrent individuals and intravitreal methotrexate is effective in those with isolated ocular recurrences (Hoang-Xuan et al. 2015; Nguyen et al. 2016).

Radiation

Ocular radiotherapy had been widely used for the treatment of PIOL, as most lymphomas are extremely radiosensitive (Margolis et al. 1980). The main targets of ocular irradiation with prophylactic CNS treatment include control of intraocular lymphoma, maintenance of vision, and prevention of CNS involvement (Margolis et al. 1980). As mentioned above, the low-dose stereotactic external beam radiotherapy constitutes the recommended topical PVRL treatment in cases with ocular involvement only (Chan et al. 2011). Berenbom et al. (2007) evaluated a total of 21 eyes of 12 patients with PIOL and found that radiotherapy is highly effective with acceptable complications in these patients. Taking into consideration the risks of intravitreal chemotherapy, especially after repetitive injections, radiotherapy is probably the most appropriate first-line therapeutic approach in the majority of cases. Topical brain radiotherapy is usually the first-line treatment for primary CNS lymphoma. If vitreoretinal lymphoma presents after the CNS lymphoma treatment, whole eye radiotherapy may be considered (Sagoo et al. 2014). Integral brain and ocular radiotherapy are recommended in patients with concurrent CNS involvement that did not respond to systemic chemotherapy, and are not fit for more aggressive therapeutic approaches (Berenbom et al. 2007). Complications of high-dose low fractionation CNS radiotherapy include neurotoxicity (DeAngelis 2001) cataract, radiation retinopathy, optic neuropathy, glaucoma, corneal defects, and dry eye (Berenbom et al. 2007). Currently, it remains controversial whether intravitreal medications or eye radiation should be considered as a first-line treatment (Kalogeropoulos et al. 2019; Berenbom et al. 2007).

Systemic Chemotherapy

The most paramount benefit of systemic chemotherapy in VRL is the concurrent treatment of brain involvement, which must be suspected in all PCNSL patients (Jahnke et al. 2007). The efficacy of systemic chemotherapy for the treatment of VRL relies on complex intraocular pharmacokinetics that has not been yet completely clarified. The blood-aqueous barrier and the blood-retinal barrier constitute obstacles to the drug diffusion from the blood into ocular tissues. The retinal pigment epithelium (RPE) and the retinal vessels are two distinct sites of the blood-retinal barrier representing the intraocular counterpart of the blood-brain barrier ((Raviola 1977; Cunha-Vaz 1979). The blood-aqueous barrier is technically a hindrance among the blood and ocular tissues, as well as fluids posterior to the iris plane (Freddo 2001). As can be inferred, the drug penetration through these barriers depends on various drug features, such as lipid solubility. The most significant advantage of systemic chemotherapy in patients with VRL is the concurrent treatment of brain involvement that must be suspected in all individuals with PCNSL (Jahnke et al. 2006). Methotrexate, cytarabine, isosfamide, and trofosfamide have been successfully employed to treat VRL in oral or intravenous formulations, as they can bypass the blood-ocular barriers and accumulate in cytostatic levels (Strauchen et al. 1989; de Smet et al. 1996; Jahnke et al. 2006). The majority of studies in the current literature have assessed the role of high-dose methotrexate for VRL.

High-dose methotrexate can achieve a response rate of up to 72% when used alone and up to 94–100% in combinations and is therefore considered the most active drug (Bessell et al. 2007; Sandor et al. 1998; Tempescul et al. 2011). Batchelor et al. (2003) showed that a subset of patients with intraocular lymphoma (IOL) may achieve sustained remission when treated with high-dose intravenous methotrexate alone. Intravenous high-dose cytarabine has also been administered with variable success rates (Strauchen et al. 1989; Baumann et al. 1986; Peterson et al. 1993). Blood-brain barrier disruption which is used in treating PCNSL, as it increases brain penetration of chemotherapy, may cause non-sight threatening maculopathy in almost two-thirds of patients (Vicuna-Kojchen et al. 2008). Soussain et al. (1996, 2008) conducted a large multicenter retrospective analysis to evaluate the efficacy of intensive chemotherapy followed by hematopoietic stem cell rescue in patients with PCNSL or intraocular lymphoma after the failure of treatment with high-dose methotrexate. Their results showed that intensive chemotherapy with thiotepa, busulfan, and cyclophosphamide, followed by hematopoietic stem cell rescue is promising in chemosensitive patients up to 65 years of age with refractory and recurrent PCNSL (Soussain et al. 1996, 2008). A similar study by Abrey et al. (2003) evaluated the administration of intensive chemotherapy with methotrexate followed by high-dose chemotherapy with autologous stem-cell transplantation in 14 patients with newly diagnosed PCNSL, of whom 2 had ocular disease. One patient responded completely, while the second achieved a temporal complete remission after chemotherapy, but had a relapse before the stem cell rescue. The combination of intrathecal methotrexate and cytarabine has been used for recurrent VRL with PCNSL (Mason and Fischer 2003). Finally, it has been found that the administration of oral trofosphamide or intravenous iraphosphamide may provide 6–18 months of progression-free survival (Jahnke et al. 2004; Jahnke et al. 2009; Janhke et al. 2005).

Chemoradiation

Chemoradiation or chemoradiotherapy refers to the combined administration of both chemotherapy and radiotherapy as an anticancer treatment (Rallis et al. 2021). Particularly, chemotherapy after prior radiotherapy of the eyes and brain in patients with PIOL was mainly used as a salvage treatment, but may also be considered as a primary therapeutic approach (Plowman et al. 1993; Plowman et al. 2002; Woolf et al. 2012). Ocular and CNS radiation can be combined with intrathecal (Valluri et al. 1995; Char et al. 1988) or systemic chemotherapy. Isobe et al. (2006) investigated the clinical manifestations and treatment outcome of 15 PIOL patients. All patients were treated with radiation therapy, and 10 of them also received chemotherapy. Thirteen patients achieved a complete remission. The two-year disease-free and overall survival rates were 58% and 74%, respectively. Ferreri et al. (2002) reported that chemotherapy, when combined with ocular irradiation, led to more efficient control of ocular disease, which appears to be linked with survival. Although multimodality treatment contributes to better disease control, it has been associated with cognitive neurotoxicity, especially in older patients (Nasir and DeAngelis 2000). Therefore, according to the British Committee for Standards in Haematology, a multimodality approach should be individually adjusted to each patient depending on the disease severity, degree, and patient's performance status (Marcus et al. 2012).

Ocular Chemotherapy

Intravitreal drug administration is not hindered by the ocular barriers providing more efficient intraocular delivery of hydrophilic medications (e.g., methotrexate). At the same time, there is reduced toxicity compared to systemic treatment (Grimm et al. 2008).

- Intravitreal methotrexate

Intravitreal administration of 0.4 mg methotrexate in 0.1 mL has been reported to achieve local tumor control in patients with relapsed PIOL (de Smet et al. 1996; de Smet et al. 1999; Wang et al. 2006) in ocular relapse of PCNSL (Kim et al. 2009), and as a primary treatment combined with intrathecal chemotherapy (Sou et al. 2008), systemic chemotherapy (Fishburne et al. 1997), or sub-Tenon steroid injections (Velez et al. 2002a, b). After intravitreal injection of methotrexate, the therapeutic doses are maintained for over 5 days and de Smet et al. (1996, 1999) reported the results of their 10-year experience of using intravitreal methotrexate as a first-line treatment for vitreoretinal involvement of PCNSL. Clinical remission was achieved after 6.4 (3.4) (2–16) of intravitreal methotrexate injections (mean (SD) (range)), with 95% of eyes needing 13 injections or less to eliminate malignant cells. Intraocular recurrence was recorded in none of the patients. One of the most common side effects was corneal epitheliopathy, typically after the third injection, which subsided as the interval among injections increased. Other complications include cataract, vitreous hemorrhage, maculopathy, and optic nerve atrophy (Smith et al. 2002). Moreover, repeated injections may lead to drug resistance (Sen et al. 2008), whereas intravitreal chemotherapy does not prevent CNS disease or involvement of the fellow eye (Smith et al. 2002).

- Intravitreal rituximab

Based on its efficacy in systemic lymphoma, rituximab (1 mg in 0.1 mL), an anti-CD20 monoclonal antibody, can be administered intravitreally (Itty and Pulido 2009; Kim et al. 2006). Although the initial response appeared to be favorable with PIOL clearance, subsequent relapse had to be treated with radiation and intravitreal methotrexate (Itty and Pulido 2009). Interestingly, it has been found that statins interfere with both detection of CD20 and anti-lymphoma activity of rituximab, leading to delayed diagnosis, postponing effective treatment, or impairing anti-lymphoma activity of rituximab (Hashida et al. 2014).

Emerging Treatments

The use of high-dose chemotherapy and autologous stem cell transplantation in cases with refractory or recurrent VRL or PCNSL has been reported in a restricted number of studies showing favorable outcomes (Illerhaus et al. 2006). However, it has not yet been clarified if this approach should be employed as consolidation in primary VRL treatment. Thalidomide-related agents, such as lenalidomide, have been reported to be well-tolerated with durable responses in relapsed or refractory aggressive non-Hodgkin's lymphoma (Witzig et al. 2011). Similarly, pomalidomide has been shown to be efficient in CNS lymphoma experimental models ((Li et al. 2013). Ibrutinib is a Bruton's tyrosine kinase (BTK) and HCK tyrosine kinase protein inhibitor that has been previously used in cases of chronic lymphocytic leukemia (CLL) (Byrd et al. 2013). Both pathways are associated with the

MYD88 L265P mutation, which is implicated in several vitreoretinal lymphoma cases (Yang et al. 2016). Ibrutinib has been approved for the treatment of Waldenström's macroglobulinemia. Taking into consideration that there is a similar MYD88 mutation in both Waldenström's and VRL could orientate further clinical trials evaluating the efficacy of ibrutinib in vitreoretinal lymphoma.

However, it is crucial to highlight that the concurrent administration of ibrutinib and high-dose steroids has been associated with invasive fungal and pneumocystis infections (Ahn et al. 2016; Okamoto et al. 2016).

Vitrectomy may also have therapeutic potential but there are currently insufficient data to support that this treatment could be employed as a monotherapy, but may be used as an additional approach in eyes with increased cellular load (Iaccheri et al. 2017).

9 Secondary Intraocular Lymphoma

Secondary intraocular lymphoma (SIOL), which derives from systemic lymphomas, is a much less frequent type of intraocular lymphoma (IOL) (Kalogeropoulos et al. 2019).

Therefore, the clinical features, clinical course, and prognosis of SIOL have not been sufficiently investigated.

More specifically, intraocular lymphoma secondary to disseminated, systemic disease is typically restricted to the choroid (Fredrick et al. 1989; Gündüz et al. 1999; Verbraeken et al. 1997). Secondary lymphomatous ocular pathology defined mainly by retinal infiltration without uveal involvement has been only rarely reported (Lewis and Clark 1975; Parikh et al. 2005). Less frequently, systemic lymphoma may manifest with findings from the anterior segment such as iris infiltration or pseudohypopyon (Duker et al. 1987; Shakin et al. 1988; Tranos et al. 2002). Other uncommon features of SIOL include optic disc edema (Pinna et al. 2003) and serous macular detachment (Pilon et al. 2005). The most frequent systemic lymphoma subtype implicating the choroid is diffuse large B-cell lymphoma (DLBCL) (Parikh et al. 2005; Duker et al. 1987; Fredrick et al. 1989; Hattenhauer and Pach 1996), followed by multiple myeloma (Knapp et al. 1987; Orellana and Friedman 1981), extramedullary plasmacytoma (Honavar et al. 2001), lymphoplasmocytic lymphoma/immunocytoma (including Waldenström's macroglobulinemia) (Pinna et al. 2003; Pilon et al. 2005; Orellana and Friedman 1981; Sen et al. 2004), marginal zone B-cell lymphoma (Kalogeropoulos et al. 2018), and B-cell chronic lymphocytic leukemia (Coupland et al. 2001; Jahnke et al. 2005; Kincaid and Green 1983). Quite rarely, secondary intraocular involvement may be associated with orbital Burkitt's lymphoma (Feman et al. 1969; Payne et al. 1971; Wysenbeek et al. 1987) and intravascular lymphoma (Mudhar et al. 2007; Elner et al. 1986).

10 Conclusion

It is crucial not only to diagnose VRL but to treat it. There are gaps between clinical manifestations, accurate diagnosis, and wholesome treatment. In the future, multidisciplinary approach aims to bridging these gaps and find whether the earlier systematic intervention would be necessary for VRL cases with occult CNS disease.

Compliance with Ethical Standards No conflicts of interest.

References

Abrey LE, Moskowitz CH, Mason WP, Crump M, Stewart D, Forsyth P, Paleologos N, Correa DD, Anderson ND, Caron D, Zelenetz A, Nimer SD, DeAngelis LM (2003) Intensive methotrexate and cytarabine followed by high-dose chemotherapy with autologous stem-cell rescue in patients with newly diagnosed primary CNS lymphoma: an intent-to-treat analysis. J Clin Oncol 21(22):4151–4156. https://doi.org/10.1200/JCO.2003.05.024

Abu Samra K, Oray M, Ebrahimiadib N, Lee S, Anesi S, Foster CS (2018) Intraocular lymphoma: descriptive data of 26 patients including Clinico-pathologic features, vitreous findings, and treatment outcomes. Ocul Immunol Inflamm 26(3):347–352. https://doi.org/10.1080/09273948.2016.1193206

Ahn IE, Jerussi T, Farooqui M, Tian X, Wiestner A, Gea-Banacloche J (2016) Atypical Pneumocystis jirovecii pneumonia in previously untreated patients with CLL on single-agent ibrutinib. Blood 128(15):1940–1943. https://doi.org/10.1182/blood-2016-06-722991

Akpek EK, Ahmed I, Hochberg FH, Soheilian M, Dryja TP, Jakobiec FA, Foster CS (1999a) Intraocular-central nervous system lymphoma: clinical features, diagnosis, and outcomes. Ophthalmology 106(9):1805–1810. https://doi.org/10.1016/S0161-6420(99)90341-X

Akpek EK, Maca SM, Christen WG, Foster CS (1999b) Elevated vitreous interleukin-10 level is not diagnostic of intraocular-central nervous system lymphoma. Ophthalmology 106(12):2291–2295. https://doi.org/10.1016/s0161-6420(99)90528-6

Baehring JM, Androudi S, Longtine JJ, Betensky RA, Sklar J, Foster CS, Hochberg FH (2005) Analysis of clonal immunoglobulin heavy chain rearrangements in ocular lymphoma. Cancer 104(3):591–597. https://doi.org/10.1002/cncr.21191

Balamurugan S, Das D, Hasanreisoglu M, Toy BC, Akhter M, Anuradha VK, Anthony E, Gurnani B, Kaur K (2020) Interleukins and cytokine biomarkers in uveitis. Indian J Ophthalmol 68(9):1750–1763. https://doi.org/10.4103/ijo.IJO_564_20

Bardenstein DS (1998) Intraocular lymphoma. Cancer Control J Moffitt Cancer Center 5(4):317–325. https://doi.org/10.1177/107327489800500403

Barry RJ, Tasiopoulou A, Murray PI, Patel PJ, Sagoo MS, Denniston AK, Keane PA (2018) Characteristic optical coherence tomography findings in patients with primary vitreoretinal lymphoma: a novel aid to early diagnosis. Br J Ophthalmol 102(10):1362–1366. https://doi.org/10.1136/bjophthalmol-2017-311612

Batchelor TT, Kolak G, Ciordia R, Foster CS, Henson JW (2003) High-dose methotrexate for intraocular lymphoma. Clin Cancer Res 9(2):711–715

Baumann MA, Ritch PS, Hande KR, Williams GA, Topping TM, Anderson T (1986) Treatment of intraocular lymphoma with high-dose Ara-C. Cancer 57(7):1273–1275. https://doi.org/10.1002/1097-0142(19860401)57:7<1273::aid-cncr2820570702>3.0.co;2-c

Benjamin D, Park CD, Sharma V (1994) Human B cell interleukin 10. Leuk Lymphoma 12(3–4):205–210. https://doi.org/10.3109/10428199409059591

Berenbom A, Davila RM, Lin HS, Harbour JW (2007) Treatment outcomes for primary intraocular lymphoma: implications for external beam radiotherapy. Eye (Lond) 21(9):1198–1201. https://doi.org/10.1038/sj.eye.6702437

Bessell EM, Hoang-Xuan K, Ferreri AJ, Reni M (2007) Primary central nervous system lymphoma: biological aspects and controversies in management. Eur J Cancer (Oxford, England: 1990) 43(7):1141–1152. https://doi.org/10.1016/j.ejca.2006.12.011

Bonzheim I, Giese S, Deuter C, Süsskind D, Zierhut M, Waizel M, Szurman P, Federmann B, Schmidt J, Quintanilla-Martinez L, Coupland SE, Bartz-Schmidt KU, Fend F (2015) High frequency of MYD88 mutations in vitreoretinal B-cell lymphoma: a valuable tool to improve diagnostic yield of vitreous aspirates. Blood 126(1):76–79. https://doi.org/10.1182/blood-2015-01-620518

Brown SM, Jampol LM, Cantrill HL (1994) Intraocular lymphoma presenting as retinal vasculitis. Surv Ophthalmol 39(2):133–140. https://doi.org/10.1016/0039-6257(94)90158-9

Byrd JC, Furman RR, Coutre SE, Flinn IW, Burger JA, Blum KA, Grant B, Sharman JP, Coleman M, Wierda WG, Jones JA, Zhao W, Heerema NA, Johnson AJ, Sukbuntherng J, Chang BY, Clow F, Hedrick E, Buggy JJ, James DF, . . ., O'Brien S (2013) Targeting BTK with ibrutinib in relapsed chronic lymphocytic leukemia. N Engl J Med 369(1):32–42. https://doi.org/10.1056/NEJMoa1215637

Cao X, Shen D, Callanan DG, Mochizuki M, Chan CC (2011) Diagnosis of systemic metastatic retinal lymphoma. Acta Ophthalmol 89(2):e149–e154. https://doi.org/10.1111/j.1755-3768.2009.01797.x

Carbonell D, Mahajan S, Chee SP, Sobolewska B, Agrawal R, Bülow T, Gupta V, Jones NP, Accorinti M, Agarwal M, Batchelor T, Biswas J, Cimino L, tenDam-van Loon NH, de-la-Torre A, Frenkel S, Pe'er J, Kramer M, Miserocchi E, Mochizuki M, . . . Study Group for Vitreoretinal Lymphoma Diagnostics (2021) Consensus recommendations for the diagnosis of vitreoretinal lymphoma. Ocul Immunol Inflamm 29(3):507–520. https://doi.org/10.1080/09273948.2021.1878233

Casady M, Faia L, Nazemzadeh M, Nussenblatt R, Chan CC, Sen HN (2014) Fundus autofluorescence patterns in primary intraocular lymphoma. Retina (Philadelphia, Pa.) 34(2):366–372. https://doi.org/10.1097/IAE.0b013e31829977fa

Cassoux N, Merle-Beral H, Leblond V, Bodaghi B, Miléa D, Gerber S, Fardeau C, Reux I, Xuan KH, Chan CC, LeHoang P (2000) Ocular and central nervous system lymphoma: clinical features and diagnosis. Ocul Immunol Inflamm 8(4):243–250. https://doi.org/10.1076/ocii.8.4.243.646

Cassoux N, Giron A, Bodaghi B, Tran TH, Baudet S, Davy F, Chan CC, Lehoang P, Merle-Béral H (2007) IL-10 measurement in aqueous humor for screening patients with suspicion of primary intraocular lymphoma. Invest Ophthalmol Vis Sci 48(7):3253–3259. https://doi.org/10.1167/iovs.06-0031

Caya JG, Clowry LJ, Wollenberg NJ, Aaberg TM, Tieu TM (1985) The clinicopathologic significance of vitreous fluid cytology examinations in a series of 38 patients. Diagn Cytopathol 1(4):267–271. https://doi.org/10.1002/dc.2840010404

Chan CC (2003) Molecular pathology of primary intraocular lymphoma. Trans Am Ophthalmol Soc 101:275–292

Chan CC, Gonzalez JA (2007) Primary intraocular lymphoma. World Scientific Publishing Co Pte Ltd, Hackensack

Chan CC, Wallace DJ (2004) Intraocular lymphoma: update on diagnosis and management. Cancer Control J Moffitt Cancer Center 11(5):285–295. https://doi.org/10.1177/107327480401100502

Chan CC, Whitcup SM, Solomon D, Nussenblatt RB (1995) Interleukin-10 in the vitreous of patients with primary intraocular lymphoma. Am J Ophthalmol 120(5):671–673. https://doi.org/10.1016/s0002-9394(14)72217-2

Chan CC, Buggage RR, Nussenblatt RB (2002) Intraocular lymphoma. Curr Opin Ophthalmol 13(6):411–418. https://doi.org/10.1097/00055735-200212000-00012

Chan CC, Rubenstein JL, Coupland SE, Davis JL, Harbour JW, Johnston PB, Cassoux N, Touitou V, Smith JR, Batchelor TT, Pulido JS (2011) Primary vitreoretinal lymphoma: a report from an international primary central nervous system lymphoma collaborative group symposium. Oncologist 16(11):1589–1599. https://doi.org/10.1634/theoncologist.2011-0210

Char DH, Ljung BM, Deschênes J, Miller TR (1988) Intraocular lymphoma: immunological and cytological analysis. Br J Ophthalmol 72(12):905–911. https://doi.org/10.1136/bjo.72.12.905

Char DH, Kemlitz AE, Miller T (2005) Intraocular biopsy. Ophthalmol Clin N Am 18(1):177. https://doi.org/10.1016/j.ohc.2004.08.007

Chen W, Iida S, Louie DC, Dalla-Favera R, Chaganti RS (1998) Heterologous promoters fused to BCL6 by chromosomal translocations affecting band 3q27 cause its deregulated expression during B-cell differentiation. Blood 91(2):603–607

Chen X, Hu Y, Wenru S, Yang S, Wang X, Zhang P, Hong X, Liang C, Qian Z, Li Z, Tao Y, Huang H, Liang D (2021) Diagnostic value of genetic mutation analysis and mutation profiling of CfDNA in intraocular fluid for vitreoretinal lymphoma. MedRxiv. https://doi.org/10.1101/2021.10.18.21265102

Choi JY, Kafkala C, Foster CS (2006) Primary intraocular lymphoma: a review. Semin Ophthalmol 21(3):125–133. https://doi.org/10.1080/08820530500350498

Cimino L, Chan CC, Shen D, Masini L, Ilariucci F, Masetti M, Asioli S, Sartori A, Cappuccini L (2009) Ocular involvement in nasal natural killer T-cell lymphoma. Int Ophthalmol 29(4): 275–279. https://doi.org/10.1007/s10792-008-9220-8

Coupland SE, Damato B (2008) Understanding intraocular lymphomas. Clin Exp Ophthalmol 36(6):564–578. https://doi.org/10.1111/j.1442-9071.2008.01843.x

Coupland SE, Foss HD, Bechrakis NE, Hummel M, Stein H (2001) Secondary ocular involvement in systemic "memory" B-cell lymphocytic leukemia. Ophthalmology 108(7):1289–1295. https://doi.org/10.1016/s0161-6420(01)00594-2

Coupland SE, Hummel M, Stein H (2002) Ocular adnexal lymphomas: five case presentations and a review of the literature. Surv Ophthalmol 47(5):470–490. https://doi.org/10.1016/s0039-6257(02)00337-5

Coupland SE, Bechrakis NE, Anastassiou G, Foerster AM, Heiligenhaus A, Pleyer U, Hummel M, Stein H (2003) Evaluation of vitrectomy specimens and chorioretinal biopsies in the diagnosis of primary intraocular lymphoma in patients with masquerade syndrome. Graefes Arch Clin Exp Ophthalmol/Albrecht von Graefes Archiv fur klinische und experimentelle Ophthalmologie 241(10):860–870. https://doi.org/10.1007/s00417-003-0749-y

Coupland SE, Heimann H, Bechrakis NE (2004) Primary intraocular lymphoma: a review of the clinical, histopathological and molecular biological features. Graefes Arch Clin Exp Ophthalmol/Albrecht von Graefes Archiv fur klinische und experimentelle Ophthalmologie 242(11):901–913. https://doi.org/10.1007/s00417-004-0973-0

Coupland SE, Anastassiou G, Bornfeld N, Hummel M, Stein H (2005a) Primary intraocular lymphoma of T-cell type: report of a case and review of the literature. Graefes Arch Clin Exp Ophthalmol/Albrecht von Graefes Archiv fur klinische und experimentelle Ophthalmologie 243(3):189–197. https://doi.org/10.1007/s00417-004-0890-2

Coupland SE, Loddenkemper C, Smith JR, Braziel RM, Charlotte F, Anagnostopoulos I, Stein H (2005b) Expression of immunoglobulin transcription factors in primary intraocular lymphoma and primary central nervous system lymphoma. Invest Ophthalmol Vis Sci 46(11):3957–3964. https://doi.org/10.1167/iovs.05-0318

Coupland SE, Hummel M, Stein H, Willerding G, Jahnke K, Stoltenburg-Didinger G (2005c) Demonstration of identical clonal derivation in a case of "oculocerebral" lymphoma. Br J Ophthalmol 89(2):238–239. https://doi.org/10.1136/bjo.2004.047001

Cunha-Vaz J (1979) The blood-ocular barriers. Surv Ophthalmol 23(5):279–296. https://doi.org/10.1016/0039-6257(79)90158-9

Davis JL (2004) Diagnosis of intraocular lymphoma. Ocul Immunol Inflamm 12(1):7–16. https://doi.org/10.1076/ocii.12.1.7.28072

Davis JL (2013) Intraocular lymphoma: a clinical perspective. Eye (Lond) 27(2):153–162. https://doi.org/10.1038/eye.2012.250

Davis JL, Solomon D, Nussenblatt RB, Palestine AG, Chan CC (1992) Immunocytochemical staining of vitreous cells. Indications, techniques, and results. Ophthalmology 99(2):250–256. https://doi.org/10.1016/s0161-6420(92)31984-0

Davis JL, Viciana AL, Ruiz P (1997) Diagnosis of intraocular lymphoma by flow cytometry. Am J Ophthalmol 124(3):362–372. https://doi.org/10.1016/s0002-9394(14)70828-1

Davis JL, Miller DM, Ruiz P (2005) Diagnostic testing of vitrectomy specimens. Am J Ophthalmol 140(5):822–829. https://doi.org/10.1016/j.ajo.2005.05.032

de Smet MD, Stark-Vancs V, Kohler DR, Smith J, Wittes R, Nussenblatt RB (1996) Intraocular levels of methotrexate after intravenous administration. Am J Ophthalmol 121(4):442–444. https://doi.org/10.1016/s0002-9394(14)70444-1

de Smet MD, Vancs VS, Kohler D, Solomon D, Chan CC (1999) Intravitreal chemotherapy for the treatment of recurrent intraocular lymphoma. Br J Ophthalmol 83(4):448–451. https://doi.org/10.1136/bjo.83.4.448

Deák GG, Goldstein DA, Zhou M, Fawzi AA, Jampol LM (2019) Vertical Hyperreflective lesions on optical coherence tomography in vitreoretinal lymphoma. JAMA Ophthalmol 137(2):194–198. https://doi.org/10.1001/jamaophthalmol.2018.5835

DeAngelis LM (2001) Primary central nervous system lymphomas. Curr Treat Options in Oncol 2(4):309–318. https://doi.org/10.1007/s11864-001-0024-6

Duker JS, Shields JA, Ross M (1987) Intraocular large cell lymphoma presenting as massive thickening of the uveal tract. Retina (Philadelphia, Pa.) 7(1):41–45

Dworkin LL, Gibler TM, Van Gelder RN (2002) Real-time quantitative polymerase chain reaction diagnosis of infectious posterior uveitis. Arch Ophthalmol (Chicago, Ill.: 1960) 120(11):1534–1539. https://doi.org/10.1001/archopht.120.11.1534

Elner VM, Hidayat AA, Charles NC, Davitz MA, Smith ME, Burgess D, Dawson N (1986) Neoplastic angioendotheliomatosis. A variant of malignant lymphoma immunohistochemical and ultrastructural observations of three cases. Ophthalmology 93(9):1237–1245

Faia LJ, Chan CC (2009) Primary intraocular lymphoma. Arch Pathol Lab Med 133(8):1228–1232. https://doi.org/10.5858/133.8.1228

Fardeau C, Lee CP, Merle-Béral H, Cassoux N, Bodaghi B, Davi F, Lehoang P (2009) Retinal fluorescein, indocyanine green angiography, and optic coherence tomography in non-Hodgkin primary intraocular lymphoma. Am J Ophthalmol 147(5):886–894.e1. https://doi.org/10.1016/j.ajo.2008.12.025

Farkas T, Harbour JW, Dávila RM (2004) Cytologic diagnosis of intraocular lymphoma in vitreous aspirates. Acta Cytol 48(4):487–491. https://doi.org/10.1159/000326409

Feman SS, Niwayama G, Hepler RS, Foos RY (1969) "Burkitt tumor" with intraocular involvement. Surv Ophthalmol 14(2):106–111

Fend F, Ferreri AJ, Coupland SE (2016) How we diagnose and treat vitreoretinal lymphoma. Br J Haematol 173(5):680–692. https://doi.org/10.1111/bjh.14025

Fernández-Rodríguez C, Bellosillo B, García-García M, Sánchez-González B, Gimeno E, Vela MC, Serrano S, Besses C, Salar A (2014) MYD88 (L265P) mutation is an independent prognostic factor for outcome in patients with diffuse large B-cell lymphoma. Leukemia 28(10):2104–2106. https://doi.org/10.1038/leu.2014.184

Ferreri AJ, Blay JY, Reni M, Pasini F, Gubkin A, Tirelli U, Calderoni A, Zucca E, Cortelazzo S, Chassagne C, Tinguely M, Borisch B, Berger F, Ponzoni M, Cavalli F, International Extranodal Lymphoma Study Group (IELSG) (2002) Relevance of intraocular involvement in the management of primary central nervous system lymphomas. Ann Oncol 13(4):531–538. https://doi.org/10.1093/annonc/mdf080

Fishburne BC, Wilson DJ, Rosenbaum JT, Neuwelt EA (1997) Intravitreal methotrexate as an adjunctive treatment of intraocular lymphoma. Arch Ophthalmol (Chicago, Ill.: 1960) 115(9):1152–1156. https://doi.org/10.1001/archopht.1997.01100160322009

Fonkem E, Lok E, Robison D, Gautam S, Wong ET (2014) The natural history of intravascular lymphomatosis. Cancer Med 3(4):1010–1024. https://doi.org/10.1002/cam4.269

Freddo TF (2001) Shifting the paradigm of the blood-aqueous barrier. Exp Eye Res 73(5):581–592. https://doi.org/10.1006/exer.2001.1056

Fredrick DR, Char DH, Ljung BM, Brinton DA (1989) Solitary intraocular lymphoma as an initial presentation of widespread disease. Arch Ophthalmol (Chicago, Ill.: 1960) 107(3):395–397. https://doi.org/10.1001/archopht.1989.01070010405034

Freeman LN, Schachat AP, Knox DL, Michels RG, Green WR (1987) Clinical features, laboratory investigations, and survival in ocular reticulum cell sarcoma. Ophthalmology 94(12): 1631–1639. https://doi.org/10.1016/s0161-6420(87)33256-7

Frenkel S, Hendler K, Siegal T, Shalom E, Pe'er J (2008) Intravitreal methotrexate for treating vitreoretinal lymphoma: 10 years of experience. Br J Ophthalmol 92(3):383–388. https://doi.org/10.1136/bjo.2007.127928

Goldey SH, Stern GA, Oblon DJ, Mendenhall NP, Smith LJ, Duque RE (1989) Immunophenotypic characterization of an unusual T-cell lymphoma presenting as anterior uveitis. A clinicopathologic case report. Arch Ophthalmol (Chicago, Ill.: 1960) 107(9):1349–1353. https://doi.org/10.1001/archopht.1989.01070020419047

Gonzales JA, Chan CC (2007) Biopsy techniques and yields in diagnosing primary intraocular lymphoma. Int Ophthalmol 27(4):241–250. https://doi.org/10.1007/s10792-007-9065-6

Gorochov G, Parizot C, Bodaghi B, Raphael D, Cassoux N, Merle-Beral H, Debré P, LeHoang P (2003) Characterization of vitreous B-cell infiltrates in patients with primary ocular lymphoma, using CDR3 size polymorphism analysis of antibody transcripts. Invest Ophthalmol Vis Sci 44(12):5235–5241. https://doi.org/10.1167/iovs.03-0035

Goto H, Murase K, Usui M (2006) Case of spontaneous regression of intraocular lymphoma demonstrated by subretinal biopsy. Nippon Ganka Gakkai Zasshi 110(3):226–231

Grimm SA, McCannel CA, Omuro AM, Ferreri AJ, Blay JY, Neuwelt EA, Siegal T, Batchelor T, Jahnke K, Shenkier TN, Hall AJ, Graus F, Herrlinger U, Schiff D, Raizer J, Rubenstein J, Laperriere N, Thiel E, Doolittle N, Iwamoto FM, . . ., Abrey LE (2008) Primary CNS lymphoma with intraocular involvement: international PCNSL collaborative group report. Neurology 71(17):1355–1360. https://doi.org/10.1212/01.wnl.0000327672.04729.8c

Gündüz K, Shields JA, Shields CL, Eagle RC Jr, Diniz W, Mercado G, Chang W (1999) Transscleral choroidal biopsy in the diagnosis of choroidal lymphoma. Surv Ophthalmol 43(6):551–555. https://doi.org/10.1016/s0039-6257(99)00041-7

Gupta V, Gupta A, Rao NA (2007) Intraocular tuberculosis--an update. Surv Ophthalmol 52(6): 561–587. https://doi.org/10.1016/j.survophthal.2007.08.015

Haldorsen IS, Krossnes BK, Aarseth JH, Scheie D, Johannesen TB, Mella O, Espeland A (2007) Increasing incidence and continued dismal outcome of primary central nervous system lymphoma in Norway 1989-2003: time trends in a 15-year national survey. Cancer 110(8): 1803–1814. https://doi.org/10.1002/cncr.22989

Hashida N, Nakai K, Saitoh N, Nishida K (2014) Association between ocular findings and preventive therapy with onset of central nervous system involvement in patients with primary vitreoretinal lymphoma. Graefes Arch Clin Exp Ophthalmol/Albrecht von Graefes Archiv fur klinische und experimentelle Ophthalmologie 252(4):687–693. https://doi.org/10.1007/s00417-014-2584-8

Hattenhauer MG, Pach JM (1996) Ocular lymphoma in a patient with chronic lymphocytic leukemia. Am J Ophthalmol 122(2):266–268. https://doi.org/10.1016/s0002-9394(14)72022-7

Higashide T, Takahira M, Okumura H, Torisaki M, Sakurai M, Shirao Y, Sugiyama K (2004) Bilaterally identical monoclonality in a case of primary intraocular lymphoma. Am J Ophthalmol 138(2):306–308. https://doi.org/10.1016/j.ajo.2004.03.00

Hoang-Xuan K, Bessell E, Bromberg J, Hottinger AF, Preusser M, Rudà R, Schlegel U, Siegal T, Soussain C, Abacioglu U, Cassoux N, Deckert M, Dirven CM, Ferreri AJ, Graus F, Henriksson R, Herrlinger U, Taphoorn M, Soffietti R, Weller M., . . ., European Association for Neuro-Oncology Task Force on Primary CNS Lymphoma (2015) Diagnosis and treatment of primary CNS lymphoma in immunocompetent patients: guidelines from the European Association for Neuro-Oncology. Lancet Oncol, 16(7):e322–e332. https://doi.org/10.1016/S1470-2045(15)00076-5

Hoffman PM, McKelvie P, Hall AJ, Stawell RJ, Santamaria JD (2003) Intraocular lymphoma: a series of 14 patients with clinicopathological features and treatment outcomes. Eye (Lond) 17(4):513–521. https://doi.org/10.1038/sj.eye.6700378

Honavar SG, Shields JA, Shields CL, Demirci H, Ehya H (2001) Extramedullary plasmacytoma confined to the choroid. Am J Ophthalmol 131(2):277–278. https://doi.org/10.1016/s0002-9394(00)00706-6

Hong JT, Chae JB, Lee JY, Kim JG, Yoon YH (2011) Ocular involvement in patients with primary CNS lymphoma. J Neuro-Oncol 102(1):139–145. https://doi.org/10.1007/s11060-010-0303-9

Hormigo A, Abrey L, Heinemann MH, DeAngelis LM (2004) Ocular presentation of primary central nervous system lymphoma: diagnosis and treatment. Br J Haematol 126(2):202–208. https://doi.org/10.1111/j.1365-2141.2004.05028.x

Hunt MA, Jahnke K, Murillo TP, Neuwelt EA (2006) Distinguishing primary central nervous system lymphoma from other central nervous system diseases: a neurosurgical perspective on diagnostic dilemmas and approaches. Neurosurg Focus 21(5):E3. https://doi.org/10.3171/foc.2006.21.5.4

Iaccheri B, Fiore T, Cerquaglia A, Lupidi M, Cagini C (2017) Transient therapeutic effect of vitrectomy in primary intraocular lymphoma. Int Ophthalmol 37(6):1333–1335. https://doi.org/10.1007/s10792-016-0405-2

Illerhaus G, Marks R, Ihorst G, Guttenberger R, Ostertag C, Derigs G, Frickhofen N, Feuerhake F, Volk B, Finke J (2006) High-dose chemotherapy with autologous stem-cell transplantation and hyperfractionated radiotherapy as first-line treatment of primary CNS lymphoma. J Clin Oncol 24(24):3865–3870. https://doi.org/10.1200/JCO.2006.06.2117

Intzedy L, Teoh SC, Hogan A, Mangwana S, Mayer EJ, Dick AD, Pawade J (2008) Cytopathological analysis of vitreous in intraocular lymphoma. Eye (Lond) 22(2):289–293. https://doi.org/10.1038/sj.eye.6702965

Ishida T, Ohno-Matsui K, Kaneko Y, Tobita H, Shimada N, Takase H, Mochizuki M (2010) Fundus autofluorescence patterns in eyes with primary intraocular lymphoma. Retina (Philadelphia, Pa.) 30(1):23–32. https://doi.org/10.1097/IAE.0b013e3181b408a2

Isobe K, Ejima Y, Tokumaru S, Shikama N, Suzuki G, Takemoto M, Tsuchida E, Nomura M, Shibamoto Y, Hayabuchi N (2006) Treatment of primary intraocular lymphoma with radiation therapy: a multi-institutional survey in Japan. Leuk Lymphoma 47(9):1800–1805. https://doi.org/10.1080/10428190600632881

Itty S, Pulido JS (2009) Rituximab for intraocular lymphoma. Retina (Philadelphia, Pa.) 29(2):129–132. https://doi.org/10.1097/IAE.0b013e318192f574

Jahnke K, Bechrakis NE, Coupland SE, Schmittel A, Foerster MH, Fischer L, Thiel E, Korfel A (2004) Treatment of primary intraocular lymphoma with oral trofosfamide: report of two cases and review of the literature. Graefes Arch Clin Exp Ophthalmol/Albrecht von Graefes Archiv fur klinische und experimentelle Ophthalmologie 242(9):771–776. https://doi.org/10.1007/s00417-004-0947-2

Jahnke K, Wagner T, Bechrakis NE, Willerding G, Coupland SE, Fischer L, Thiel E, Korfel A (2005) Pharmacokinetics and efficacy of ifosfamide or trofosfamide in patients with intraocular lymphoma. Ann Oncol 16(12):1974–1978. https://doi.org/10.1093/annonc/mdi409

Jahnke K, Korfel A, Komm J, Bechrakis NE, Stein H, Thiel E, Coupland SE (2006) Intraocular lymphoma 2000-2005: results of a retrospective multicentre trial. Graefes Arch Clin Exp Ophthalmol/Albrecht von Graefes Archiv fur klinische und experimentelle Ophthalmologie 244(6):663–669. https://doi.org/10.1007/s00417-005-0138-9

Jahnke K, Thiel E, Abrey LE, Neuwelt EA, Korfel A (2007) Diagnosis and management of primary intraocular lymphoma: an update. Clin Ophthalmol (Auckland, NZ) 1(3):247–258

Jahnke K, Thiel E, Bechrakis NE, Willerding G, Kraemer DF, Fischer L, Korfel A (2009) Ifosfamide or trofosfamide in patients with intraocular lymphoma. J Neuro-Oncol 93(2): 213–217. https://doi.org/10.1007/s11060-008-9761-8

Johnston RL, Tufail A, Lightman S, Luthert PJ, Pavesio CE, Cooling RJ, Charteris D (2004) Retinal and choroidal biopsies are helpful in unclear uveitis of suspected infectious or malignant origin. Ophthalmology 111(3):522–528. https://doi.org/10.1016/j.ophtha.2002.10.002

Kalogeropoulos D, Papoudou-Bai A, Kanavaros P, Kalogeropoulos C (2018) Ocular adnexal marginal zone lymphoma of mucosa-associated lymphoid tissue. Clin Exp Med 18(2): 151–163. https://doi.org/10.1007/s10238-017-0474-1

Kalogeropoulos D, Vartholomatos G, Mitra A, Elaraoud I, Ch'ng SW, Zikou A, Papoudou-Bai A, Moschos MM, Kanavaros P, Kalogeropoulos C (2019) Primary vitreoretinal lymphoma. Saudi J Ophthalmol 33(1):66–80. https://doi.org/10.1016/j.sjopt.2018.12.008

Karma A, von Willebrand EO, Tommila PV, Paetau AE, Oskala PS, Immonen IJ (2007) Primary intraocular lymphoma: improving the diagnostic procedure. Ophthalmology 114(7):1372–1377. https://doi.org/10.1016/j.ophtha.2006.11.009

Kase S, Namba K, Jin XH, Kubota KC, Ishida S (2012) Spontaneous regression of intraocular lymphoma. Ophthalmology 119(5):1083–1084. https://doi.org/10.1016/j.ophtha.2011.12.011

Katai N, Kuroiwa S, Fujimori K, Yoshimura N (1997) Diagnosis of intraocular lymphoma by polymerase chain reaction. Graefes Arch Clin Exp Ophthalmol/Albrecht von Graefes Archiv fur klinische und experimentelle Ophthalmologie 235(7):431–436. https://doi.org/10.1007/BF00947062

Kim H, Csaky KG, Chan CC, Bungay PM, Lutz RJ, Dedrick RL, Yuan P, Rosenberg J, Grillo-Lopez AJ, Wilson WH, Robinson MR (2006) The pharmacokinetics of rituximab following an intravitreal injection. Exp Eye Res 82(5):760–766. https://doi.org/10.1016/j.exer.2005.09.018

Kim E, Kim C, Lee J, Cho Y (2009) A case of primary intraocular lymphoma treated by intravitreal methotrexate. Korean J Ophthalmol 23(3):210–214. https://doi.org/10.3341/kjo.2009.23.3.210

Kimura K, Usui Y, Goto H, Japanese Intraocular Lymphoma Study Group (2012) Clinical features and diagnostic significance of the intraocular fluid of 217 patients with intraocular lymphoma. Jpn J Ophthalmol 56(4):383–389. https://doi.org/10.1007/s10384-012-0150-7

Kincaid MC, Green WR (1983) Ocular and orbital involvement in leukemia. Surv Ophthalmol 27(4):211–232. https://doi.org/10.1016/0039-6257(83)90123-6

Kirmani MH, Thomas EL, Rao NA, Laborde RP (1987) Intraocular reticulum cell sarcoma: diagnosis by choroidal biopsy. Br J Ophthalmol 71(10):748–752. https://doi.org/10.1136/bjo.71.10.748

Kitazawa K, Sotozono C, Koizumi N, Nagata K, Inatomi T, Sasaki H, Kinoshita S (2017) Safety of anterior chamber paracentesis using a 30-gauge needle integrated with a specially designed disposable pipette. Br J Ophthalmol 101(5):548–550. https://doi.org/10.1136/bjophthalmol-2016-309650

Knapp AJ, Gartner S, Henkind P (1987) Multiple myeloma and its ocular manifestations. Surv Ophthalmol 31(5):343–351. https://doi.org/10.1016/0039-6257(87)90119-6

Lavine JA, Singh AD, Sharma S, Baynes K, Lowder CY, Srivastava SK (2019) Ultra-Widefield multimodal imaging of primary vitreoretinal lymphoma. Retina (Philadelphia, Pa.) 39(10): 1861–1871. https://doi.org/10.1097/IAE.0000000000002260

Lee J, Goldstein DA (2020) The diverse multi-modal imaging findings of recurrent primary vitreoretinal lymphoma. Am J Ophthalmol Case Rep 20:100936. https://doi.org/10.1016/j.ajoc.2020.100936

Levasseur SD, Wittenberg LA, White VA (2013) Vitreoretinal lymphoma: a 20-year review of incidence, clinical and cytologic features, treatment, and outcomes. JAMA Ophthalmol 131(1): 50–55. https://doi.org/10.1001/jamaophthalmol.2013.569

Levy-Clarke GA, Byrnes GA, Buggage RR, Shen DF, Filie AC, Caruso RC, Nussenblatt RB, Chan CC (2001) Primary intraocular lymphoma diagnosed by fine needle aspiration biopsy of a subretinal lesion. Retina (Philadelphia, Pa) 21(3):281–284. https://doi.org/10.1097/00006982-200106000-00023

Lewis RA, Clark RB (1975) Infiltrative retinopathy in systemic lymphoma. Am J Ophthalmol 79(1):48–52. https://doi.org/10.1016/0002-9394(75)90454-7

Li Z, Qiu Y, Personett D, Huang P, Edenfield B, Katz J, Babusis D, Tang Y, Shirely MA, Moghaddam MF, Copland JA, Tun HW (2013) Pomalidomide shows significant therapeutic activity against CNS lymphoma with a major impact on the tumor microenvironment in murine models. PLoS One 8(8):e71754. https://doi.org/10.1371/journal.pone.0071754

Liu TY, Ibrahim M, Bittencourt M, Sepah YJ, Do DV, Nguyen QD (2012) Retinal optical coherence tomography manifestations of intraocular lymphoma. J Ophthalmic Inflamm Infect 2(4):215–218. https://doi.org/10.1007/s12348-012-0072-z

Ljung BM, Char D, Miller TR, Deschenes J (1988) Intraocular lymphoma. Cytologic diagnosis and the role of immunologic markers. Acta Cytol 32(6):840–847

Llorenç V, Fuster C, Alba-Linero C, Moll-Udina A, Serrano A, Solé M, Sainz de la Maza M, Aldecoa I, Adán A (2019) Clinical features of primary and systemic metastatic intraocular lymphomas in Spanish patients. J Ophthalmol 2019:6327041. https://doi.org/10.1155/2019/6327041

Lobo A, Lightman S (2003) Vitreous aspiration needle tap in the diagnosis of intraocular inflammation. Ophthalmology 110(3):595–599. https://doi.org/10.1016/S0161-6420(02)01895-X

Lv L, Liu Y (2021) Clinical application of liquid biopsy in non-Hodgkin lymphoma. Front Oncol 11:658234. https://doi.org/10.3389/fonc.2021.658234

Mantopoulos D, Cebulla CM (2015) Multimodal imaging of spontaneously shifting primary vitreoretinal lymphoma. Ocular Oncol Pathol 1(4):237–240. https://doi.org/10.1159/000374121

Marchese A, Miserocchi E, Giuffrè C, Cicinelli MV, Querques G, Bandello F, Modorati G (2019) Aurora borealis and string of pearls in vitreoretinal lymphoma: patterns of vitreous haze. Br J Ophthalmol 103(11):1656–1659. https://doi.org/10.1136/bjophthalmol-2018-313491

Marcus R, Hodson D, Coupland S et al (2012) British Committee for Standards in Haematology. Guidelines on the diagnosis and management of adult patients with primary CNS lymphoma (PCNSL) and primary intra-ocular lymphoma (PIOL). http://www.bcshguidelines.com/documents/PCNSL__bcsh_2007.pdf. Accessed June 2012

Margolis L, Fraser R, Lichter A, Char DH (1980) The role of radiation therapy in the management of ocular reticulum cell sarcoma. Cancer 45(4):688–692. https://doi.org/10.1002/1097-0142(19800215)45:4<688::aid-cncr2820450412>3.0.co;2-f

Mason JO, Fischer DH (2003) Intrathecal chemotherapy for recurrent central nervous system intraocular lymphoma. Ophthalmology 110(6):1241–1244. https://doi.org/10.1016/S0161-6420(03)00268-9

Mastropasqua R, Thaung C, Pavesio C, Lightman S, Westcott M, Okhravi N, Aylward W, Charteris D, da Cruz L (2015) The role of Chorioretinal biopsy in the diagnosis of intraocular lymphoma. Am J Ophthalmol 160(6):1127–1132.e1. https://doi.org/10.1016/j.ajo.2015.08.033

Mead GM, Bleehen NM, Gregor A, Bullimore J, Shirley D, Rampling RP, Trevor J, Glaser MG, Lantos P, Ironside JW, Moss TH, Brada M, Whaley JB, Stenning SP (2000) A medical research council randomized trial in patients with primary cerebral non-Hodgkin lymphoma: cerebral radiotherapy with and without cyclophosphamide, doxorubicin, vincristine, and prednisone chemotherapy. Cancer 89(6):1359–1370

Merle-Béral H, Davi F, Cassoux N, Baudet S, Colin C, Gourdet T, Bodaghi B, LeHoang P (2004) Biological diagnosis of primary intraocular lymphoma. Br J Haematol 124(4):469–473. https://doi.org/10.1046/j.1365-2141.2003.04800.x

Meunier J, Lumbroso-Le Rouic L, Vincent-Salomon A, Dendale R, Asselain B, Arnaud P, Fourquet A, Desjardins L, Plancher C, Validire P, Chaoui D, Lévy C, Decaudin D (2004) Ophthalmologic and intraocular non-Hodgkin's lymphoma: a large single centre study of initial characteristics, natural history, and prognostic factors. Hematol Oncol 22(4):143–158. https://doi.org/10.1002/hon.741

Mochizuki M, Singh AD (2009) Epidemiology and clinical features of intraocular lymphoma. Ocul Immunol Inflamm 17(2):69–72. https://doi.org/10.1080/09273940902957305

Mudhar HS, Sethuraman C, Khan MD, Jan SU (2007) Intracular, pan-uveal intravascular large B-cell lymphoma associated with choroidal infarction and choroidal tri-lineage extramedullary haemtopoiesis. Histopathology 51(2):275–279. https://doi.org/10.1111/j.1365-2559.2007.02750.x

Narasimhan S, Joshi M, Parameswaran S, Rishi P, Khetan V, Ganesan S, Biswas J, Sundaram N, Sreenivasan J, Verma S, Krishnamurthy V, Subramanian K (2020) MYD88 L265P mutation in intraocular lymphoma: a potential diagnostic marker. Indian J Ophthalmol 68(10):2160–2165. https://doi.org/10.4103/ijo.IJO_1712_19

Nasir S, DeAngelis LM (2000) Update on the management of primary CNS lymphoma. Oncology (Williston Park) 14(2):228–244

Nelson DF, Martz KL, Bonner H, Nelson JS, Newall J, Kerman HD, Thomson JW, Murray KJ (1992) Non-Hodgkin's lymphoma of the brain: can high dose, large volume radiation therapy improve survival? Report on a prospective trial by the Radiation Therapy Oncology Group (RTOG): RTOG 8315. Int J Radiat Oncol Biol Phys 23(1):9–17. https://doi.org/10.1016/0360-3016(92)90538-s

Nguyen DT, Houillier C, Choquet S, Cassoux N, Soussain C, Le Cossec C, Legarf-Tavernier M, Costopoulos M, LeHoang P, Bodaghi B, Omuro A, Hoang-Xuan K, Touitou V (2016) Primary Oculocerebral lymphoma: MTX Polychemotherapy alone on intraocular disease control. Ophthalmology 123(9):2047–2050. https://doi.org/10.1016/j.ophtha.2016.03.043

O'Neill BP, Wang CH, O'Fallon JR, Colgan JD, Earle JD, Krigel RL, Brown LD, McGinnis WL (1999) Primary central nervous system non-Hodgkin's lymphoma (PCNSL): survival advantages with combined initial therapy? A final report of the North Central Cancer Treatment Group (NCCTG) Study 86-72-52. Int J Radiat Oncol Biol Phys 43(3):559–563. https://doi.org/10.1016/s0360-3016(98)00450-7

Okamoto K, Proia LA, Demarais PL (2016) Disseminated Cryptococcal disease in a patient with chronic lymphocytic leukemia on Ibrutinib. Case Rep Infect Dis 2016:4642831. https://doi.org/10.1155/2016/4642831

Orellana J, Friedman AH (1981) Ocular manifestations of multiple myeloma, Waldenström's macroglobulinemia and benign monoclonal gammopathy. Surv Ophthalmol 26(3):157–169. https://doi.org/10.1016/0039-6257(81)90065-5

Palexas GN, Green WR, Goldberg MF, Ding Y (1995) Diagnostic pars plana vitrectomy report of a 21-year retrospective study. Trans Am Ophthalmol Soc 93:281–314

Parikh AH, Khan SH, Wright JD Jr, Oh KT (2005) Systemic non-Hodgkin's lymphoma simulating primary intraocular lymphoma. Am J Ophthalmol 139(3):573–574. https://doi.org/10.1016/j.ajo.2004.09.047

Parver LM, Font RL (1979) Malignant lymphoma of the retina and brain. Initial diagnosis by cytologic examination of vitreous aspirate. Arch Ophthalmol (Chicago, Ill : 1960) 97(8): 1505–1507. https://doi.org/10.1001/archopht.1979.01020020167016

Pavan PR, Oteiza EE, Margo CE (1995) Ocular lymphoma diagnosed by internal subretinal pigment epithelium biopsy. Arch Ophthalmol (Chicago, Ill.: 1960) 113(10):1233–1234. https://doi.org/10.1001/archopht.1995.01100100021014

Payne T, Karp LA, Zimmerman LE (1971) Intraocular involvement in Burkitt's lymphoma. Arch Ophthalmol (Chicago, Ill.: 1960) 85(3):295–298. https://doi.org/10.1001/archopht.1971.00990050297009

Pe'er J, Hochberg FH, Foster CS (2009) Clinical review: treatment of vitreoretinal lymphoma. Ocul Immunol Inflamm 17(5):299–306. https://doi.org/10.3109/09273940903370755

Peterson K, Gordon KB, Heinemann MH, DeAngelis LM (1993) The clinical spectrum of ocular lymphoma. Cancer 72(3):843–849. https://doi.org/10.1002/1097-0142(19930801)72:3<843::aid-cncr2820720333>3.0.co;2-#

Phan RT, Dalla-Favera R (2004) The BCL6 proto-oncogene suppresses p53 expression in germinal-centre B cells. Nature 432(7017):635–639. https://doi.org/10.1038/nature03147

Pilon AF, Rhee PS, Messner LV (2005) Bilateral, persistent serous macular detachments with Waldenström's macroglobulinemia. Optom Vis Sci 82(7):573–578. https://doi.org/10.1097/01.opx.0000171333.64652.a4
Pinna A, Dore S, Dore F, Salis A, Carta F (2003) Bilateral optic disc swelling as the presenting sign of Waldenström's macroglobulinaemia. Acta Ophthalmol Scand 81(4):413–415. https://doi.org/10.1034/j.1600-0420.2003.00106.x
Plowman PN, Montefiore DS, Lightman S (1993) Multiagent chemotherapy in the salvage cure of ocular lymphoma relapsing after radiotherapy. Clin Oncol (R Coll Radiol) 5(5):315–316. https://doi.org/10.1016/s0936-6555(05)80909-4
Plowman PN, Taylor A, Jackson AS, Lightman S, Pavesio C (2002) Optimal therapy of primary ocular lymphoma. Clin Oncol (R Coll Radiol) 14(6):521. https://doi.org/10.1053/clon.2002.0154
Puente XS, Pinyol M, Quesada V, Conde L, Ordóñez GR, Villamor N, Escaramis G, Jares P, Beà S, González-Díaz M, Bassaganyas L, Baumann T, Juan M, López-Guerra M, Colomer D, Tubío JM, López C, Navarro A, Tornador C, Aymerich M, … Campo E (2011) Whole-genome sequencing identifies recurrent mutations in chronic lymphocytic leukaemia. Nature 475(7354): 101–105. https://doi.org/10.1038/nature10113
Pulido JS, Salomao DR, Frederick LA, Viswanatha DS (2015) MyD-88 L265P mutations are present in some cases of vitreoretinal lymphoma. Retina (Philadelphia, Pa.) 35(4):624–627. https://doi.org/10.1097/IAE.0000000000000589
Raja H, Salomão DR, Viswanatha DS, Pulido JS (2016) Prevalence of MYD88 L265P mutation in histologically proven, diffuse large B-cell vitreoretinal lymphoma, Retina (Philadelphia, Pa.). 36(3):624–628. https://doi.org/10.1097/IAE.0000000000000996
Rajagopal R, Harbour JW (2011) Diagnostic testing and treatment choices in primary vitreoretinal lymphoma. Retina (Philadelphia, Pa.) 31(3):435–440. https://doi.org/10.1097/IAE.0b013e31820a6743
Rallis KS, Lai Yau TH, Sideris M (2021) Chemoradiotherapy in cancer treatment: rationale and clinical applications. Anticancer Res 41(1):1–7. https://doi.org/10.21873/anticanres.14746
Raparia K, Chang CC, Chévez-Barrios P (2009) Intraocular lymphoma: diagnostic approach and immunophenotypic findings in vitrectomy specimens. Arch Pathol Lab Med 133(8):1233–1237. https://doi.org/10.5858/133.8.1233
Raviola G (1977) The structural basis of the blood-ocular barriers. Exp Eye Res 25(Suppl):27–63. https://doi.org/10.1016/s0014-4835(77)80009-2
Sagoo MS, Mehta H, Swampillai AJ, Cohen VM, Amin SZ, Plowman PN, Lightman S (2014) Primary intraocular lymphoma. Surv Ophthalmol 59(5):503–516. https://doi.org/10.1016/j.survophthal.2013.12.001
Salomão DR, Pulido JS, Johnston PB, Canal-Fontcuberta I, Feldman AL (2013) Vitreoretinal presentation of secondary large B-cell lymphoma in patients with systemic lymphoma. JAMA Ophthalmol 131(9):1151–1158. https://doi.org/10.1001/jamaophthalmol.2013.334
Sandor V, Stark-Vancs V, Pearson D, Nussenblat R, Whitcup SM, Brouwers P, Patronas N, Heiss J, Jaffe E, deSmet M, Kohler D, Simon R, Wittes R (1998) Phase II trial of chemotherapy alone for primary CNS and intraocular lymphoma. J Clin Oncol 16(9):3000–3006. https://doi.org/10.1200/JCO.1998.16.9.3000
Schneider C, Pasqualucci L, Dalla-Favera R (2011) Molecular pathogenesis of diffuse large B-cell lymphoma. Semin Diagn Pathol 28(2):167–177. https://doi.org/10.1053/j.semdp.2011.04.001
Schultz C, Scott C, Sherman W, Donahue B, Fields J, Murray K, Fisher B, Abrams R, Meis-Kindblom J (1996) Preirradiation chemotherapy with cyclophosphamide, doxorubicin, vincristine, and dexamethasone for primary CNS lymphomas: initial report of radiation therapy oncology group protocol 88-06. J Clin Oncol 14(2):556–564. https://doi.org/10.1200/JCO.1996.14.2.556
Sen HN, Chan CC, Caruso RC, Fariss RN, Nussenblatt RB, Buggage RR (2004) Waldenström's macroglobulinemia-associated retinopathy. Ophthalmology 111(3):535–539. https://doi.org/10.1016/j.ophtha.2003.05.036

Sen HN, Chan CC, Byrnes G, Fariss RN, Nussenblatt RB, Buggage RR (2008) Intravitreal methotrexate resistance in a patient with primary intraocular lymphoma. Ocul Immunol Inflamm 16(1):29–33. https://doi.org/10.1080/09273940801899764

Sen HN, Bodaghi B, Hoang PL, Nussenblatt R (2009) Primary intraocular lymphoma: diagnosis and differential diagnosis. Ocul Immunol Inflamm 17(3):133–141. https://doi.org/10.1080/09273940903108544

Shakin EP, Augsburger JJ, Eagle RC Jr, Ehya H, Shields JA, Fischer D, Koepsell DG (1988) Multiple myeloma involving the iris. Arch Ophthalmol (Chicago, Ill.: 1960) 106(4):524–526. https://doi.org/10.1001/archopht.1988.01060130570039

Shen DF, Zhuang Z, LeHoang P, Böni R, Zheng S, Nussenblatt RB, Chan CC (1998) Utility of microdissection and polymerase chain reaction for the detection of immunoglobulin gene rearrangement and translocation in primary intraocular lymphoma. Ophthalmology 105(9): 1664–1669. https://doi.org/10.1016/S0161-6420(98)99036-4

Shields CL, Manalac J, Das C, Saktanasate J, Shields JA (2015) Review of spectral domain enhanced depth imaging optical coherence tomography of tumors of the choroid. Indian J Ophthalmol 63(2):117–121. https://doi.org/10.4103/0301-4738.154377

Sjö LD (2009) Ophthalmic lymphoma: epidemiology and pathogenesis. Acta Ophthalmol 87 Thesis 1:1–20. https://doi.org/10.1111/j.1755-3768.2008.01478.x

Smith JR, Rosenbaum JT, Wilson DJ, Doolittle ND, Siegal T, Neuwelt EA, Pe'er J (2002) Role of intravitreal methotrexate in the management of primary central nervous system lymphoma with ocular involvement. Ophthalmology 109(9):1709–1716. https://doi.org/10.1016/s0161-6420(02)01125-9

Sobrin L, Dubovy SR, Davis JL, Murray TG (2005) Isolated, bilateral intraocular lymphoma in a 15-year-old girl. Retina (Philadelphia, Pa.) 25(3):370–373. https://doi.org/10.1097/00006982-200504000-00021

Sou R, Ohguro N, Maeda T, Saishin Y, Tano Y (2008) Treatment of primary intraocular lymphoma with intravitreal methotrexate. Jpn J Ophthalmol 52(3):167–174. https://doi.org/10.1007/s10384-008-0519-9

Soussain C, Merle-Béral H, Reux I, Sutton L, Fardeau C, Gerber S, Ben Othman T, Binet JL, Lehoang P, Leblond V (1996) A single-center study of 11 patients with intraocular lymphoma treated with conventional chemotherapy followed by high-dose chemotherapy and autologous bone marrow transplantation in 5 cases. Leuk Lymphoma 23(3–4):339–345. https://doi.org/10.3109/10428199609054837

Soussain C, Hoang-Xuan K, Taillandier L, Fourme E, Choquet S, Witz F, Casasnovas O, Dupriez B, Souleau B, Taksin AL, Gisselbrecht C, Jaccard A, Omuro A, Sanson M, Janvier M, Kolb B, Zini JM, Leblond V, Société Française de Greffe de Moëlle Osseuse-Thérapie Cellulaire (2008) Intensive chemotherapy followed by hematopoietic stem-cell rescue for refractory and recurrent primary CNS and intraocular lymphoma: Société Française de Greffe de Moëlle Osseuse-Thérapie Cellulaire. J Clin Oncol 26(15):2512–2518. https://doi.org/10.1200/JCO.2007.13.5533

Stefanovic A, Davis J, Murray T, Markoe A, Lossos IS (2010) Treatment of isolated primary intraocular lymphoma with high-dose methotrexate-based chemotherapy and binocular radiation therapy: a single-institution experience. Br J Haematol 151(1):103–106. https://doi.org/10.1111/j.1365-2141.2010.08321.x

Strauchen JA, Dalton J, Friedman AH (1989) Chemotherapy in the management of intraocular lymphoma. Cancer 63(10):1918–1921. https://doi.org/10.1002/1097-0142(19890515)63:10<1918::aid-cncr2820631008>3.0.co;2-d

Sugita S, Takase H, Sugamoto Y, Arai A, Miura O, Mochizuki M (2009) Diagnosis of intraocular lymphoma by polymerase chain reaction analysis and cytokine profiling of the vitreous fluid. Jpn J Ophthalmol 53(3):209–214. https://doi.org/10.1007/s10384-009-0662-y

Swerdlow SH, Kuzu I, Dogan A, Dirnhofer S, Chan JK, Sander B, Ott G, Xerri L, Quintanilla-Martinez L, Campo E (2016a) The many faces of small B cell lymphomas with plasmacytic differentiation and the contribution of MYD88 testing. Virchows Arch Int J Pathol 468(3): 259–275. https://doi.org/10.1007/s00428-015-1858-9

Swerdlow SH, Campo E, Pileri SA, Harris NL, Stein H, Siebert R, Advani R, Ghielmini M, Salles GA, Zelenetz AD, Jaffe ES (2016b) The 2016 revision of the World Health Organization classification of lymphoid neoplasms. Blood 127(20):2375–2390. https://doi.org/10.1182/blood-2016-01-643569

Takase H, Arai A, Iwasaki Y, Imai A, Nagao T, Kawagishi M, Ishida T, Mochizuki M (2022) Challenges in the diagnosis and management of vitreoretinal lymphoma – Clinical and basic approaches. Prog Retinal Eye Res:101053. Advance online publication. https://doi.org/10.1016/j.preteyeres.2022.101053

Takeda A, Yoshikawa H, Fukuhara T, Hikita S, Hijioka K, Otomo T, Arita R, Hisatomi T, Kimura K, Yoshida S, Kawano Y, Sonoda KH, Ishibashi T (2015) Distinct profiles of soluble cytokine receptors between B-cell vitreoretinal lymphoma and uveitis. Invest Ophthalmol Vis Sci 56(12):7516–7523. https://doi.org/10.1167/iovs.15-17465

Tan SZ, Steeples LR, Chhabra R, Jones NP (2018) An unusual case report of primary vitreoretinal lymphoma. BMC Ophthalmol 18(Suppl 1):223. https://doi.org/10.1186/s12886-018-0860-9

Tanaka R, Kaburaki T, Taoka K, Karakawa A, Tsuji H, Nishikawa M, Yatomi Y, Shinozaki-Ushiku A, Ushiku T, Araki F (2021) More accurate diagnosis of vitreoretinal lymphoma using a combination of diagnostic test results: a prospective observational study. Ocul Immunol Inflamm 30:1–7. Advance online publication. https://doi.org/10.1080/09273948.2021.1873394

Tempescul A, Pradier O, Marianowski-Cochard C, Ianotto JC, Berthou C (2011) Combined therapy associating systemic platinum-based chemotherapy and local radiotherapy into the treatment of primary intraocular lymphoma. Ann Hematol 90(9):1117–1118. https://doi.org/10.1007/s00277-010-1139-2

Tiu Teo HM, Çiftçi S, Elner VM, Demirci H (2019) Systemic anti-CD20 (rituximab) as primary treatment for symptomatic primary uveal lymphoma. Am J Ophthalmol Case Rep 15:100484. https://doi.org/10.1016/j.ajoc.2019.100484

Tranos PG, Andreou PS, Wickremasinghe SS, Brazier JD (2002) Pseudohypopyon as a feature of multiple myeloma. Arch Ophthalmol (Chicago, Ill.: 1960) 120(1):87–88

Treon SP, Xu L, Yang G, Zhou Y, Liu X, Cao Y, Sheehy P, Manning RJ, Patterson CJ, Tripsas C, Arcaini L, Pinkus GS, Rodig SJ, Sohani AR, Harris NL, Laramie JM, Skifter DA, Lincoln SE, Hunter ZR (2012) MYD88 L265P somatic mutation in Waldenström's macroglobulinemia. N Engl J Med 367(9):826–833. https://doi.org/10.1056/NEJMoa1200710

Trudeau M, Shepherd FA, Blackstein ME, Gospodarowicz M, Fitzpatrick P, Moffatt KP (1988) Intraocular lymphoma: report of three cases and review of the literature. Am J Clin Oncol 11(2): 126–130

Tuaillon N, Chan CC (2001) Molecular analysis of primary central nervous system and primary intraocular lymphomas. Curr Mol Med 1(2):259–272. https://doi.org/10.2174/1566524013363915

Usui Y, Wakabayashi Y, Okunuki Y, Kimura K, Tajima K, Matsuda R, Ueda S, Ma J, Nagai T, Mori H, Kezuka T, Nagao T, Goto H (2012) Immune mediators in vitreous fluids from patients with vitreoretinal B-cell lymphoma. Invest Ophthalmol Vis Sci 53(9):5395–5402. https://doi.org/10.1167/iovs.11-8719

Valentincic NV, de Groot-Mijnes JD, Kraut A, Koroscc P, Hawlina M, Rothova A (2011) Intraocular and serum cytokine profiles in patients with intermediate uveitis. Mol Vis 17:2003–2010

Valluri S, Moorthy RS, Khan A, Rao NA (1995) Combination treatment of intraocular lymphoma. Retina (Philadelphia, Pa.) 15(2):125–129. https://doi.org/10.1097/00006982-199515020-00006

van Dongen JJ, Langerak AW, Brüggemann M, Evans PA, Hummel M, Lavender FL, Delabesse E, Davi F, Schuuring E, García-Sanz R, van Krieken JH, Droese J, González D, Bastard C, White HE, Spaargaren M, González M, Parreira A, Smith JL, Morgan GJ, . . ., Macintyre EA (2003) Design and standardization of PCR primers and protocols for detection of clonal immunoglobulin and T-cell receptor gene recombinations in suspect lymphoproliferations: report of the

BIOMED-2 Concerted Action BMH4-CT98-3936. Leukemia 17(12):2257–2317. https://doi.org/10.1038/sj.leu.2403202

Velez G, Boldt HC, Whitcup SM, Nussenblatt RB, Robinson MR (2002a) Local methotrexate and dexamethasone phosphate for the treatment of recurrent primary intraocular lymphoma. Ophthalmic Surg Lasers 33(4):329–333

Velez G, Chan CC, Csaky KG (2002b) Fluorescein angiographic findings in primary intraocular lymphoma. Retina (Philadelphia, Pa) 22(1):37–43. https://doi.org/10.1097/00006982-200202000-00007

Verbraeken HE, Hanssens M, Priem H, Lafaut BA, De Laey JJ (1997) Ocular non-Hodgkin's lymphoma: a clinical study of nine cases. Br J Ophthalmol 81(1):31–36. https://doi.org/10.1136/bjo.81.1.31

Vicuna-Kojchen J, Frenkel S, Siegal T, Shalom E, Chowers I, Pe'er J (2008) Maculopathy in patients with primary CNS lymphoma treated with chemotherapy in conjunction with blood-brain barrier disruption. Br J Ophthalmol 92(2):231–235. https://doi.org/10.1136/bjo.2007.130062

Wallace DJ, Shen D, Reed GF, Miyanaga M, Mochizuki M, Sen HN, Dahr SS, Buggage RR, Nussenblatt RB, Chan CC (2006) Detection of the bcl-2 t(14;18) translocation and proto-oncogene expression in primary intraocular lymphoma. Invest Ophthalmol Vis Sci 47(7):2750–2756. https://doi.org/10.1167/iovs.05-1312

Wang JK, Yang CM, Lin CP, Shan YD, Lo AY, Tien HF (2006) An Asian patient with intraocular lymphoma treated by intravitreal methotrexate. Jpn J Ophthalmol 50(5):474–478. https://doi.org/10.1007/s10384-005-0327-4

Wang Y, Shen D, Wang VM, Sen HN, Chan CC (2011) Molecular biomarkers for the diagnosis of primary vitreoretinal lymphoma. Int J Mol Sci 12(9):5684–5697. https://doi.org/10.3390/ijms12095684

Wang JQ, Jeelall YS, Ferguson LL, Horikawa K (2014) Toll-like receptors and cancer: MYD88 mutation and inflammation. Front Immunol 5:367. https://doi.org/10.3389/fimmu.2014.00367

Whitcup SM, de Smet MD, Rubin BI, Palestine AG, Martin DF, Burnier M Jr, Chan CC, Nussenblatt RB (1993) Intraocular lymphoma. Clinical and histopathologic diagnosis. Ophthalmology 100(9):1399–1406. https://doi.org/10.1016/s0161-6420(93)31469-7

Whitcup SM, Stark-Vancs V, Wittes RE, Solomon D, Podgor MJ, Nussenblatt RB, Chan CC (1997) Association of interleukin 10 in the vitreous and cerebrospinal fluid and primary central nervous system lymphoma. Arch Ophthalmol (Chicago, Ill.: 1960) 115(9):1157–1160. https://doi.org/10.1001/archopht.1997.01100160327010

Whitcup SM, Chan CC, Buggage RR, Nussenblatt RB, Byrnes GA, Rubin BI (2000) Improving the diagnostic yield of vitrectomy for intraocular lymphoma. Arch Ophthalmol (Chicago, Ill.: 1960) 118(3):446

White VA, Gascoyne RD, Paton KE (1999) Use of the polymerase chain reaction to detect B- and T-cell gene rearrangements in vitreous specimens from patients with intraocular lymphoma. Arch Ophthalmol (Chicago, Ill.: 1960) 117(6):761–765. https://doi.org/10.1001/archopht.117.6.761

Wilson DJ, Braziel R, Rosenbaum JT (1992) Intraocular lymphoma. Immunopathologic analysis of vitreous biopsy specimens. Arch Ophthalmol (Chicago, Ill.: 1960) 110(10):1455–1458. https://doi.org/10.1001/archopht.1992.01080220117032

Witzig TE, Vose JM, Zinzani PL, Reeder CB, Buckstein R, Polikoff JA, Bouabdallah R, Haioun C, Tilly H, Guo P, Pietronigro D, Ervin-Haynes AL, Czuczman MS (2011) An international phase II trial of single-agent lenalidomide for relapsed or refractory aggressive B-cell non-Hodgkin's lymphoma. Ann Oncol 22(7):1622–1627. https://doi.org/10.1093/annonc/mdq626

Wolf LA, Reed GF, Buggage RR, Nussenblatt RB, Chan CC (2003) Vitreous cytokine levels. Ophthalmology 110(8):1671–1672. https://doi.org/10.1016/S0161-6420(03)00811-X

Woolf DK, Ahmed M, Plowman PN (2012) Primary lymphoma of the ocular adnexa (orbital lymphoma) and primary intraocular lymphoma. Clin Oncol (R Coll Radiol) 24(5):339–344. https://doi.org/10.1016/j.clon.2012.03.001

Wysenbeek YS, Nissenkorn I, Cohen S, Ben-Sira I, Stark B, Zaizov R (1987) Intraocular involvement of Burkitt's lymphoma in a Bedouin child. Pediatr Hematol Oncol 4(4):309–314. https://doi.org/10.3109/08880018709141283

Yang G, Buhrlage SJ, Tan L, Liu X, Chen J, Xu L, Tsakmaklis N, Chen JG, Patterson CJ, Brown JR, Castillo JJ, Zhang W, Zhang X, Liu S, Cohen P, Hunter ZR, Gray N, Treon SP (2016) HCK is a survival determinant transactivated by mutated MYD88, and a direct target of ibrutinib. Blood 127(25):3237–3252. https://doi.org/10.1182/blood-2016-01-695098

Yonese I, Takase H, Yoshimori M, Onozawa E, Tsuzura A, Miki T, Mochizuki M, Miura O, Arai A (2019) CD79B mutations in primary vitreoretinal lymphoma: diagnostic and prognostic potential. Eur J Haematol 102(2):191–196. https://doi.org/10.1111/ejh.13191

Zaldivar RA, Martin DF, Holden JT, Grossniklaus HE (2004) Primary intraocular lymphoma: clinical, cytologic, and flow cytometric analysis. Ophthalmology 111(9):1762–1767. https://doi.org/10.1016/j.ophtha.2004.03.021

Zamora DO, Riviere M, Choi D, Pan Y, Planck SR, Rosenbaum JT, David LL, Smith JR (2007) Proteomic profiling of human retinal and choroidal endothelial cells reveals molecular heterogeneity related to tissue of origin. Mol Vis 13:2058–2065

Bleeding in the Setting of Lymphoma, with a Focus on Waldenström Macroglobulinaemia

Simone A. Brysland, Muhammad Gohar Maqbool, Dipti Talaulikar, and Elizabeth E. Gardiner

Abstract

Bleeding symptoms are common in haematological malignancies, with 26% experiencing minor to moderate bleeding and 6% experiencing major or fatal haemorrhages. However, bleeding is underreported and remains poorly understood in this patient cohort. This book chapter will discuss the known contributors to bleeding in haematological malignancies, including possible platelet dysfunction, common therapies that increase bleeding risk, limitations of tests evaluating bleeding in the diagnostic laboratory and future directions for research in this field. This book chapter will particularly focus on Waldenström Macroglobulinaemia (WM), a rare, low-grade, incurable B-cell lymphoma, which offers interesting insights into bleeding in haematological malignancies, as it is characterised by 1) bone marrow (BM) and organ infiltration by proliferating lymphoplasmacytoid B cells, and 2) immunoglobulin (Ig) M hypersecretion. Interestingly, bleeding has been observed in WM and other haematological malignancy patients who do not have thrombocytopenia, acquired von Willebrand syndrome (AVWS), disseminated intravascular coagulation (DIC), blood hyperviscosity, cryoglobulinaemia or hyperfibrinolysis, implying that other potential contributors exist. One alternative contributor is a possible acquired platelet lesion, caused by disruption to the BM microenvironment and megakaryocyte (MK)-mediated platelet production by

---

Authors Dipti Talaulikar and Elizabeth E. Gardiner have equally contributed to this chapter.

S. A. Brysland · E. E. Gardiner (✉)
Australian Cancer Research Foundation Department of Cancer Biology and Therapeutics, John Curtin School of Medical Research, The Australian National University, Canberra, ACT, Australia
e-mail: elizabeth.gardiner@anu.edu.au

M. G. Maqbool · D. Talaulikar
School of Medicine and Psychology, College of Health and Medicine, Australian National University, Canberra, ACT, Australia

© The Author(s), under exclusive license to Springer Nature Switzerland AG 2022
Interdisciplinary Cancer Research, https://doi.org/10.1007/16833_2022_98
Published online: 30 December 2022

infiltrating malignant cells, or by acquisition of common driver mutations in the megakaryocytes (MKs). The resulting platelets could have altered adhesion, activation and aggregation responses due to the loss of important surface receptors, such as glycoprotein (GP) Ibα and GPVI; or altered intra-platelet signalling due to the acquisition of mutant constitutively active signalling proteins. Additionally, bleeding is exacerbated in haematological malignancies by standard and novel therapies, and the treatment of comorbidities with anticoagulants and antiplatelets further increases this risk. Therefore, it is important to better understand the molecular mechanisms behind bleeding in haematological malignancies, to stratify patients based on bleeding risk, aid in clinical decisions surrounding therapy, minimise bleeding morbidity and improve patients' quality of life.

Keywords

Haemostasis · Hyperviscosity · Platelets · Receptors · Thrombocytopenia

1 Introduction

Haematological malignancies are characterised by the overproliferation of malignant lymphoid or myeloid haematopoietic cells, at various stages of maturation, that accumulate in specific regions of the body (bone marrow [BM], peripheral blood or lymph nodes). In the United States, one person is diagnosed with a haematological malignancy every 3 minutes, with over 1.5 million people currently living with active disease or in remission, highlighting the scale at which these malignancies impact the general population. Out of these patients, 26% will experience mild to moderate bleeding, such as bruising and chronic nose bleeds, and 6% will experience major and sometimes fatal intracranial or gastrointestinal (GI) haemorrhages. Therefore, bleeding represents an underappreciated and poorly understood lifelong morbidity in patients with haematological malignancies. Additionally, standard and novel therapies, such as chemotherapy and Bruton's tyrosine kinase (BTK) inhibitors, carry an attendant bleeding risk, which can be potentially life threatening. It is critical to understand bleeding at a molecular level in haematological malignancies, to enable the stratification of patients for bleeding risk and aid clinicians in making confidently informed decisions surrounding therapy, which minimise bleeding morbidity and improve quality of life. Waldenström Macroglobulinaemia (WM) is a rare, low-grade, indolent, B-cell lymphoma, which offers an interesting angle on bleeding in haematological malignancies and will be focused on throughout this book chapter, as it is both characterised by the infiltration of the BM and other organs by malignant lymphoplasmacytoid cells, and the hypersecretion of haemostasis-inhibiting immunoglobulin (Ig) Ms.

2 Haematological Malignancies, with a Focus on Waldenström Macroglobulinaemia

2.1 The Genetic Background and Clinical Features of Haematological Malignancies

Haematological malignancies are characterised by the overproliferation of malignant haematopoietic cells leading to constitutional symptoms, cytopenias and organomegaly. The proliferative features of these malignancies can be explained by specific driver mutations which constitutively activate important intracellular signalling pathways and promote malignant cell survival and proliferation. Some examples of these driver mutations include the JAK2^{V617F} mutation in 95% of polycythaemia vera (PV) myeloproliferative neoplasm (MPN) patients (Rumi et al. 2014), the BCR-ABL translocation in 90% of chronic myeloid leukaemia (CML) patients (Sattler and Griffin 2003), the PML-RARα translocation in 98% of acute promyelocytic lymphoma (APL) patients (Liquori et al. 2020) and the BRAF^{V600E} mutation in 80% of hairy cell leukaemia (HCL) patients (Falini et al. 2016). Other haematological malignancies, such as chronic lymphocytic leukaemia (CLL), acute myeloid leukaemia (AML) and myelodysplastic syndromes (MDS), have a more complex genomic landscape and a highly prevalent driver mutation is usually not identified (Gaidano and Rossi 2017).

2.2 Waldenström Macroglobulinaemia

Waldenström Macroglobulinaemia (WM) is a rare, low-grade, indolent, B-cell lymphoma, characterised by the accumulation of IgM-secreting malignant lymphoplasmacytoid cells in the BM and other organs (Talaulikar et al. 2017). WM is a subtype of lymphoplasmacytic lymphoma, making up 5% of non-Hodgkin lymphomas (NHLs) and occurring at a rate of 3/1,000,000 cases per year (Talaulikar et al. 2017). WM is an incurable disease with a median overall survival of 10–12 years (Castillo et al. 2014).

WM is preceded by two sequential pre-malignant asymptomatic states. The first, IgM monoclonal gammopathy of undetermined significance (MGUS), is characterised by an intermediate IgM level (1.5–30 g/L) and low BM involvement (<10%). IgM-MGUS progresses into smouldering WM (SWM) at a rate of ~1.5% per year or 10% in 5 years (Kyle et al. 2009; Maqbool et al. 2020). SWM is characterised by higher IgM levels and/or high BM infiltration (>10%). SWM progresses into symptomatic WM at a rate of ~12% per year, or 55% in 5 years (Maqbool et al. 2020; Bustoros et al. 2019), thus requiring treatment intervention.

WM symptoms, which indicate need for treatment, include anaemia (in 72%), constitutional symptoms (in 58%), IgM-related symptoms (blood hyperviscosity, neuropathy, in 54%), organomegaly (in 17%) and mild bleeding (Buske et al. 2018). These symptoms are a direct consequence of the lymphoma, which develops as a result of a specific genetic landscape.

WM pathology can be explained by several distinct driver mutations found in the malignant lymphoplasmacytoid B cells. The major driver mutation is the gain-of-function leucine-265 to proline (L265P) mutation in the myeloid differentiation primary response (*MYD*) 88 gene, identified in over 90% of patients (Treon et al. 2012). Myd88 is a Toll-like receptor (TLR) and Interleukin-1 receptor (IL-1R) adaptor protein, which plays a crucial role in immune cell signalling (Fig. 1). Following receptor-ligand binding, Myd88 complexes with Interleukin 1 receptor-associated kinase (IRAK) 1/4, which enables the nuclear translocation of the nuclear factor (NF)-κB transcription factor (Fig. 1) (Deguine and Barton 2014). Myd88^{L265P} does not require TLR or IL-1R receptor homodimerisation and complexes with IRAK1/4 faster than Myd88^{WT}, resulting in a constitutively active Myd88-NF-κB pathway and increased lymphoplasmacytoid B-cell proliferation, survival and IgM secretion (Fig. 1) (Yu et al. 2021). *MYD88*^{L265P} is not specific to WM, as it also occurs in 60% of IgM-MGUS and 30% of diffuse large B-cell lymphoma (DLBCL) patients (Landgren and Tageja 2014; Jimenez et al. 2013; Varettoni et al. 2013; Xu et al. 2013). *MYD88*^{L265P} is insufficient in driving oncogenesis, but likely acts an early oncogenic event requiring additional mutations for the development of pre-malignant IgM-MGUS into malignant WM (Jimenez et al. 2013). Aside from MYD88^{L265P}, several other somatic NF-κB activating mutations have also been identified (*TBL1XR1, PTPN13, MALT1, BCL10, NFKB2, NFKBIB, NFKBIZ* and *UDRL1F*), particularly in *MYD88*^{WT} patients, highlighting the importance of the Myd88-NF-κB pathway in WM pathology (Hunter et al. 2018).

The next most abundant driver mutations in WM occur in the C-X-C chemokine receptor (*CXCR*) 4 gene in 30% of WM patients (Hunter et al. 2014;

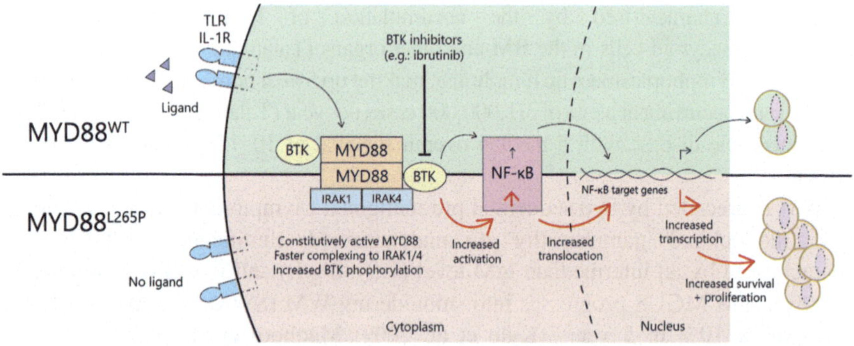

Fig. 1 B-cell development in MYD88^{WT} healthy individuals and MYD88^{L265P} WM patients. In B-cell malignancy patients, somatic mutations and epigenetic dysregulation occur in the B lymphocytes. The most common driver mutation in WM is MYD88^{L265P}, which occurs in over 90% of patients. MYD88^{L265P} complexes to IRAK1/4 faster than MYD88^{WT}, in response to low or absent TLR or IL-1R stimulation. This increases BTK phosphorylation, NF-κB activation, nuclear translocation and target gene transcription, which increases lymphoplasmacytoid cell survival, proliferation and IgM secretion. *WM* Waldenström Macroglobulinaemia, *MYD88* Myeloid differentiation primary response 88, *TLR* Toll-like receptor, *IL-1R* Interleukin-1 receptor, *IRAK* IL-1 receptor-associated kinase, *BTK* Bruton's tyrosine kinase, *NF-κB* nuclear factor kappa-light-chain-enhancer of activated B cells, *IgM* Immunoglobulin M

Poulain et al. 2016; Treon et al. 2015). CXCR4 is involved in the homing of cells to the BM by engaging the stromal-derived factor (SDF)-1 chemokine. The common *CXCR4*^{S338X} frameshift gain-of-function mutation prevents CXCR4 internalisation following SDF-1 stimulation (Dotta et al. 2011), which results in a constitutively active pathway, aberrant lymphoplasmacytoid cell infiltration into the BM, and increased proliferation and survival (Cao et al. 2015; Roccaro et al. 2014). CXCR4 is also overexpressed in other haematological malignancies, such as AML (Brault et al. 2014), acute lymphoblastic leukaemia (ALL) (Crazzolara et al. 2001), CLL (Han et al. 2014) and MDS (Zhang et al. 2013), and has been associated with poor prognosis and resistance to therapy.

The driver mutations of WM and other haematological malignancies can explain some of the proliferative and marrow-infiltrating features of the diseases. However, many additional mutations are required for oncogenesis, including mutations in epigenetic regulation (*DNMT3A*, *KMT2D*, *KMT2C* and *KDM6A*) and DNA damage repair genes (*TP53*, *ATM* and *TRRAP*) (Hunter et al. 2018; Roos-Weil et al. 2020). Despite the seminal discoveries in the haematological malignancy field this millennium, one feature that still remains poorly understood is the bleeding phenotype, which will be discussed further in this book chapter.

3 Haematological Malignancy Patients Can Present with a Bleeding Phenotype

Easy bruising and mild mucosal bleeding are reported as common symptoms in WM, however few studies have described their prevalence. Of the four studies available, bleeding was observed in 17% of patients ($n = 615$), with minimal description of severity (Gertz et al. 2004; Merchionne et al. 2011; Zangari et al. 2007; Garcia-Sanz et al. 2001). Bleeding is also a common symptom in other pre-treated haematological malignancies. Any-grade bleeding was observed in 26% of patients with haematological malignancies, of which 6% were fatal ($n = 469$) (Cartoni et al. 2009). Major haemorrhages were observed in 9.1% of CLL patients ($n = 24{,}198$) (Georgantopoulos et al. 2019) and 2.4% of APL patients (n = 792) (Choudhry and DeLoughery 2012), mild bleeding was observed in 40–70% of ALL patients at diagnosis (Bick 1992), 21.3% of CML patients ($n = 43\ 0$) (Savage et al. 1997) and 3–53% of MDS patients, of which 14–24% were fatal ($n = 728$) (Franchini et al. 2013; Matsuda et al. 2005). Several pathophysiological features of WM and other haematological malignancies are known to contribute to their bleeding phenotype and will be discussed below.

3.1 Thrombocytopenia

Thrombocytopenia is characterised by a platelet count below 150×10^9/L. In haematological malignancies associated with marrow infiltration, such as WM, mantle cell lymphoma (MCL) and marginal zone lymphoma (MZL), thrombocytopenia is predominantly caused by the overcrowding of the BM and platelet-producing resident megakaryocytes (MKs) by malignant cells, resulting in impaired platelet production. In 1–5% of CLL patients, thrombocytopenia can be immune-mediated, whereby non-malignant, self-reactive B lymphocytes produce IgG autoantibodies leading to platelet destruction (Fattizzo and Barcellini 2020). Thrombocytopenia is also occasionally caused by the sequestration of platelets into the spleen. Thrombocytopenia is observed in 18% of WM patients (Buske et al. 2018), but is an unlikely cause of bleeding, as bleeding also occurs in patients with normal platelet counts (150–400 × 10^9 platelets/L) and WM platelet counts rarely drop below the threshold of 20 × 10^9/L requiring prophylactic platelet transfusion intervention.

3.2 Acquired von Willebrand Syndrome

Acquired von Willebrand syndrome (AVWS) is characterised by the loss of high-molecular-weight von Willebrand factor (VWF) multimers from circulation. AVWS always occurs in conjunction with an underlying disorder, of which 48% are lymphoproliferative, such as WM, CLL, multiple myeloma (MM) and MGUS; and 15% are myeloproliferative (Federici et al. 2000, 2013). In haematological malignancies, AVWS is caused by one or more of the following factors (Mital 2016): (1) the binding of autoantibodies to VWF functional or non-functional domains, resulting in VWF inhibition or the rapid clearance of autoantibody-VWF-Factor VIII (FVIII) immune complexes by the reticuloendoethlial system, as seen in MM and MGUS (Dicke et al. 2016); (2) the absorption of VWF-FVIII complexes onto the surface of the malignant cells and their subsequent clearance, as seen in MM, NHLs and HCL (Richard et al. 1990; Massimo and Pier 2020); and (3) increased fluid shear stress and VWF multimer unfolding and proteolysis by ADAMTS-13, as seen in cardiovascular diseases (Mehta et al. 2019; Tsai et al. 1994). AVWS occurs in 6% of WM patients, resulting in mucocutaneous and GI bleeding. In these patients, VWF/FVIII antigen levels and VWF activity below 50% of the healthy mean were correlated with high IgM levels (>30 g/L) (Castillo et al. 2019). Universal AVWS treatments involve increasing VWF multimer levels with desmopressin and/or FVIII/VWF concentrate transfusions. In WM specifically, plasmapheresis is used to reduce IgM paraprotein (autoantibody) levels (Tiede et al. 2011).

3.3 Disseminated Intravascular Coagulation

Disseminated intravascular coagulation (DIC) is an acquired, systemic, thrombotic disorder characterised by excessive tissue factor (TF) release, extrinsic blood coagulation activation, micro-thrombus formation throughout the vasculature, organ dysfunction, thrombocytopenia and bleeding (Ikezoe 2021). Thrombocytopenia is caused by the excess consumption of platelets in the micro-thrombi. In haematological malignancies, TF is aberrantly presented on the surface of the malignant cells; or released by endothelial cells which are stimulated or damaged by the excessive production of interleukin (IL)-1β, tumour necrosis factor (TNF)-α and IL-6 (Franchini et al. 2010). DIC is most common in APL, occurring in 77% of patients (Chang et al. 2012), but is also observed in 17% of non-APL AML patients (Uchiumi et al. 2007) and 11% of NHL patients (Chi and Ikezoe 2015). DIC has not been reported in WM. Patients with DIC can have significant bleeding due to thrombocytopenia and consumptive coagulopathy. Management of DIC requires aggressive control of the underlying disease and additional platelet or plasma product transfusion may be required.

3.4 Blood Hyperviscosity

Blood hyperviscosity is characterised by an increase in serum viscosity and the subsequent increase in rheological drag and hypoperfusion (reduced blood flow) of tissues. Blood hyperviscosity occurs predominantly in IgM paraproteinemias (disease state characterised by the overproduction of antibodies), and rarely in IgG paraproteinemias. Affected haematological malignancies include WM and some cases of MM, PV and leukaemias. In WM, blood hyperviscosity is caused by the increase in large, 970-kDa, positively charged pentameric IgM paraproteins, of which 80% are intravascular and electrostatically bind and aggregate red blood cells (Stone 2009; Gertz et al. 2000). Hyperviscosity occurs in 13% of WM patients when IgM levels surpass 50–60 g/L, resulting in the tearing of small venules, spontaneous mucocutaneous bleeding, retinal haemorrhages and neurological symptoms (Stone 2009). Hyperviscosity likely also contributes to bleeding by perturbing the normal function of the endothelium and its ability to activate platelets and the coagulation system, due to the increase in circulating proteins (antibodies) in the vasculature. Treatments involve reducing IgM levels with plasmapheresis or reducing the number of IgM-secreting malignant cells with chemotherapy or targeted therapies.

3.5 Cryoglobulinaemia

Cryoglobulinaemia is characterised by the precipitation and aggregation of immunoglobulins below 37 °C (cryoglobulins), which is reversible with heating (Stone 2009). Cryoglobulinaemia can occur in conjunction with MGUS and B-cell malignancies, such as WM, MM and CLL. As much as 70% of cryoglobulinaemic WM patients have Type II Mixed Cryoglobulinaemia, where the monoclonal IgM paraproteins bind to the Fc portion of polyclonal IgG antibodies to form immune complex aggregates (Stone 2009). Cryoglobulinaemia is not concentration-dependent. Cryoglobulins are detected in 15% of WM patients, of which only 5% are symptomatic (Merlini et al. 2003). Symptoms are caused by the vascular obstruction of blood vessels by cryoglobulin aggregates, resulting in physical tearing, mild mucocutaneous bleeding, purpura and inflammation (systemic vasculitis, glomerulonephritis) (Stone 2009). Cryoglobulinaemia is treated identically to blood hyperviscosity.

3.6 Haemostasis-Inhibiting Paraproteins

Haematological malignancy patients display a deranged haemostatic system, highlighted by a bleeding diathesis. Studies have demonstrated an interaction between the VWF-FVIII complex and the paraproteinaemic Igs in WM, MGUS, MM, other lymphomas, CLL and amyloidosis (Loftus and Arnold 1994; Taher et al. 2003; Mayerhofer et al. 2009; McKelvey and Kwaan 1972; Castaldi and Penny 1970; Mazurier et al. 1980; Endo et al. 2000). This interaction was proven to be inhibitory in some cases, causing AVWS. In 9% of amyloidosis patients, acquired Factor (F) X deficiencies are caused by the absorption of FX onto overproduced, misfolded antibody light chains (light chain fibrils) in the spleen and liver (Dejhansathit and Suvannasankha 2019). Liver failure in these patients can also result in reduced FX production (Dejhansathit and Suvannasankha 2019). In WM, IgM cryoglobulins have been shown to inhibit granulocyte and erythroid progenitor cell growth in vitro, thus possibly contributing to the pancytopenias commonly observed in this patient cohort (Zago-Novaretti et al. 1994). As MKs and their progenitors also reside in the BM microenvironment and fall within the same myeloid lineage as the erythroid and granulocyte progenitors, it is possible that the IgM paraprotein also suppresses MK maturation in the BM, thus likely altering platelet production, quality and functionality. Direct inhibitory interactions between the IgM paraprotein and platelets have been reported in WM patients (Varticovski et al. 1987; Owen et al. 2001), with speculations that the IgM coats platelets and reduces adhesion and aggregation responses (van Breugel et al. 1992).

3.7 Hyperfibrinolysis

Hyperfibrinolysis is characterised by the excessive lysis of fibrin and clot dissolution, which results in bleeding. Fibrinolysis occurs when tissue plasminogen activator (t-PA) and urokinase-type plasminogen activator (u-PA) convert plasminogen into plasmin, which cleaves fibrin into its degradation products. t-PA and u-PA are inhibited by plasminogen activator inhibitor (PAI)-1, and plasmin is inhibited by α2-antiplasmin (α2-AP).

Increased fibrinolytic activity has predominantly been observed in APL, but also to a lesser extent in amyloidosis. In APL, malignant cells express Annexin II on their surface, which increases t-PA-mediated plasmin generation 60-fold (Cesarman et al. 1994). To add to this, increased levels of fibrinolytic activators t-PA and u-PA, and decreased levels fibrinolytic inhibitors PAI-1 and α2-AP also contribute to a hyperfibrinolytic system in APL. In pre-treated amyloidosis patients, bleeding complications were associated with low overall fibrinolysis times (euglobulin lysis times), increased plasmin generation due to elevated levels of u-PA expressed on malignant cells in the BM (Uchiba et al. 2009), and low levels of the plasmin inhibitor α2-AP, which were corrected with treatment with an antifibrinolytic drug (Franchini et al. 2021; Meyer and Williams 1985). Fibrinolysis has only been evaluated in a small NHL study ($n = 30$), where no change was reported (Franchini et al. 2021; Zurborn et al. 1990).

In summary, thrombocytopenia, AVWS, DIC, blood hyperviscosity, cryoglobulinaemia and hyperfibrinolysis represent known causes of bleeding in WM and other haematological malignancies. Bleeding is almost certainly multifactorial by nature, and bleeding risk needs to be evaluated with this in mind. Currently, bleeding parameters are evaluated in isolation and are only flagged if the patient falls below a pre-established healthy threshold. However, it is unlikely that an individual will bleed with isolated thrombocytopenia and a platelet count above 20×10^9/L. On the other hand, consider a patient with a drop in platelet count from 350 to 180×10^9/L (a drastic change that would not be immediately picked up by clinicians), in conjunction with intermediately, but not alarmingly, increased IgM levels, and slightly deranged VWF multimers and fibrinolysis. This patient would potentially be at a much higher risk of bleeding than a simply thrombocytopenic patient, despite all tests returning relatively normal results. Therefore, all bleeding parameters need to be closely evaluated together using a scoring matrix to accurately predict bleeding risk.

Additionally, bleeding has also been observed in the absence of these sequelae (Merchionne et al. 2011; Haider et al. 2013), implying that other potential causes exist. One possibility is that patients possess an underlying platelet lesion, which manifests itself as bleeding complications following trauma (surgery, infection or chemotherapy) rather than chronic bleeding symptoms (Zangari et al. 2009; Luu et al. 2018). This possible platelet lesion and its contribution to the WM bleeding phenotype will be discussed below.

4 Platelet Dysfunction Possibly Contributes to Bleeding in Haematological Malignancies

4.1 Vascular Disturbances in the Bone Marrow Disrupt Platelet Production

Platelets are most notably known for their role in haemostasis, the cessation of bleeding (Furie and Furie 2008). Platelets are produced by MKs in the exquisitely controlled BM microenvironment (Kaushansky 2015). This microenvironment is composed of haematopoietic, fibroblast, mesenchymal, osteoblast, osteoclast, adipocyte and endothelial cells, which enrich the space with extracellular matrix (ECM) proteins, cytokines, chemokines, growth factors and metabolites (Fig. 2). Together, these promote hematopoietic cell maintenance, migration, proliferation and differentiation (Mendelson and Frenette 2014).

During thrombopoiesis (platelet production), hematopoietic stem cells (HSCs) migrate from the endosteal niche to the vascular niche in the BM, where they undergo several lineage commitment steps to become MKs (Fig. 2) (Noetzli et al. 2019; Stone et al. 2022). These MKs gradually acquire surface receptors (αIIbβ3, followed by glycoprotein (GP) Ibα, then GPVI) (Lagrue-Lak-Hal et al. 2001), undergo endomitosis (DNA replication without cell division) and extend proplatelets into nearby sinusoidal blood vessels (Fig. 2). These proplatelets bud off from the MK membrane and become mature circulating platelets, replete with surface receptors, granules, RNAs, Golgi apparatuses, ribosomes and mitochondria of MK origin (Fig. 2) (Rondina and Weyrich 2015). In healthy individuals, platelet counts are tightly regulated and stable, but range from 150 to 400 × 10^9/L (Grozovsky et al. 2015).

WM and other haematological malignancies are characterised by malignant cell homing, infiltration and overcrowding of the BM (Fig. 2), which is amplified by the gain-of-function mutations to the *CXCR4* gene. The WM BM microenvironment has been shown to contribute to WM pathogenesis by promoting lymphoplasmacytoid cell survival and proliferation (Jalali and Ansell 2018). This is enabled by: (1) increased mast cell numbers and proliferative CD40-CD40L interactions with the malignant lymphoplasmacytoid cells (Tournilhac et al. 2006), (2) increased IL-6, granulocyte colony-stimulating factor (G-CSF), C-C motif ligand (CCL)-5 and IL-21 levels resulting in increased malignant cell proliferation and IgM secretion (Elsawa et al. 2011; Hodge et al. 2012), (3) overexpression of programmed death-ligand (PD-L)-1 by monocytes resulting in compromised T-cell function, immune evasion and proliferation by malignant cells (Jalali et al. 2018), and (4) increased glutathione metabolism by malignant cells to protect from increases in BM oxidative stress (Jalali et al. 2020) caused by increased lipid peroxidation (Jalali et al. 2021).

The BM microenvironment is also disrupted in other haematological malignancies, which contributes to disease progression and therapy resistance (Méndez-Ferrer et al. 2020). In MDS, there is increased apoptosis of BM mesenchymal stem cells (MSCs) (Fig. 2), which results in increased IL-1β, IL-6 and tumour necrosis factor (TNF)-α production, increased inflammation and malignant cell propagation and engraftment to the BM (Flores-Figueroa et al. 2002). In T-cell ALL, a loss of osteoblasts is associated with impaired HSC function and haematopoiesis, and accelerated disease progression (Fig. 2) (Hawkins et al. 2016). In CLL and AML, extracellular vesicles (EVs) produced by malignant cells cause a reduction in the production of HSC-supporting-factors by BM stromal cells, which alters natural killer T-cell function and accelerates disease progression (Kumar et al. 2018; Szczepanski et al. 2011; Paggetti et al. 2015). Finally, in DLBCL and CLL, neutrophils support malignant cell survival via contact-dependent signalling through B-cell-activating factor and a proliferation-inducing ligand; and by stimulating stromal cells to secrete CCL2 and IL-8 to recruit additional neutrophils, promote their survival, trigger neutrophil extracellular trap (NET) formation and activate the NF-κB pathway (Grégoire et al. 2015; Sangaletti et al. 2014).

Such drastic changes to the usually intricately controlled BM microenvironment, particularly physical changes to cell distributions and vasculature due to

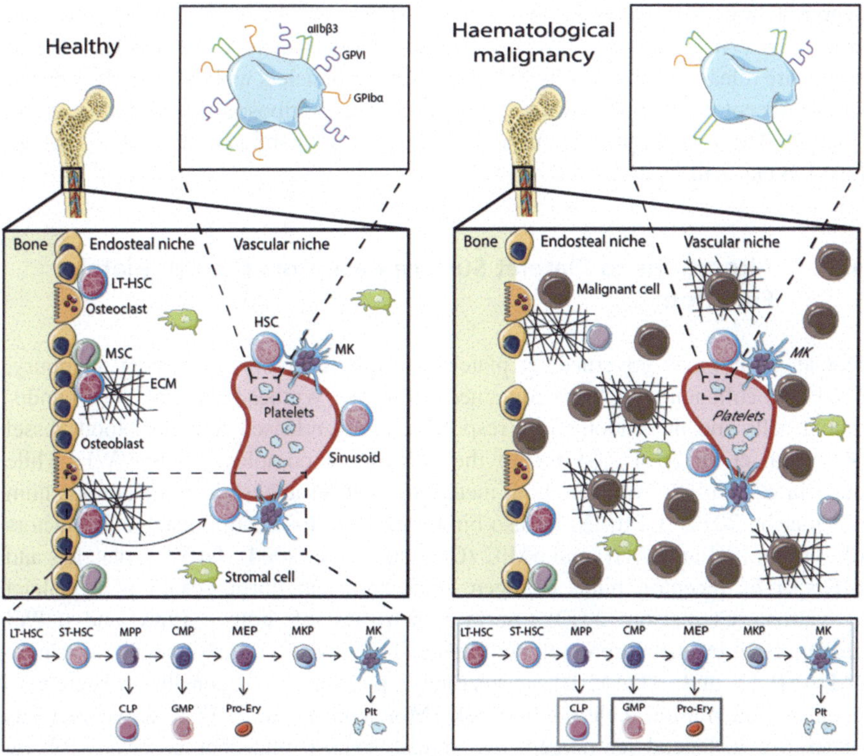

Fig. 2 The bone marrow microenvironment in health and in haematological malignancies. In healthy individuals, osteoblasts, osteoclasts, quiescent LT-HSCs, MSCs and ECM proteins reside in the endosteal niche within the BM. LT-HSCs migrate to sinusoids in the vascular niche and undergo multiple lineage commitment steps to become MKs. MKs mature, undergo endomitosis, extend proplatelets into the sinusoids, which break off and become circulating platelets. Platelets obtain many surface receptors, including αIIbβ3, GPIbα and GPVI, as well as granules, RNAs, Golgi apparatuses, ribosomes and mitochondria, all of MK origin. In haematological malignancies, malignant cells infiltrate and overcrowd the BM. As a result, there is physical disruption to distribution of hematopoietic cells, ECM proteins and vasculature in the BM microenvironment. There is a loss of osteoblasts, osteoclasts and MSCs, and an interference in stromal cell functions, which are all important in maintaining HSC populations. In WM specifically, IgM paraproteins inhibit GMP and Pro-Ery growth (dark grey box). CLPs are likely to be deranged as WM is a lymphoproliferative neoplasm (medium grey box). BM overcrowding and excess IgM binding may interfere with the function of other haematopoietic BM cells, including HSCs and MKs (light grey box). Stress-induced accelerated MK maturation and platelet production (in italics) results in the loss of platelet surface receptors, such as GPIbα and GPVI, usually acquired late in MK maturation. *LT-HSC* long-term haematopoietic stem cell, *ST-HSC* short-term HSC, *MPP* multipotent progenitor, *CLP* common lymphoid progenitor, *CMP* common myeloid progenitor, *GMP* granulocyte-macrophage progenitor, *MEP* megakaryocyte erythroid progenitor, *Ery* erythrocyte, *MKP* megakaryocyte progenitor, *MK* megakaryocyte, *Plt* platelet, *MSC* mesenchymal stem cell, *ECM* extracellular matrix protein, *WM* Waldenström Macroglobulinaemia, *GP* glycoprotein

overcrowding, and increases in oxidative stress, are likely to disturb MK and progenitor cell functions (Fig. 2). Possible outcomes include the accelerated and premature release of platelets by MKs, resulting in platelets with potentially reduced surface receptors, granules, RNAs, Golgi apparatuses, ribosomes and mitochondria (Fig. 2). The loss of platelet surface receptors is of particular interest to platelet function and will be discussed below.

4.2 Alterations to Platelet Surface Receptors Disrupt Platelet Function

Platelet function is governed by platelet surface receptors. Following an injury, VWF and collagen, which are contained within the Weibel-Palade bodies of endothelial cells and the endothelium respectively, are released into the blood vessel (Peyvandi et al. 2011). GPIbα, of the GPIb-IX-V complex, binds VWF, while integrin α2β1 binds collagen, both mediating platelet adhesion to the endothelium (Clemetson 2012). GPIb-IX-V also binds collagen, P-selectin, coagulation factors (thrombin, FXI and FXII) and αMβ2 (Canobbio et al. 2004). GPVI, a platelet- and MK-specific receptor, binds collagen, fibrinogen and fibrin, resulting in platelet activation (Clemetson 2012). Surface densities of both GPIbα and GPVI are regulated by metalloproteolytic cleavage by a disintegrin and metalloproteinase (ADAM)-17 and ADAM-10, respectively, possibly to desensitise platelets to agonists and minimise thrombotic risk (Montague et al. 2018). Active platelets undergo cytoskeletal rearrangements, spread, expose phosphatidylserine (PS) on their surface and release α and dense granules. These granules contain platelet agonists, pro-coagulant and pro-inflammatory factors and receptors, including adenosine 5′-diphosphate (ADP), serotonin, thromboxane A_2 and P-selectin. These are all involved in attracting additional platelets and immune cells to the site of the injury and promoting platelet adhesion to the endothelium, platelet aggregation and vasoconstriction (Andre 2004; McNicol and Israels 1999). Integrin αIIbβ3, another platelet- and MK-specific receptor, is activated and binds fibrinogen, as well as VWF and fibronectin, bridging adjacent platelets together and enabling a platelet plug to form at the injured site (Clemetson 2012).

Simultaneously, the blood coagulation system is activated by tissue factor (TF), contained within the endothelium and released during injury. Prothrombin is cleaved into thrombin by activated FX (FXa) and fibrin is generated to secure the platelet plug at the injured site (Dahlback 2000). Thrombin generation is accelerated by the negatively charged PS-coated active platelet surface. FXa and FVa assemble to form the prothrombinase complex and accelerates thrombin generation 278,000-fold compared to FXa alone (Nesheim et al. 1979). Thrombin generation is also heavily influenced by the surface levels of GPIbα, GPVI and αIIbβ3, as blockade, cleavage or knockdown of these receptors experimentally, results in a reduction in total thrombin produced (Mammadova-Bach et al. 2015; Li et al. 2000; Maurer et al. 2013).

Reduced platelet surface receptor numbers and densities have been associated with impaired haemostasis and bleeding in several pathologies, including in mechanocirculatory support (Muthiah et al. 2016) and trauma patients (Vulliamy et al. 2020). In CLL, low levels of GPVI and GPIbα have been observed on platelets alongside normal soluble cleaved GPVI ectodomain fragments, implying that the low-surface receptor levels are due to a platelet production issue in the BM rather than metalloproteollytic cleavage of receptors in the periphery (Qiao et al. 2013). The observed low GPVI levels in CLL are consistent with an impaired GPVI-mediated collagen activation response (Kamel et al. 2015). CLL pathology shares some similarities with WM, as it involves the infiltration of the BM by malignant B cells. The surface receptor levels observed in CLL patients may support the hypothesis of a stress-induced, accelerated platelet production pathway, which releases platelets into the bloodstream with normal levels of αIIbβ3, but with inadequate GPIbα and GPVI possibly due to accelerated MK maturation. Platelet surface receptor levels have not yet been evaluated in treatment-naïve WM patients but are suspected to be similar to those observed in CLL. Additionally, in WM, the suspected IgM-mediated steric hindrance of ligand access to platelet receptors could also impact platelet function and haemostasis by impairing platelet function and thrombin generation.

4.3 Possible Megakaryocyte Mutations and Alterations to Intra-Platelet Signalling Disrupt Platelet Function

Following platelet-receptor interactions, intra-platelet signalling is required for correct platelet function. Certain haematological malignancies are characterised by driver mutations that are known to occur in the malignant cells, such as the $MYD88^{L265P}$ and CXCR4 mutations in WM. These driver mutations result in the constitutive activation of important kinases and the disruption of signalling pathways shared by most haematopoietic cell types. Interestingly, the L265P mutation within MYD88 has been detected in normal precursor as well as mature B lymphocytes from patients with lymphoma (Rodriguez et al. 2022). The authors suggested that this mutation was a preneoplastic event and that progression to malignant lymphoma may be fundamentally driven by both the cellular origin of the $MYD88^{L265P}$ mutation and the emergence of cooperating genetic alterations (Rodriguez et al. 2022). Three possible mechanisms exist which would enable these driver mutations to be passed on to the MKs and potentially be evident and functionally relevant in platelets.

The first, if the mutations originated in an early hematopoietic progenitor, shared by the lymphoid and myeloid lineages, and were passed on to the MKs during specific lineage commitment steps. The cell of origin of the driver mutations in WM and other haematological malignancies, including DLBCL and T-cell lymphomas, remains unknown. It is therefore plausible that the mutations arise in the HSCs. This is supported by the discovery of the pre-malignant state known as clonal haematopoiesis of indeterminate potential (CHIP), which is characterised by HSCs bearing somatic mutations, particularly in the *DNMT3A*, *TET2*, *ASXL1*, *JAK2*, *TP53*, *PPM1D* and *SF3B1* genes, giving rise to mutated blood cells and associated with a ten-fold increased risk of developing a haematological malignancy (1%/year) and an increased risk of cardiovascular disease (Jaiswal et al. 2014; Jaiswal and Libby 2020). One such haematological malignancy is AML, where the well-defined mutant HSCs give rise to a myeloid blast incapable of terminal differentiation due to a specific mutational and epigenetic landscape (Shlush et al. 2017). CHIP also increases the risk of NHL development (Jaiswal and Libby 2020), which includes WM. The genes commonly mutated in CHIP are often involved in epigenetic regulation, of which *DNMT3A*, *TET2* and *TP53* are also commonly mutated in WM (Wang et al. 2021). It is possible that the MYD88 and CXCR4 mutations also occur in the pre-malignant CHIP HSCs, however they likely occur at a lower frequency in the general population compared to the commonly mutated epigenetic regulators, explaining why MYD88 and CXCR4 mutations were not found in a large proportion of the CHIP patients analysed.

The second mechanism enabling the haematological malignancy driver mutations to occur in the MKs is via MK emperipolesis of malignant cells. Emperipolesis is characterised by the engulfment of cells without causing cell death, resulting in the transfer of membrane, proteins and nucleic acids. Emperipolesis of malignant cells by MKs has been reported in several haematological malignancies, including Hodgkin's lymphoma (HL), AML, CML, MDS, MPNs and NHLs (Rastogi et al. 2014), but not in WM specifically.

The third mechanism consists of the MK-mediated uptake of EVs in the BM. EVs are released by all cell types, and contain proteins, lipids, nucleic acids and carbohydrates. Serum from haematological malignancy patients, including WM, HL, MM, AML and some MPNs, exhibited elevated levels of EVs compared to healthy individuals. These EVs displayed malignancy-specific markers on their surface (Caivano et al. 2015), and analysis of miRNA transcripts in WM EVs identified unique markers of diseases progression (Bouyssou et al. 2018). EVs isolated from the supernatants of WM cell lines and patient BM aspirates were found to contain MYD88^{L265P} (Mancek-Keber et al. 2018). These EVs were taken up by MYD88^{WT} recipient cells and activated the NF-κB pathway (Mancek-Keber et al. 2018). When injected into mice, these EVs triggered the development of a WM-like pathology, as seen in MYD88^{L265P} transgenic mice, with increased globulin production, recruitment of additional mast cells to the BM, but no BM infiltration (Mancek-Keber et al. 2018). The evaluation of EVs carrying CXCR4 mRNA mutants has not yet been assessed in WM. In summary, MKs can acquire mutant mRNA transcripts from EVs that are derived from malignant cells, which can alter their signalling pathways and overall functions.

MKs and platelets possess functional Myd88 and CXCR4 (Burkhart et al. 2012). In wild-type platelets, Myd88 is required for the downstream signalling of TLRs following stimulation by high concentrations of pathogen- or danger-associated molecular patterns, resulting in NF-κB activation, platelet-neutrophil interactions and neutrophil extracellular trap (NET) formation (Vallance et al. 2017). The gain-of-function MYD88^{L265P} mutation in platelets would result in excessive Myd88 and

NF-κB signalling in response to low or absent TLR stimulation (Vallance et al. 2017). In sepsis patients, this hyperactivated pathway causes excessive neutrophil activation, NET formation, tissue damage and bleeding (Denning et al. 2019). Additionally, in wild-type platelets, membrane-bound CXCR4 and SDF-1 contained within platelet α granules (Burkhart et al. 2012) are required for MK localisation in the BM and thrombopoiesis (Niswander et al. 2014). The gain-of-function $CXCR4^{S338X}$ mutation in platelets would result in excessive CXCR4 signalling following SDF-1 stimulation due to an inability to internalise CXCR4 (Dotta et al. 2011), which could result in aberrant MK and platelet migration and platelet production. Both MYD88 and CXCR4 mutations in platelets would therefore impair intra-platelet signalling and could alter platelet aggregation, thromboxane A_2 production and granule secretion (Chatterjee et al. 2015), thus potentially contributing to a bleeding phenotype.

Furthermore, novel B-cell malignancy therapies target B-cell signalling pathways to reduce malignant lymphoplasmacytoid cell numbers and disease burden. However, as platelets share many signalling pathways with B lymphocytes, these therapies result in platelet off-target effects and bleeding symptoms, which will be discussed further.

5 Standard B-cell Haematological Malignancy Therapies Exacerbate the Bleeding Phenotype

Many haematological malignancies, such as WM and CLL, are incurable diseases, with therapies aiming to improve symptoms and prolong remissions and survival, rather than aiming to cure. In WM, decisions surrounding therapy are based on patient symptoms, their diagnostic laboratory profile and clinical trial or drug availability (Castillo et al. 2020). For asymptomatic patients, no treatment is required (Treon 2015). For symptomatic patients, first-line therapies include rituximab in combination with alkylating chemotherapies or BTK inhibitors (ibrutinib, acalabrutinib, zanubrutinib, tirabrutinib) and second-line therapies include BTK inhibitors or proteasome inhibitors (bortezomib, carfilzomib) (Castillo et al. 2020). These treatments are also commonly used to treat many other B-cell malignancies, such as CLL, MCL and MZL. Most of these treatments are associated with an increased bleeding risk, which will be discussed below.

5.1 Rituximab

Rituximab is an anti-CD20 monoclonal antibody that depletes both healthy and malignant B lymphocytes and subsequently reduces disease burden in B-cell malignancies, and IgM production in WM specifically (Salles et al. 2017). In WM, rituximab monotherapy is associated with a >25% increase in IgM, known as an IgM flare, which increases the risk of hyperviscosity and IgM-related bleeding (Ghobrial et al. 2004). Rituximab-associated acute thrombocytopenia due to cytokine release syndrome and complement activation has been documented in WM, MCL, HCL and CLL, which also increases bleeding risk due to low platelet counts (Ram et al. 2009; Jaglowski and Byrd 2010). These risks are reduced when rituximab is used in combination therapy.

5.2 Alkylating Chemotherapies

Alkylating chemotherapies, such as bendamustine and cyclophosphamide, kill rapidly dividing cells by inducing double-stranded DNA breaks, causing cell-cycle arrest, preventing proliferation and resulting in cell death (Cheson and Leoni 2011). In haematological malignancies, rapidly dividing cells include malignant cells, but also HSCs and progenitor cells. These therapies therefore reduce disease burden, but also induce cytopenias, notably thrombocytopenia, which increases bleeding risk.

5.3 Bruton's Tyrosine Kinase Inhibitors

BTK inhibitors covalently bind to the active site of BTK, thus blocking its enzymatic activity (Burger and Buggy 2013). Three BTK inhibitors (ibrutinib, acalabrutinib and zanubrutinib) have been approved by the U.S. Food and Drug Administration for the treatment of B-cell malignancies, including WM, CLL, activated B cell-like (ABC)-DLBCL, MCL and MZL. These inhibitors are effective therapeutic options because they block the B-cell receptor (BCR) signalling pathway and the activation of nuclear factor (NF)-κB, nuclear factor of activated T cells and mitogen-activated protein kinases, which usually enable healthy and malignant B-cell development, survival and proliferation (Fig. 3) (Estupiñán et al. 2021). In WM, there is constitutive activation of BTK, particularly in patients harbouring the MYD88^{L265P} mutation (Hunter et al. 2018), rendering the lymphoplasmacytoid cells more susceptible to death following BTK inhibition. BTK and the BCR signalling pathway are also overexpressed in other haematological malignancies (Pal Singh et al. 2018), including ABC-DLBCL, MCL (Cinar et al. 2013) and CLL (Singh et al. 2017). For example, ABC-DLBCL patients harbour the *MYD88*^{L265P} mutation (in 30%) and *CARD11* mutations (in 10%) which both constitutively activate the NF-κB pathway; and 20% of patients display mutations in the ITAM (immunoreceptor tyrosine-based activation motif) of *CD79A* and *CD79B*, which aberrantly activate BTK, spleen tyrosine kinase, phosphoinositide 3-kinase (PI3K) and protein kinase C (PKC)-β. The various BTK inhibitors differ in their specificity to BTK, with the first-generation inhibitor ibrutinib being less specific to BTK and inhibiting other tyrosine kinases, such as TEC, IL-2-inducible T-cell kinase (ITK), TXK and epidermal growth factor receptor, compared to the other second-generation inhibitors (Estupiñán et al. 2021). Despite the deep and durable responses achieved with BTK inhibitors in the treatment of B-cell malignancies, these drugs are associated with significant bleeding off-target effects.

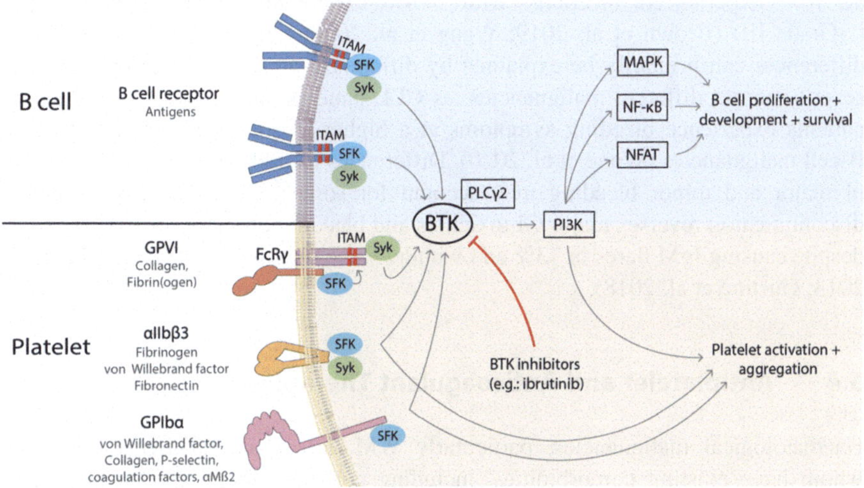

Fig. 3 Common signalling pathways in platelets and B lymphocytes involve Bruton's tyrosine kinase. BTK is required for signalling in platelets and B lymphocytes. Following ligand binding to the major platelet receptors (αIIbβ3, GPVI and GPIbα of the GPIb-IX-V complex) and the BCR, there is receptor clustering, phosphorylation of ITAMs, recruitment of SFKs, phosphorylation of Syk, activation of PI3K, engagement of BTK and phosphorylation of PLCγ. In platelets, this leads to activation and aggregation responses. In B lymphocytes, there is further activation of NF-κB, MAPKs and NFAT, which leads to B lymphocyte proliferation, development and survival. GPVI and BCR signalling are more susceptible to BTK inhibition as these receptors rely more on BTK pathways compared to αIIbβ3 and GPIbα. *BCR* B-cell receptor BTK: Bruton's tyrosine kinase, *GP* glycoprotein, *ITAM* immunoreceptor tyrosine-based activation motif, *SFK* Src-family kinases, *Syk* spleen tyrosine kinase, *PI3K* phosphoinositide 3-kinase, *PLCγ* phospholipase C γ, *NF-κB* nuclear factor κB, *NFAT* nuclear factor of activated T cells, *MAPK* mitogen-activated protein kinase

The activation of BTK and other tyrosine kinases is required in the downstream signalling following platelet GPVI, C-type lectin-like receptor (CLEC)-2, αIIbβ3 and GPIb-IX-V ligand binding (Fig. 3) (Liu et al. 2006; Li et al. 2010). This enables enhanced platelet adhesion to the endothelium in flowing blood and enhanced platelet aggregation at the site of injury (Fig. 3). Importantly, there is redundancy in the platelet BTK pathway, as X-linked agammaglobulinemia mice and patients, with loss-of-function mutations to the *BTK* gene, do not display a bleeding propensity, most likely due to compensatory signalling by Tec and other kinases (Atkinson et al. 2003; Quek et al. 1998). As anticipated, BTK inhibitors have been shown to inhibit platelet function (Levade et al. 2014; Bye et al. 2017; Tullemans et al. 2018), likely aided by their broad off-target kinase spectrum. Platelets from ibrutinib-treated, but not zanubrutinib-treated, WM patients displayed reduced levels of GPIb-IX-V and αIIbβ3 and impaired platelet aggregation responses (Dobie et al. 2019). Consistent with these findings, 23% of WM patients receiving ibrutinib experienced bleeding symptoms (Treon et al. 2015; Abeykoon et al. 2020; Ali et al. 2017; Dimopoulos et al. 2017, 2018; Treon et al. 2018). This is slightly lower than what is observed in the entire B-cell malignancy cohort, of which

32–40% experienced bleeding, with 3–8% experiencing major haemorrhage (>Grade III) (Brown et al. 2019; Wang et al. 2020; Shatzel et al. 2017). These differences can primarily be explained by differences in disease pathogenesis and severity across different malignancies, as CLL patients and relapsed or refractory patients experience bleeding symptoms at a higher frequency compared to other B-cell malignancies (Wang et al. 2020). Differences in study design and definitions of major and minor bleeding may account for some of the disparity. Ibrutinib discontinuation reverses major GI toxicities and bleeding observed in WM patients, despite causing IgM flares in 73% and withdrawal symptoms in 19% (Castillo et al. 2018; Gustine et al. 2018).

5.4 Antiplatelet and Anticoagulant Therapies

Haematological malignancies, particularly WM, affect older patients, many of whom have existing comorbidities, including coronary artery disease and atrial fibrillation. These patients require antiplatelet or anticoagulation therapies, such as Aspirin, Clopidogrel, Warfarin or direct oral anticoagulants, which increase their bleeding risk.

The BTK inhibitor ibrutinib increases the risk of atrial fibrillation by 10% (Treon et al. 2015; Abeykoon et al. 2020; Ali et al. 2017; Dimopoulos et al. 2017, 2018; Treon et al. 2018; Gustine et al. 2016; Fradley et al. 2019), thus requiring anticoagulation therapy. The concurrent use of these therapeutics increases the risk of bleeding. In one study, major bleeding occurred in 3.7% of B-cell malignancy patients receiving ibrutinib monotherapy compared with 5.1% receiving ibrutinib in combination with antiplatelet reagents or anticoagulants (Brown et al. 2019).

Taken together, haematological malignancy treatments increase bleeding propensity, which is further amplified by the treatment of existing comorbidities. These patients require specialised treatment plans, which balance existing disease-related bleeding propensity with treatment-induced bleeding risk. It is therefore important to broaden our knowledge on the molecular basis of bleeding pathogenesis, to accurately predict bleeding risk, make the best clinical decisions surrounding therapy, minimise bleeding morbidity and improve patient quality of life.

6 Evaluating the Bleeding Phenotype Patients with Haematological Malignancy in the Diagnostic Laboratory

Several laboratory tests currently exist to evaluate blood coagulation, platelet function and overall haemostasis in patients displaying a bleeding diasthesis, such as haematological malignancy patients.

6.1 Laboratory Tests Evaluating Blood Coagulation

Coagulation tests measuring the prothrombin time (PT) and activated partial thromboplastin time (aPTT) are routinely performed on patients experiencing bleeding. Several variables are known to influence these tests, including phlebotomy technique, anticoagulant volume (which changes based on haematocrit), sample handling (mixing, centrifugation, transport), patient age, gender, physiological state (pregnancy, post-surgery) and medications (antiplatelets, anticoagulants, anti-inflammatories).

Coagulation tests measuring VWF antigen and cofactor binding assays are also performed in haematological malignancy patients when AVWS is suspected. The VWF antigen assay uses monoclonal antibodies to measure the total amount of VWF. The VWF ristocetin cofactor binding assay evaluates VWF activity by unfolding VWF with the ristocetin antibiotic, which then binds donor platelet GPIbα and causes platelet agglutination. This assay is sensitive to several variables, including patient VWF levels (must exceed 20 U/dL), the source of ristocetin and donor platelets and the presence of VWF A1 domain mutations (which cause poor ristocetin-VWF binding and VWF unfolding) (Sharma and Haberichter 2019).

6.2 Laboratory Tests Evaluating Platelet Function

Platelet function tests evaluate platelet responses to physiological agonists, such as collagen, ADP and thrombin. Established diagnostic platelet function tests that are performed on patients displaying a bleeding diathesis include light transmission aggregometry (LTA) and the platelet function analyser (PFA-100), with the Multiplate platelet function analyser representing an emerging tool (Vinholt 2019). In addition to the PT and aPTT pre-analytical variables shared by the platelet function tests, these tests are also influenced by the platelet count (which generally must exceed 150×10^9/L) and bilirubin and lipid concentrations (which increase plasma turbidity, affecting LTA measurements). The PFA-100 is the simplest test to perform in the laboratory and is more physiological than the LTA due to the higher shear rates used. The LTA is also disadvantaged by its requirement for large samples volumes, while the PFA-100 cannot evaluate platelet secretion defects. Finally, all three tests lack sensitivity to small changes in platelet receptor levels and function (Paniccia et al. 2015).

6.3 Laboratory Tests Evaluating Blood Coagulation and Platelet Function

Viscoelastic tests, including thromboelastography (TEG) and rotational thromboelastometry (ROTEM), are used to evaluate overall haemostasis. Both TEG and ROTEM are capable of evaluating clot formation and lysis rates, clot size and firmness and platelet contributions to clotting parameters. These tests are widely used in transfusion and acute bleeding settings, and show some utility in predicting bleeding and diagnosing hyperfibrinolysis in haematological malignancies (Walsh et al. 2020). However, no studies on WM patients exist. TEG and ROTEM have not been standardised in thrombocytopenic patients, which may be relevant to many haematological malignancy cohorts and should be taken into consideration.

6.4 Considerations and Future Directions

There are several challenges associated with evaluating platelet function in the diagnostic laboratory. Firstly, there are no standardised tests that evaluate platelet function in thrombocytopenic patients. Secondly, platelet surface receptor levels are not routinely evaluated in the diagnostic laboratory, despite these determining platelet function and being reduced in haematological malignancies (Luu et al. 2018). Finally, no tests accurately replicate the physiological conditions of flowing blood in the vasculature.

Several additional techniques exist that could be incorporated into the diagnostic laboratory to further evaluate platelet and MK function. Platelet flow cytometry can be used in thrombocytopenic patients to evaluate the levels of platelet surface receptors and markers of activation. Enzyme-linked immunosorbent assays can be used to quantify shed platelet receptor ectodomains (Al-Tamimi et al. 2012). Microfluidic systems can be used to measure platelet adhesion to immobilised substrates, such as collagen, fibrinogen or fibrin, under physiological pulsatile flowing conditions (Lui et al. 2019; Mangin et al. 2020; de Witt et al. 2014). Fluorescence resonance energy transfer assays can be used to evaluate changes to thrombin generation potential, particularly in paraproteinemic settings. Finally, genetic analyses on fluorescence-activated cell sorted MKs can be performed.

7 Conclusions

Bleeding is an underreported symptom in haematological malignancies, which causes significant patient morbidity and sometimes mortality. Some patients experience bleeding in the absence of thrombocytopenia, AVWS, DIC, blood hyperviscosity, cryoglobulinaemia and hyperfibrinolysis, implying that other mechanisms of bleeding exist. One possibility is that patients with haematological malignancy possess an underlying platelet lesion. Platelets could be characterised by reduced surface receptor levels due to premature release by stressed MKs in an overcrowded BM; or by the acquisition of mutated constitutively active signalling proteins. With the ever-evolving treatment landscape of haematological malignancies, which incorporate therapies like BTK inhibitors that carry an attendant bleeding risk, it is particularly important to understand the molecular mechanisms behind bleeding, to stratify patients for bleeding risk, aid in clinical personalised decisions surrounding therapy and improve quality of life. Research techniques to do this involve existing clinical diagnostic tools (PT, aPTT, LTA, PFA-100, ROTEM) complemented by more sensitive research-based methods (platelet flow cytometry, thrombin generation assays, platelet adhesion and spreading assays, MK genetic analysis). Additional research is required to better understand and predict bleeding in haematological malignancy patients, with emerging research techniques and scoring matrices offering promising first steps.

References

Abeykoon JP, Zanwar S, Ansell SM, Gertz MA, Kumar S, Manske M et al (2020) Ibrutinib monotherapy outside of clinical trial setting in Waldenstrom macroglobulinaemia: practice patterns, toxicities and outcomes. Br J Haematol 188(3):394–403

Ali N, Malik F, Jafri SIM, Naglak M, Sundermeyer M, Pickens PV (2017) Analysis of efficacy and tolerability of Bruton tyrosine kinase inhibitor ibrutinib in various B-cell malignancies in the general community: a single-center experience. Clin Lymphoma Myeloma Leuk 17S:S53–S61

Al-Tamimi M, Arthur JF, Gardiner EE, Andrews RK (2012) Focusing on plasma glycoprotein VI. Thromb Haemost 107(4):648–655

Andre P (2004) P-selectin in haemostasis. Br J Haematol 126(3):298–306

Atkinson BT, Ellmeier W, Watson SP (2003) Tec regulates platelet activation by GPVI in the absence of Btk. Blood 102(10):3592–3599

Bick RL (1992) Coagulation abnormalities in malignancy: a review. Semin Thromb Hemost 18(4): 353–372

Bouyssou JM, Liu C-J, Bustoros M, Sklavenitis-Pistofidis R, Aljawai Y, Manier S et al (2018) Profiling of circulating exosomal miRNAs in patients with Waldenström Macroglobulinemia. PLoS One 13(10):e0204589

Brault L, Rovó A, Decker S, Dierks C, Tzankov A, Schwaller J (2014) CXCR4-SERINE339 regulates cellular adhesion, retention and mobilization, and is a marker for poor prognosis in acute myeloid leukemia. Leukemia 28(3):566–576

Brown JR, Moslehi J, Ewer MS, O'Brien SM, Ghia P, Cymbalista F et al (2019) Incidence of and risk factors for major haemorrhage in patients treated with ibrutinib: an integrated analysis. Br J Haematol 184(4):558–569

Burger JA, Buggy JJ (2013) Bruton tyrosine kinase inhibitor ibrutinib (PCI-32765). Leuk Lymphoma 54(11):2385–2391

Burkhart JM, Vaudel M, Gambaryan S, Radau S, Walter U, Martens L et al (2012) The first comprehensive and quantitative analysis of human platelet protein composition allows the comparative analysis of structural and functional pathways. Blood 120(15):e73–e82

Buske C, Sadullah S, Kastritis E, Tedeschi A, Garcia-Sanz R, Bolkun L et al (2018) Treatment and outcome patterns in European patients with Waldenstrom's macroglobulinaemia: a large, observational, retrospective chart review. Lancet Haematol 5(7):e299–e309

Bustoros M, Sklavenitis-Pistofidis R, Kapoor P, Liu CJ, Kastritis E, Zanwar S et al (2019) Progression risk stratification of asymptomatic Waldenström Macroglobulinemia. J Clin Oncol 37(16):1403–1411

Bye AP, Unsworth AJ, Desborough MJ, Hildyard CAT, Appleby N, Bruce D et al (2017) Severe platelet dysfunction in NHL patients receiving ibrutinib is absent in patients receiving acalabrutinib. Blood Adv 1(26):2610–2623

Caivano A, Laurenzana I, De Luca L, La Rocca F, Simeon V, Trino S et al (2015) High serum levels of extracellular vesicles expressing malignancy-related markers are released in patients with various types of hematological neoplastic disorders. Tumour Biol 36(12):9739–9752

Canobbio I, Balduini C, Torti M (2004) Signalling through the platelet glycoprotein Ib-V–IX complex. Cell Signal 16(12):1329–1344

Cao Y, Hunter ZR, Liu X, Xu L, Yang G, Chen J et al (2015) The WHIM-like CXCR4(S338X) somatic mutation activates AKT and ERK, and promotes resistance to ibrutinib and other agents used in the treatment of Waldenstrom's Macroglobulinemia. Leukemia 29(1):169–176

Cartoni C, Niscola P, Breccia M, Brunetti G, D'Elia GM, Giovannini M et al (2009) Hemorrhagic complications in patients with advanced hematological malignancies followed at home: an Italian experience. Leuk Lymphoma 50(3):387–391

Castaldi PA, Penny R (1970) A macroglobulin with inhibitory activity against coagulation factor VIII. Blood 35(3):370–376

Castillo JJ, Olszewski AJ, Cronin AM, Hunter ZR, Treon SP (2014) Survival trends in Waldenström macroglobulinemia: an analysis of the surveillance, epidemiology and end results database. Blood 123(25):3999–4000

Castillo JJ, Gustine JN, Meid K, Dubeau T, Severns P, Treon SP (2018) Ibrutinib withdrawal symptoms in patients with Waldenstrom macroglobulinemia. Haematologica 103(7):e307–ee10

Castillo JJ, Gustine JN, Meid K, Dubeau T, Severns P, Xu L et al (2019) Low levels of von Willebrand markers associate with high serum IgM levels and improve with response to therapy, in patients with Waldenström macroglobulinaemia. Br J Haematol 184(6):1011–1014

Castillo JJ, Advani RH, Branagan AR, Buske C, Dimopoulos MA, D'Sa S et al (2020) Consensus treatment recommendations from the tenth International Workshop for Waldenström Macroglobulinaemia. Lancet Haematol 7(11):e827–ee37

Cesarman GM, Guevara CA, Hajjar KA (1994) An endothelial cell receptor for plasminogen/tissue plasminogen activator (t-PA). II. Annexin II-mediated enhancement of t-PA-dependent plasminogen activation. J Biol Chem 269(33):21198–21203

Chang H, Kuo M-C, Shih L-Y, Dunn P, Wang P-N, Wu J-H et al (2012) Clinical bleeding events and laboratory coagulation profiles in acute promyelocytic leukemia. Eur J Haematol 88(4): 321–328

Chatterjee M, Rath D, Gawaz M (2015) Role of chemokine receptors CXCR4 and CXCR7 for platelet function. Biochem Soc Trans 43(4):720–726

Cheson BD, Leoni L (2011) Bendamustine: mechanism of action and clinical data. Clin Adv Hematol Oncol 9(8 Suppl 19):1–11

Chi S, Ikezoe T (2015) Disseminated intravascular coagulation in non-Hodgkin lymphoma. Int J Hematol 102(4):413–419

Choudhry A, DeLoughery TG (2012) Bleeding and thrombosis in acute promyelocytic leukemia. Am J Hematol 87(6):596–603

Cinar M, Hamedani F, Mo Z, Cinar B, Amin HM, Alkan S (2013) Bruton tyrosine kinase is commonly overexpressed in mantle cell lymphoma and its attenuation by Ibrutinib induces apoptosis. Leuk Res 37(10):1271–1277

Clemetson KJ (2012) Platelets and primary haemostasis. Thromb Res 129(3):220–224

Crazzolara R, Kreczy A, Mann G, Heitger A, Eibl G, Fink F-M et al (2001) High expression of the chemokine receptor CXCR4 predicts extramedullary organ infiltration in childhood acute lymphoblastic leukaemia. Br J Haematol 115(3):545–553

Dahlback B (2000) Blood coagulation. Lancet 355(9215):1627–1632

de Witt SM, Swieringa F, Cavill R, Lamers MM, van Kruchten R, Mastenbroek T et al (2014) Identification of platelet function defects by multi-parameter assessment of thrombus formation. Nat Commun 5:e4257

Deguine J, Barton GM (2014) MyD88: a central player in innate immune signaling. F1000Prime Rep 6:97

Dejhansathit S, Suvannasankha A (2019) Acquired factor X deficiency in patients with primary light chain amyloidosis. J Investig Med High Impact Case Rep 7:2324709619832332

Denning NL, Aziz M, Gurien SD, Wang P (2019) DAMPs and NETs in sepsis. Front Immunol 10:2536

Dicke C, Schneppenheim S, Holstein K, Spath B, Bokemeyer C, Dittmer R et al (2016) Distinct mechanisms account for acquired von Willebrand syndrome in plasma cell dyscrasias. Ann Hematol 95(6):945–957

Dimopoulos MA, Trotman J, Tedeschi A, Matous JV, Macdonald D, Tam C et al (2017) Ibrutinib for patients with rituximab-refractory Waldenstrom's macroglobulinaemia (iNNOVATE): an open-label substudy of an international, multicentre, phase 3 trial. Lancet Oncol 18(2):241–250

Dimopoulos MA, Tedeschi A, Trotman J, Garcia-Sanz R, Macdonald D, Leblond V et al (2018) Phase 3 trial of Ibrutinib plus Rituximab in Waldenstrom's Macroglobulinemia. N Engl J Med 378(25):2399–2410

Dobie G, Kuriri FA, Omar MMA, Alanazi F, Gazwani AM, Tang CPS et al (2019) Ibrutinib, but not zanubrutinib, induces platelet receptor shedding of GPIb-IX-V complex and integrin αIIbβ3 in mice and humans. Blood Adv 3(24):4298–4311

Dotta L, Tassone L, Badolato R (2011) Clinical and genetic features of Warts, Hypogammaglobulinemia, Infections and Myelokathexis (WHIM) syndrome. Curr Mol Med 11(4):317–325

Elsawa SF, Novak AJ, Ziesmer SC, Almada LL, Hodge LS, Grote DM et al (2011) Comprehensive analysis of tumor microenvironment cytokines in Waldenstrom macroglobulinemia identifies CCL5 as a novel modulator of IL-6 activity. Blood 118(20):5540–5549

Endo T, Yatomi Y, Amemiya N, Mizoe A, Iino M, Ozawa H et al (2000) Antibody studies of factor VIII inhibitor in a case with Waldenstrom's macroglobulinemia. Am J Hematol 63(3):145–148

Estupiñán HY, Berglöf A, Zain R, Smith CIE (2021) Comparative analysis of BTK inhibitors and mechanisms underlying adverse effects. Front Cell Dev Biol 9(331):630942

Falini B, Martelli MP, Tiacci E (2016) BRAF V600E mutation in hairy cell leukemia: from bench to bedside. Blood 128(15):1918–1927

Fattizzo B, Barcellini W (2020) Autoimmune Cytopenias in chronic lymphocytic leukemia: focus on molecular aspects. Front Oncol 9:1435

Federici AB, Rand JH, Bucciarelli P, Budde U, van Genderen PJ, Mohri H et al (2000) Acquired von Willebrand syndrome: data from an international registry. Thromb Haemost 84(2):345–349

Federici AB, Budde U, Castaman G, Rand JH, Tiede A (2013) Current diagnostic and therapeutic approaches to patients with acquired von Willebrand syndrome: a 2013 update. Semin Thromb Hemost 39(2):191–201

Flores-Figueroa E, Gutiérrez-Espíndola G, Montesinos JJ, Arana-Trejo RM, Mayani H (2002) In vitro characterization of hematopoietic microenvironment cells from patients with myelodysplastic syndrome. Leuk Res 26(7):677–686

Fradley MG, Gliksman M, Emole J, Viganego F, Rhea I, Welter-Frost A et al (2019) Rates and risk of atrial arrhythmias in patients treated with ibrutinib compared with cytotoxic chemotherapy. Am J Cardiol 124(4):539–544

Franchini M, Di Minno MN, Coppola A (2010) Disseminated intravascular coagulation in hematologic malignancies. Semin Thromb Hemost 36(4):388–403

Franchini M, Frattini F, Crestani S, Bonfanti C (2013) Bleeding complications in patients with hematologic malignancies. Semin Thromb Hemost 39(1):94–100

Franchini M, Zaffanello M, Mannucci PM (2021) Bleeding disorders in primary fibrinolysis. Int J Mol Sci 22(13):7027

Furie B, Furie BC (2008) Mechanisms of thrombus formation. N Engl J Med 359(9):938–949

Gaidano G, Rossi D (2017) The mutational landscape of chronic lymphocytic leukemia and its impact on prognosis and treatment. Hematology Am Soc Hematol Educ Program 2017(1): 329–337

Garcia-Sanz R, Montoto S, Torrequebrada A, de Coca AG, Petit J, Sureda A et al (2001) Waldenstrom macroglobulinaemia: presenting features and outcome in a series with 217 cases. Br J Haematol 115(3):575–582

Georgantopoulos P, Yang H, Norris LB, Bennett CL (2019) Major hemorrhage in chronic lymphocytic leukemia patients in the US Veterans Health Administration system in the pre-ibrutinib era: incidence and risk factors. Cancer Med 8(5):2233–2240

Gertz MA, Fonseca R, Rajkumar SV (2000) Waldenstrom's macroglobulinemia. Oncologist 5(1): 63–67

Gertz MA, Merlini G, Treon SP (2004) Amyloidosis and Waldenstrom's macroglobulinemia. Hematology Am Soc Hematol Educ Program 1:257–282

Ghobrial IM, Fonseca R, Greipp PR, Blood E, Rue M, Vesole DH et al (2004) Initial immunoglobulin M 'flare' after rituximab therapy in patients diagnosed with Waldenstrom macroglobulinemia: an Eastern Cooperative Oncology Group Study. Cancer 101(11):2593–2598

Grégoire M, Guilloton F, Pangault C, Mourcin F, Sok P, Latour M et al (2015) Neutrophils trigger a NF-κB dependent polarization of tumor-supportive stromal cells in germinal center B-cell lymphomas. Oncotarget 6(18):16471–16487

Grozovsky R, Giannini S, Falet H, Hoffmeister KM (2015) Regulating billions of blood platelets: glycans and beyond. Blood 126(16):1877–1884

Gustine JN, Meid K, Dubeau TE, Treon SP, Castillo JJ (2016) Atrial fibrillation associated with ibrutinib in Waldenstrom macroglobulinemia. Am J Hematol 91(6):E312–E313

Gustine JN, Meid K, Dubeau T, Severns P, Hunter ZR, Guang Y et al (2018) Ibrutinib discontinuation in Waldenstrom macroglobulinemia: etiologies, outcomes, and IgM rebound. Am J Hematol 93(4):511–517

Haider S, Latif T, Hochhausler A, Lucas F, Abdel KN (2013) Waldenstrom's macroglobulinemia and peripheral neuropathy, organomegaly, endocrinopathy, monoclonal gammopathy, and skin changes with a bleeding diathesis and rash. Case Rep Oncol Med 2013:890864

Han T-T, Fan L, Li J-Y, Xu W (2014) Role of chemokines and their receptors in chronic lymphocytic leukemia: function in microenvironment and targeted therapy. Cancer Biol Ther 15(1):3–9

Hawkins ED, Duarte D, Akinduro O, Khorshed RA, Passaro D, Nowicka M et al (2016) T-cell acute leukaemia exhibits dynamic interactions with bone marrow microenvironments. Nature 538(7626):518–522

Hodge LS, Ziesmer SC, Yang ZZ, Secreto FJ, Gertz MA, Novak AJ et al (2012) IL-21 in the bone marrow microenvironment contributes to IgM secretion and proliferation of malignant cells in Waldenstrom macroglobulinemia. Blood 120(18):3774–3782

Hunter ZR, Xu L, Yang G, Zhou Y, Liu X, Cao Y et al (2014) The genomic landscape of Waldenstrom macroglobulinemia is characterized by highly recurring MYD88 and WHIM-like CXCR4 mutations, and small somatic deletions associated with B-cell lymphomagenesis. Blood 123(11):1637–1646

Hunter ZR, Xu L, Tsakmaklis N, Demos MG, Kofides A, Jimenez C et al (2018) Insights into the genomic landscape of MYD88 wild-type Waldenstrom macroglobulinemia. Blood Adv 2(21): 2937–2946

Ikezoe T (2021) Advances in the diagnosis and treatment of disseminated intravascular coagulation in haematological malignancies. Int J Hematol 113(1):34–44

Jaglowski SM, Byrd JC (2010) Rituximab in chronic lymphocytic leukemia. Semin Hematol 47(2): 156–169

Jaiswal S, Libby P (2020) Clonal haematopoiesis: connecting ageing and inflammation in cardiovascular disease. Nat Rev Cardiol 17(3):137–144

Jaiswal S, Fontanillas P, Flannick J, Manning A, Grauman PV, Mar BG et al (2014) Age-related clonal hematopoiesis associated with adverse outcomes. N Engl J Med 371(26):2488–2498

Jalali S, Ansell SM (2018) The bone marrow microenvironment in Waldenström macroglobulinemia. Hematol Oncol Clin North Am 32(5):777–786

Jalali S, Price-Troska T, Paludo J, Villasboas J, Kim H-J, Yang Z-Z et al (2018) Soluble PD-1 ligands regulate T-cell function in Waldenstrom macroglobulinemia. Blood Adv 2(15): 1985–1997

Jalali S, Shi J, Buko A, Ahsan N, Paludo J, Serres M et al (2020) Increased glutathione utilization augments tumor cell proliferation in Waldenstrom Macroglobulinemia. Redox Biol 36:101657

Jalali S, Shi J, Ahsan N, Wellik L, Serres M, Buko A et al (2021) Progression from Monoclonal gammopathy of undetermined significance of the immunoglobulin M class (IgM-MGUS) to Waldenstrom Macroglobulinemia is associated with an alteration in lipid metabolism. Redox Biol 41:101927

Jimenez C, Sebastian E, Chillon MC, Giraldo P, Mariano Hernandez J, Escalante F et al (2013) MYD88 L265P is a marker highly characteristic of, but not restricted to, Waldenstrom's macroglobulinemia. Leukemia 27(8):1722–1728

Kamel S, Horton L, Ysebaert L, Levade M, Burbury K, Tan S et al (2015) Ibrutinib inhibits collagen-mediated but not ADP-mediated platelet aggregation. Leukemia 29(4):783–787

Kaushansky K (2015) Thrombopoiesis. Semin Hematol 52(1):4–11

Kumar B, Garcia M, Weng L, Jung X, Murakami JL, Hu X et al (2018) Acute myeloid leukemia transforms the bone marrow niche into a leukemia-permissive microenvironment through exosome secretion. Leukemia 32(3):575–587

Kyle RA, Benson J, Larson D, Therneau T, Dispenzieri A, Melton Iii LJ et al (2009) IgM monoclonal gammopathy of undetermined significance and smoldering Waldenstrom's macroglobulinemia. Clin Lymphoma Myeloma 9(1):17–18

Lagrue-Lak-Hal AH, Debili N, Kingbury G, Lecut C, Le Couedic JP, Villeval JL et al (2001) Expression and function of the collagen receptor GPVI during megakaryocyte maturation. J Biol Chem 276(18):15316–15325

Landgren O, Tageja N (2014) MYD88 and beyond: novel opportunities for diagnosis, prognosis and treatment in Waldenstrom's Macroglobulinemia. Leukemia 28(9):1799–1803

Levade M, David E, Garcia C, Laurent PA, Cadot S, Michallet AS et al (2014) Ibrutinib treatment affects collagen and von Willebrand factor-dependent platelet functions. Blood 124(26): 3991–3995

Li Y, Spencer FA, Ball S, Becker RC (2000) Inhibition of platelet-dependent prothrombinase activity and thrombin generation by glycoprotein IIb/IIIa receptor-directed antagonists: potential contributing mechanism of benefit in acute coronary syndromes. J Thromb Thrombolysis 10(1):69–76

Li Z, Delaney MK, O'Brien KA, Du X (2010) Signaling during platelet adhesion and activation. Arterioscler Thromb Vasc Biol 30(12):2341–2349

Liquori A, Ibañez M, Sargas C, Sanz M, Barragán E, Cervera J (2020) Acute Promyelocytic leukemia: a constellation of molecular events around a single PML-RARA fusion gene. Cancers (Basel) 12(3):624

Liu J, Fitzgerald ME, Berndt MC, Jackson CW, Gartner TK (2006) Bruton tyrosine kinase is essential for botrocetin/VWF-induced signaling and GPIb-dependent thrombus formation *in vivo*. Blood 108(8):2596–2603

Loftus LS, Arnold WN (1994) Acquired hemophilia in a patient with myeloma. West J Med 160(2): 173–176

Lui M, Gardiner EE, Arthur JF, Pinar I, Lee WM, Ryan K et al (2019) Novel stenotic microchannels to study thrombus formation in shear gradients: influence of shear forces and human platelet-related factors. Int J Mol Sci 20(12):2967

Luu S, Gardiner EE, Andrews RK (2018) Bone marrow defects and platelet function: a focus on MDS and CLL. Cancers (Basel) 10(5):147

Mammadova-Bach E, Ollivier V, Loyau S, Schaff M, Dumont B, Favier R et al (2015) Platelet glycoprotein VI binds to polymerized fibrin and promotes thrombin generation. Blood 126(5): 683–691

Mancek-Keber M, Lainscek D, Bencina M, Chen JG, Romih R, Hunter ZR et al (2018) Extracellular vesicle-mediated transfer of constitutively active MyD88(L265P) engages MyD88(wt) and activates signaling. Blood 131(15):1720–1729

Mangin PH, Gardiner EE, Nesbitt WS, Kerrigan SW, Korin N, Lam WA et al (2020) In vitro flow based systems to study platelet function and thrombus formation: recommendations for standardization: communication from the SSC on biorheology of the ISTH. J Thromb Haemost 18(3):748–752

Maqbool MG, Tam CS, Morison IM, Simpson D, Mollee P, Schneider H et al (2020) A practical guide to laboratory investigations at diagnosis and follow up in Waldenström macroglobulinaemia: recommendations from the Medical and Scientific Advisory Group, Myeloma Australia, the pathology sub-committee of the Lymphoma and Related Diseases Registry and the Australasian Association of Clinical Biochemists monoclonal gammopathy working group. Pathology 52(2):167–178

Massimo F, Pier Mannuccio M (2020) Acquired von Willebrand syndrome: focused for hematologists. Haematologica 105(8):2032–2037

Matsuda A, Germing U, Jinnai I, Misumi M, Kuendgen A, Knipp S et al (2005) Difference in clinical features between Japanese and German patients with refractory anemia in myelodysplastic syndromes. Blood 106(8):2633–2640

Maurer E, Tang C, Schaff M, Bourdon C, Receveur N, Ravanat C et al (2013) Targeting platelet GPIbbeta reduces platelet adhesion, GPIb signaling and thrombin generation and prevents arterial thrombosis. Arterioscler Thromb Vasc Biol 33(6):1221–1229

Mayerhofer M, Haushofer A, Kyrle PA, Chott A, Mullner C, Quehenberger P et al (2009) Mechanisms underlying acquired von Willebrand syndrome associated with an IgM paraprotein. Eur J Clin Investig 39(9):833–836

Mazurier C, Parquet-Gernez A, Descamps J, Bauters F, Goudemand M (1980) Acquired von Willebrand's syndrome in the course of Waldenstrom's disease. Thromb Haemost 44(3): 115–118

McKelvey EM, Kwaan HC (1972) An IgM circulating anticoagulant with factor VIII inhibitory activity. Ann Intern Med 77(4):571–575

McNicol A, Israels SJ (1999) Platelet dense granules: structure, function and implications for haemostasis. Thromb Res 95(1):1–18

Mehta R, Athar M, Girgis S, Hassan A, Becker RC (2019) Acquired Von Willebrand Syndrome (AVWS) in cardiovascular disease: a state of the art review for clinicians. J Thromb Thrombolysis 48(1):14–26

Mendelson A, Frenette PS (2014) Hematopoietic stem cell niche maintenance during homeostasis and regeneration. Nat Med 20(8):833–846

Méndez-Ferrer S, Bonnet D, Steensma DP, Hasserjian RP, Ghobrial IM, Gribben JG et al (2020) Bone marrow niches in haematological malignancies. Nat Rev Cancer 20(5):285–298

Merchionne F, Procaccio P, Dammacco F (2011) Waldenstrom's macroglobulinemia. An overview of its clinical, biochemical, immunological and therapeutic features and our series of 121 patients collected in a single center. Crit Rev Oncol Hematol 80(1):87–99

Merlini G, Baldini L, Broglia C, Comelli M, Goldaniga M, Palladini G et al (2003) Prognostic factors in symptomatic Waldenstrom's macroglobulinemia. Semin Oncol 30(2):211–215

Meyer K, Williams EC (1985) Fibrinolysis and acquired alpha-2 plasmin inhibitor deficiency in amyloidosis. Am J Med 79(3):394–396

Mital A (2016) Acquired von Willebrand syndrome. Adv Clin Exp Med 25(6):1337–1344

Montague SJ, Andrews RK, Gardiner EE (2018) Mechanisms of receptor shedding in platelets. Blood 132(24):2535–2545

Muthiah K, Connor D, Ly K, Gardiner EE, Andrews RK, Qiao J et al (2016) Longitudinal changes in haemostatic parameters and reduced pulsatility contribute to non-surgical bleeding in patients with centrifugal continuous flow left ventricular assist devices. J Heart Lung Transplant 35(6): 745–751

Nesheim ME, Taswell JB, Mann KG (1979) The contribution of bovine Factor V and Factor Va to the activity of prothrombinase. J Biol Chem 254(21):10952–10962

Niswander LM, Fegan KH, Kingsley PD, McGrath KE, Palis J (2014) SDF-1 dynamically mediates megakaryocyte niche occupancy and thrombopoiesis at steady state and following radiation injury. Blood 124(2):277–286

Noetzli LJ, French SL, Machlus KR (2019) New insights into the differentiation of megakaryocytes from hematopoietic progenitors. Arterioscler Thromb Vasc Biol 39(7):1288–1300

Owen RG, Lubenko A, Savage J, Parapia LA, Jack AS, Morgan GJ (2001) Autoimmune thrombocytopenia in Waldenström's macroglobulinemia. Am J Hematol 66(2):116–119

Paggetti J, Haderk F, Seiffert M, Janji B, Distler U, Ammerlaan W et al (2015) Exosomes released by chronic lymphocytic leukemia cells induce the transition of stromal cells into cancer-associated fibroblasts. Blood 126(9):1106–1117

Pal Singh S, Dammeijer F, Hendriks RW (2018) Role of Bruton's tyrosine kinase in B cells and malignancies. Mol Cancer 17(1):57

Paniccia R, Priora R, Liotta AA, Abbate R (2015) Platelet function tests: a comparative review. Vasc Health Risk Manag 11:133–148

Peyvandi F, Garagiola I, Baronciani L (2011) Role of von Willebrand factor in the haemostasis. Blood Transfus 9(Suppl 2):s3–s8

Poulain S, Roumier C, Venet-Caillault A, Figeac M, Herbaux C, Marot G et al (2016) Genomic landscape of CXCR4 mutations in Waldenstrom Macroglobulinemia. Clin Cancer Res 22(6): 1480–1488

Qiao J, Schoenwaelder SM, Mason KD, Tran H, Davis AK, Kaplan ZS et al (2013) Low adhesion receptor levels on circulating platelets in patients with lymphoproliferative diseases before receiving Navitoclax (ABT-263). Blood 121(8):1479–1481

Quek LS, Bolen J, Watson SP (1998) A role for Bruton's tyrosine kinase (Btk) in platelet activation by collagen. Curr Biol 8(20):1137–1140

Ram R, Bonstein L, Gafter-Gvili A, Ben-Bassat I, Shpilberg O, Raanani P (2009) Rituximab-associated acute thrombocytopenia: an under-diagnosed phenomenon. Am J Hematol 84(4): 247–250

Rastogi V, Sharma R, Misra SR, Yadav L, Sharma V (2014) Emperipolesis – a review. J Clin Diagn Res 8(12):ZM01-ZM2

Richard C, Cuadrado MA, Prieto M, Batlle J, López Fernández MF, Rodriguez Salazar ML et al (1990) Acquired von Willebrand disease in multiple myeloma secondary to absorption of von Willebrand factor by plasma cells. Am J Hematol 35(2):114–117

Roccaro AM, Sacco A, Jimenez C, Maiso P, Moschetta M, Mishima Y et al (2014) C1013G/CXCR4 acts as a driver mutation of tumor progression and modulator of drug resistance in lymphoplasmacytic lymphoma. Blood 123(26):4120–4131

Rodriguez S, Celay J, Goicoechea I, Jimenez C, Botta C, Garcia-Barchino M-J, Garces J-J, Larrayoz M, Santos S, Alignani D, Vilas-Zornoza A, Perez C, Garate S, Sarvide S, Lopez A, Reinhardt H-C, Carrasco YR, Sanchez-Garcia I, Larrayoz M-J, Calasanz M-J, Panizo C, Prosper F, Lamo-Espinosa J-M, Motta M, Tucci A, Sacco A, Gentile M, Duarte S, Vitoria H, Geraldes C, Paiva A, Puig N, Garcia-Sanz R, Roccaro AM, Fuerte G, San JF, Miguel J-AM-C, Paiva B (2022) Sci Adv 8(3):eabl4644. https://doi.org/10.1126/sciadv.abl4644

Rondina MT, Weyrich AS (2015) Regulation of the genetic code in megakaryocytes and platelets. J Thromb Haemost 13(Suppl 1):S26–S32

Roos-Weil D, Giacopelli B, Armand M, Della-Valle V, Ghamlouch H, Decaudin C et al (2020) Identification of 2 DNA methylation subtypes of Waldenstrom macroglobulinemia with plasma and memory B-cell features. Blood 136(5):585–595

Rumi E, Pietra D, Ferretti V, Klampfl T, Harutyunyan AS, Milosevic JD et al (2014) JAK2 or CALR mutation status defines subtypes of essential thrombocythemia with substantially different clinical course and outcomes. Blood 123(10):1544–1551

Salles G, Barrett M, Foa R, Maurer J, O'Brien S, Valente N et al (2017) Rituximab in B-cell hematologic malignancies: a review of 20 years of clinical experience. Adv Ther 34(10): 2232–2273

Sangaletti S, Tripodo C, Vitali C, Portararo P, Guarnotta C, Casalini P et al (2014) Defective stromal remodeling and neutrophil extracellular traps in lymphoid tissues favor the transition from autoimmunity to lymphoma. Cancer Discov 4(1):110–129

Sattler M, Griffin JD (2003) Molecular mechanisms of transformation by the BCR-ABL oncogene. Semin Hematol 40(2 Suppl 2):4–10

Savage DG, Szydlo RM, Goldman JM (1997) Clinical features at diagnosis in 430 patients with chronic myeloid leukaemia seen at a referral centre over a 16-year period. Br J Haematol 96(1): 111–116

Sharma R, Haberichter SL (2019) New advances in the diagnosis of von Willebrand disease. Hematology 2019(1):596–600

Shatzel JJ, Olson SR, Tao DL, McCarty OJT, Danilov AV, DeLoughery TG (2017) Ibrutinib-associated bleeding: pathogenesis, management and risk reduction strategies. J Thromb Haemost 15(5):835–847

Shlush LI, Mitchell A, Heisler L, Abelson S, Ng SWK, Trotman-Grant A et al (2017) Tracing the origins of relapse in acute myeloid leukaemia to stem cells. Nature 547(7661):104–108

Singh SP, Pillai SY, de Bruijn MJW, Stadhouders R, Corneth OBJ, van den Ham HJ et al (2017) Cell lines generated from a chronic lymphocytic leukemia mouse model exhibit constitutive Btk and Akt signaling. Oncotarget 8(42):71981–71995

Stone MJ (2009) Waldenstrom's macroglobulinemia: hyperviscosity syndrome and cryoglobulinemia. Clin Lymphoma Myeloma 9(1):97–99

Stone AP, Nascimento TF, Barrachina MN (2022) The bone marrow niche from the inside out: how megakaryocytes are shaped by and shape hematopoiesis. Blood 139(4):483–491

Szczepanski MJ, Szajnik M, Welsh A, Whiteside TL, Boyiadzis M (2011) Blast-derived microvesicles in sera from patients with acute myeloid leukemia suppress natural killer cell function via membrane-associated transforming growth factor-beta1. Haematologica 96(9): 1302–1309

Taher A, Abiad R, Uthman I (2003) Coexistence of lupus anticoagulant and acquired haemophilia in a patient with monoclonal gammopathy of unknown significance. Lupus 12(11):854–856

Talaulikar D, Tam CS, Joshua D, Ho JP, Szer J, Quach H et al (2017) Treatment of patients with Waldenstrom macroglobulinaemia: clinical practice guidelines from the Myeloma Foundation of Australia Medical and Scientific Advisory Group. Intern Med J 47(1):35–49

Tiede A, Rand JH, Budde U, Ganser A, Federici AB (2011) How I treat the acquired von Willebrand syndrome. Blood 117(25):6777–6785

Tournilhac O, Santos DD, Xu L, Kutok J, Tai YT, Le Gouill S et al (2006) Mast cells in Waldenstrom's macroglobulinemia support lymphoplasmacytic cell growth through CD154/ CD40 signaling. Ann Oncol 17(8):1275–1282

Treon SP (2015) How I treat Waldenström macroglobulinemia. Blood 126(6):721–732

Treon SP, Xu L, Yang G, Zhou Y, Liu X, Cao Y et al (2012) MYD88 L265P somatic mutation in Waldenstrom's macroglobulinemia. N Engl J Med 367(9):826–833

Treon SP, Tripsas CK, Meid K, Warren D, Varma G, Green R et al (2015) Ibrutinib in previously treated Waldenstrom's macroglobulinemia. N Engl J Med 372(15):1430–1440

Treon SP, Gustine J, Meid K, Yang G, Xu L, Liu X et al (2018) Ibrutinib monotherapy in symptomatic, treatment-naive patients with Waldenstrom Macroglobulinemia. J Clin Oncol 36(27):2755–2761

Tsai HM, Sussman II, Nagel RL (1994) Shear stress enhances the proteolysis of von Willebrand factor in normal plasma. Blood 83(8):2171–2179

Tullemans BME, Heemskerk JWM, Kuijpers MJE (2018) Acquired platelet antagonism: off-target antiplatelet effects of malignancy treatment with tyrosine kinase inhibitors. J Thromb Haemost 16(9):1686–1699

Uchiba M, Imamura T, Hata H, Tatetsu H, Yonemura Y, Ueda M et al (2009) Excessive fibrinolysis in AL-amyloidosis is induced by urokinae-type plasminogen activator from bone marrow plasma cells. Amyloid 16(2):89–93

Uchiumi H, Matsushima T, Yamane A, Doki N, Irisawa H, Saitoh T et al (2007) Prevalence and clinical characteristics of acute myeloid leukemia associated with disseminated intravascular coagulation. Int J Hematol 86(2):137–142

Vallance TM, Zeuner MT, Williams HF, Widera D, Vaiyapuri S (2017) Toll-like receptor 4 Signalling and its impact on platelet function, thrombosis, and Haemostasis. Mediat Inflamm 2017: 9605894

van Breugel HF, de Groot PG, Heethaar RM, Sixma JJ (1992) Role of plasma viscosity in platelet adhesion. Blood 80(4):953–959

Varettoni M, Arcaini L, Zibellini S, Boveri E, Rattotti S, Riboni R et al (2013) Prevalence and clinical significance of the MYD88 (L265P) somatic mutation in Waldenstrom's macroglobulinemia and related lymphoid neoplasms. Blood 121(13):2522–2528

Varticovski L, Pick AI, Schattner A, Shoenfeld Y (1987) Anti-platelet and anti-DNA IgM in Waldenstrom macroglobulinemia and ITP. Am J Hematol 24(4):351–355

Vinholt PJ (2019) The role of platelets in bleeding in patients with thrombocytopenia and hematological disease. Clin Chem Lab Med (CCLM) 57(12):1808–1817

Vulliamy P, Montague SJ, Gillespie S, Chan MV, Coupland LA, Andrews RK et al (2020) Loss of GPVI and GPIbα contributes to trauma-induced platelet dysfunction in severely injured patients. Blood Adv 4(12):2623–2630

Walsh M, Kwaan H, McCauley R, Marsee M, Speybroeck J, Thomas S et al (2020) Viscoelastic testing in oncology patients (including for the diagnosis of fibrinolysis): review of existing evidence, technology comparison, and clinical utility. Transfusion 60(S6):S86–S100

Wang J, Zhao A, Zhou H, Zhu J, Niu T (2020) Risk of bleeding associated with Ibrutinib in patients with B-cell malignancies: a systematic review and meta-analysis of randomized controlled trials. Front Pharmacol 11:580622

Wang Y, Gali VL, Xu-Monette ZY, Sano D, Thomas SK, Weber DM et al (2021) Molecular and genetic biomarkers implemented from next-generation sequencing provide treatment insights in clinical practice for Waldenström macroglobulinemia. Neoplasia 23(4):361–374

Xu L, Hunter ZR, Yang G, Zhou Y, Cao Y, Liu X et al (2013) MYD88 L265P in Waldenstrom macroglobulinemia, immunoglobulin M monoclonal gammopathy, and other B-cell lymphoproliferative disorders using conventional and quantitative allele-specific polymerase chain reaction. Blood 121(11):2051–2058

Yu X, Li W, Deng Q, Liu H, Wang X, Hu H et al (2021) MYD88 L265P elicits mutation-specific ubiquitination to drive NF-kappaB activation and lymphomagenesis. Blood 137(12):1615–1627

Zago-Novaretti M, Khuri F, Miller KB, Berkman EM (1994) Waldenstrom's macroglobulinemia with an IgM paraprotein that is both a cold agglutinin and a cryoglobulin and has a suppressive effect on progenitor cell growth. Transfusion 34(10):910–914

Zangari M, Elice F, Fink L, Tricot G (2007) Hemostatic dysfunction in paraproteinemias and amyloidosis. Semin Thromb Hemost 33(4):339–349

Zangari M, Elice F, Tricot G, Fink L (2009) Bleeding disorders associated with cancer dysproteinemias. Cancer Treat Res 148:295–304

Zhang Y, Guo Q, Zhao H, Zhao D, Wu X, Zhao W et al (2013) Expression of CXCR4 is an independent prognostic factor for overall survival and progression-free survival in patients with myelodysplastic syndrome. Med Oncol 30(1):341

Zurborn KH, Duscha H, Gram J, Bruhn HD (1990) Investigations of coagulation system and fibrinolysis in patients with disseminated adenocarcinomas and Non-Hodgkin's lymphomas. Oncology 47(5):376–380

The Role of EBV and Immune Deregulation in the Development of IgG4-Related Diseases

Alessandro Ginori, Luigi Spagnoli, and Maria R. Ambrosio

Abstract

IgG4-related disease (IgG4-RD) is a recently recognized multi-organ fibro-inflammatory condition characterized by a tendency to form mass lesions in multiple sites, a dense lymphoplasmacytic infiltrate dominated by IgG4+ plasma cells, storiform fibrosis, and elevated serum IgG4 levels. Recent papers have discussed the potential role played by EBV in some cases of IgG4-RD. The aim of this review is to highlight the possible interaction between EBV-induced immune deregulation and the development of IgG4-RD, discussing also about the prognostic implications. A systematic review was conducted using studies sourced from the PubMed, Embase, and Cochrane literature databases. A total of 10 relevant publications discussing IgG4-RD associated with EBV infection were analyzed. Reactivation of EBV seems to play a role in the development of IgG4-RD and seems also to affect the prognosis of the disease. Searching EBV DNA copy number or testing EBV-encoded RNA (EBER) in situ hybridization to detect EBV in tissues could be prognostically relevant in patients with features of IgG4-RD, especially in presence of systemic or unusual manifestations. However, further investigations are required to better clarify the pathogenetic link between EBV and IgG4-RD.

Keywords

EBV · EBV lymphoproliferative disorders · IgG4 · IgG4-related diseases · Immune deregulation

A. Ginori (✉) · L. Spagnoli · M. R. Ambrosio
USL Northwest Tuscany, "Monterosso" Polyspecialistic Center, Pathology Unit, Carrara, Italy
e-mail: alessandro.ginori@gmail.com

© The Author(s), under exclusive license to Springer Nature Switzerland AG 2022

Interdisciplinary Cancer Research, https://doi.org/10.1007/16833_2022_92
Published online: 16 December 2022

1 Introduction

IgG4-related disease (IgG4-RD) is a recently recognized multi-organ fibro-inflammatory condition that mimics many malignant, infectious, and inflammatory disorders (Deshpande et al. 2012; Kamisawa et al. 2015). It is characterized by a tendency to form mass lesions in multiple sites, a dense lymphoplasmacytic infiltrate dominated by IgG4+ plasma cells, storiform fibrosis, and elevated serum IgG4 levels (the last one could be absent) (Deshpande et al. 2012). This disease was firstly described in the pancreas when the so-called autoimmune pancreatitis was associated with high levels of serum IgG4 in 2001 (Hamano et al. 2001). Nowadays IgG4-RD includes many conditions previously considered as isolated idiopathic single-organ diseases involving biliary ducts, salivary gland, periorbital tissues, kidneys, lungs, soft tissues, lymph nodes, meninges, aorta, breast, prostate, thyroid, pericardium and skin (Table 1) (Stone et al. 2012). For this reason, IgG4-RD has been considered to have many characteristics in common with sarcoidosis, another systemic immune-mediated disorder in which many different organ involvements have the same histopathological features (Deshpande et al. 2012; Kamisawa et al. 2015). Recognizing this disease is essential to ensure a proper treatment to patients and to prevent severe complications as multi-organ damages, invalidating tissue fibrosis and even death. Some parameters are useful to make a correct diagnosis or to suggest the possibility of this disease, but much remains unknown about the pathogenesis of IgG-RD. The behavior of *en vivo* IgG4 and the role that they play in the development of this disease – primary or secondary – is still debated and to date still remains a mystery (Aalberse et al. 2009; Nirula et al. 2011; Umehara et al. 2012). An intriguing new perspective about the disease concerns Epstein-Barr virus

Table 1 Most common conditions included in IgG4-RD

Disease	Site
Autoimmune pancreatitis	Pancreas
Eosinophilic angiocentric fibrosis	Orbits and upper respiratory tract
Küttner's tumor	Submandibular glands
Mikulicz's disease	Salivary and lacrimal glands
Idiopathic tubulointerstitial nephritis	Kidneys
Inflammatory pseudotumor	Orbits, lungs, kidneys
Inflammatory aortic aneurysm	Aorta
Periaortitis and periarteritis	Aorta and arteries
Multifocal fibrosclerosis	Orbits, thyroid gland, retroperitoneum, mediastinum
Retroperitoneal fibrosis (Ormond's disease)	Retroperitoneum
Fibrosing mediastinitis	Mediastinum
Sclerosing mesenteritis	Mesentery
Riedel's thyroiditis	Thyroid gland
Hypertrophic pachymeningitis	Meninges

(EBV). EBV is a member of the herpes virus family and infects nearly 95% of the adult human population worldwide (Ambrosio et al. 2015). The EBV life cycle allows the virus to access the memory B cells, escaping the immune defenses and establishing a non-pathogenetic infection during the entire life of the human host (Pattle and Farrell 2006; Thorley-Lawson and Allday 2009). It is well known that this immune deregulation may cause proliferation and transformation of lymphoid cells, leading to the so-called EBV-induced lymphoproliferative disorders (EBV-LPD) (Dojcinov et al. 2011). The development of EBV-LPD usually is the result of an immune suppression (primary or secondary), but they could develop also in a non-immunodeficiency setting (Dojcinov et al. 2010; de la Hera Magallanes et al. 2011). EBV-LPD form a spectrum of diseases that goes from benign disorders like the EBV-mucocutaneous ulcer (EBVMCU) to frank malignancies like B-cell, T-cell, and Hodgkin lymphomas. Recent papers have discussed the potential role played by EBV in some cases of IgG4-RD (Shacks et al. 1985; Cheuk et al. 2010; Takahashi et al. 2012; Wada et al. 2013; Choe et al. 2013; Takeuchi et al. 2014; Furukawa et al. 2015; Szczawinska-Poplonyk et al. 2016; Kotetsu et al. 2017; Nagata et al. 2018). The aim of this review is to summarize the clinical, histological, and pathogenetic key points of IgG4-RD and EBVLPD, highlighting the possible interaction between EBV-induced immune deregulation and the development of IgG4-RD, and discussing about the prognostic and therapeutic implications of this new scenario.

2 Clinical and Histopathological Features of IgG4-RD

2.1 Epidemiology and Clinical Manifestations

It is difficult to clearly establish the epidemiology of IgG4-RD, as the disease has started to appear in medical publications only in 2003 (Kamisawa et al. 2003a, b). The prevalence of the different manifestations also remains unclear, but autoimmune pancreatitis, sialadenitis (particularly of the submandibular gland), dacryoadenitis, and IgG4-related retroperitoneal fibrosis are the most common ones (Kamisawa et al. 2015). IgG4-RD affects typically middle-aged to elderly men (Stone et al. 2012; Kanno et al. 2012), in contrast with many autoimmune disorders that affect mainly female patients. However, the male-to-female ratio seems to be equal in the IgG4-RD of the head and neck district (orbits, salivary glands, sinuses) (Zen and Nakanuma 2010). There is no report about cases of IgG4-RD with familiarity or genetic susceptibility (Kamisawa et al. 2015). The presentation of IgG4-RD is generally subacute, with symptoms lasting for months or even years before diagnosis. The disease can progress quickly or experience alternating periods of well-being and worsening of clinical manifestations. The most common symptoms are not specific and include weight loss, fatigue, arthralgias, enthesopathies, and fever. With the progression of the disease, enlargement and loss of function of the involved organ appear with site-specific symptoms (Kamisawa et al. 2015).

2.2 Pathogenesis

Even if the pathogenesis of IgG4-RD remains largely to be solved (Takeuchi et al. 2014), it is known the role played by CD4+ T-cells in the development of this disease, activating macrophages, myofibroblasts, and fibroblasts and inducing fibrosis (Kamisawa et al. 2015). In fact, many CD4+ T-cells are present in tissues affected by IgG4-RD. Different reports have suggested a key pathogenetic function of forkhead box P3 (FOXP3)-positive regulatory T-cells (Tregs), type 1 helper T-cells (Th1) and Th2 in this disease (Okazaki et al. 2000; Ota et al. 2007; Zen et al. 2007; Takeuchi et al. 2014). Th2 cells secrete interleukins (ILs) 4, 5, and 13 and are usually implicated in fibrotic processes, but other diseases such as tuberculosis or Crohn's disease are characterized by a fibrosis induced by Th1 cells (Kamisawa et al. 2015). Some authors have reported a possible role of IL-17, transforming growth factor (TGF)-β, and periostin in IgG4-RD, due to the fibrotic induction of these molecules in some persistent chronic inflammation processes (Ohta et al. 2012, 2014). Moreover, another T-follicular helper population is supposed to induce the generation of IgG4+ plasma cells that characterize the disease (Kamisawa et al. 2015). In conclusion, the precise phenotype of CD4+ T-cells remains to be clarified. The role of IgG4 in the physiologic inflammatory process is also unclear. Perhaps they have the function of attenuating the inflammation (Kamisawa et al. 2015). For all these reasons, a model of pathogenesis of IgG4-RD argues that in some susceptible individuals, mostly elderly men, some insult, possibly a microbe, starts a tissue damage and a deregulation of the immune response. A population of CD4+ T-cells arrives at the site of the damage and induces inflammation and a fibrotic process through activation of innate immune cells and the secretion of specific cytokines. Possibly, the T-cell response is also sustained by some antigen-presenting B-cells (Khosroshahi et al. 2010, 2012). This mechanism may trigger another population of T-follicular helper cells that induce the development of secondary follicles with florid germinal centers within the regional lymph nodes and the production of plasmablasts and plasma cells secreting immunoglobulins expressing the IgG4 class (Kamisawa et al. 2015). This is only a hypothetical pathogenetic mechanism of IgG4-RD based on the available knowledge. However, many aspects of the disease are still far from being solved.

2.3 Diagnosis

Histologic examination of the involved tissue is necessary for the diagnosis (Deshpande et al. 2012; Kamisawa et al. 2015). However, also in the presence of striking histopathological features (see below for the diagnostic criteria), a clinical correlation is crucial to support the diagnosis. Imaging, and in particular PET scan, is important to show how many organs are involved and to monitor disease after therapy (Ebbo et al. 2014). Even if many IgG4-RD patients have high serum IgG4 levels, this finding is neither sensitive nor specific to make a diagnosis. Patients with multi-organ involvement can also have concentrations of serum IgG4 lower than normal (Carruthers et al. 2015) and some studies have demonstrated that high levels of serum IgG4 are also present in patients that did not have IgG4-RD or in healthy people (Ghazale et al. 2007; Tabata et al. 2009; Sadler et al. 2011; Carruthers et al. 2015). Serum IgG4 concentrations seem to be helpful in monitoring disease activity in some cases, but they should never be used alone to make therapy decisions (Kamisawa et al. 2015). In conclusion, when there is a clinical suspicion of IgG4-RD, biopsy of involved tissue is mandatory, not only to confirm the diagnosis but also to exclude other benign disorders and malignant tumors that can mimic IgG4-RD.

2.4 Histopathological Features of IgG4-RD

According to the consensus statement on the pathology of IgG4-RD (Deshpande et al. 2012), the three central histopathological features of IgG4-RD are:

- A dense lymphoplasmacytic infiltrate.
- Fibrosis, at least focally in a storiform pattern.
- Obliterative phlebitis.

Other two minor features of IgG4-RD, that alone are neither sensitive nor specific for the diagnosis, are:

- Phlebitis without obliteration of the lumen.
- Increased number of eosinophils.

A quite confident pathological diagnosis of IgG4-RD needs two of the three major histological features (mostly, the first two). However, in some organs such as lymph nodes, lung, or minor salivary glands, storiform fibrosis and obliterative phlebitis can be inconspicuous or absent (Deshpande et al. 2012).

Interestingly, in lymph nodes five histological patterns are recognized in the proper setting of IgG4-RD:

1. Multicentric Castleman's disease-like.
2. Follicular hyperplasia.
3. Angioimmunoblastic T-cell lymphoma (AITL)-like with interfollicular expansion.
4. Progressive transformation of germinal centers.
5. Nodal inflammatory pseudotumor-like.

In lung manifestations of IgG4-RD, obliterative arteritis is also seen and it is characterized by a non-necrotizing lymphoplasmacytic infiltrate with or without obliteration of the lumen (Deshpande et al. 2012). Usually, granulomata, necrosis, or dense neutrophilic infiltrate are not specific of IgG4-RD and their presence should suggest the possibility of another diagnosis (Deshpande et al. 2012; Kamisawa et al. 2015). The lymphocytic infiltrate in IgG4-RD is mostly represented by

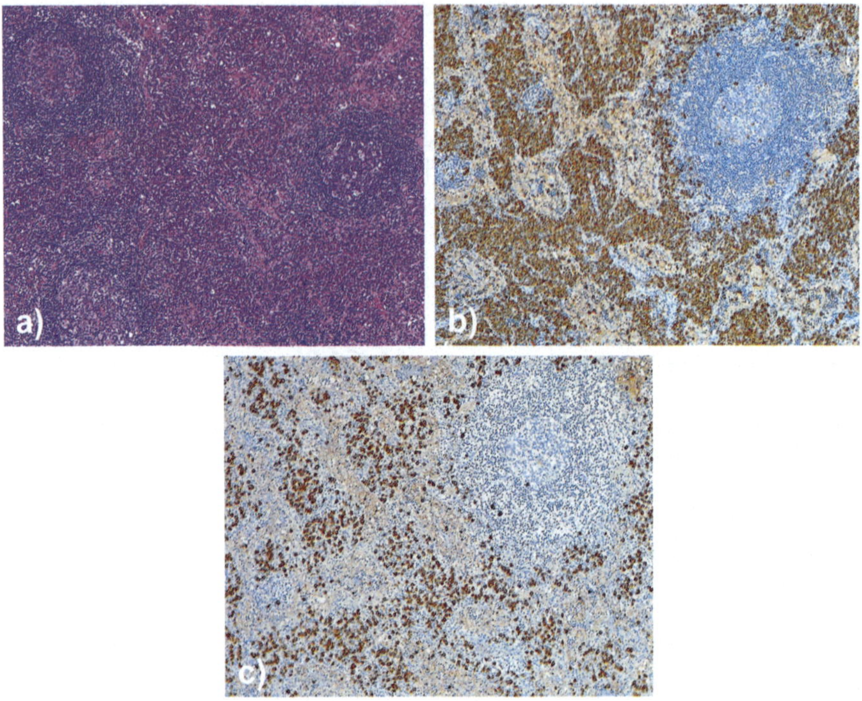

Fig. 1 IgG4-related lymphadenopathy characterized by interfollicular expansion with many plasma cells (**a**, hematoxylin and eosin stain, 100×) expressing IgG (**b**, IgG immunohistochemistry stain, 100×) and IgG4 (**c**, IgG4 immunohistochemistry stain 100×)

CD4+/FOXP3+ T-cells. IgG4 immunohistochemistry is obviously essential for the pathological diagnosis of IgG4-RD (Deshpande et al. 2012) and allows to differentiate this disease from other systemic immune disorders characterized by a dense lymphoplasmacytic infiltrate such as Castleman's disease or primary sclerosing cholangitis (Kamisawa et al. 2015). The absolute number of IgG4-positive plasma cells may vary from organ to organ mostly in relation to the different degree of fibrosis at the time of the biopsy. So, the pathology consensus statement has proposed a different diagnostic cut-off value of IgG4-positive plasma cells/high power fields (HPF) based on the involved tissue (Deshpande et al. 2012). Moreover, even if the other pathological features are present, the IgG4+/IgG+ plasma cell ratio must be >40% to ensure a histological diagnosis of IgG4-RD (Deshpande et al. 2012) (Fig. 1). However, it is correct to keep in mind that careful correlation with clinical and imaging findings is essential to arrive at a definitive diagnosis. For this reason, the consensus statement proposes three diagnostic categories based only on the histological findings (Deshpande et al. 2012):

1. *Histologically Highly Suggestive of IgG4-RD*: characterized by the presence of all the three major histological features, with the required number of IgG4-positive plasma cells/HPF and a IgG4+/IgG+ plasma cells ratio >40%
2. *Probable Histological Features of IgG4-RD*: includes cases with only one of the three major histological features (usually a dense lymphoplasmacytic infiltrate) and the required number of IgG4-positive plasma cells/HPF, needle biopsies and meningeal and cutaneous diseases due to the limited data about these manifestations
3. *Insufficient Histopathological Evidence of IgG4-RD*: includes cases that do not fulfill the criteria for the previous two categories; this histological diagnosis does not exclude an IgG4-RD completely

3 Clinical and Histopathological Features of EBV-LPD

The other "main character" of this review is EBV. EBV has a key role in many disorders and has a long history of interactions with human hosts and with other viruses. We do not have the presumption to clarify the role of EBV in any disease. We just want to try to resume some characteristics of this often quiet – but sometimes restless – roommate of human race, highlighting its potential association with IgG4-RD.

3.1 Epidemiology of EBV Infection

EBV is a member of the herpes virus family and has been discovered in 1964 by Antony Epstein and Yvonne Barr as a direct consequence of the studies about Burkitt lymphoma (BL) by Denis Burkitt (1962; Epstein et al. 1964). It infects nearly 95% of the adult human population worldwide (Ambrosio et al. 2015). In developed countries, there are two peaks of infection, the first in children aged 1–6 years and the second in adolescents and young adults aged 14–20 years (Crawford et al. 2006). In developing countries, infection occurs at a much earlier age so that 90% of children over the age of 2 have yet been infected (Hjalgrim et al. 2007). There is no difference in EBV prevalence by sex and its transmission occurs through saliva (so-called "kissing disease"). For this reason, primary EBV infection takes place in oropharynx. In adolescents and in childhood, EBV infection is asymptomatic in most of the patients, but can cause the infectious mononucleosis syndrome, a self-resolving disease characterized by a strong cellular immune response which takes the infection under control (Delgado-Eckert and Shapiro 2011). However, after the resolution of primary infection, the virus establishes a lifelong persistence in memory B-cells that usually remains clinically silent (Delecluse et al. 2007).

3.2 Structure, Life Cycle of EBV and Pathogenesis of EBV Infection

Like other herpes viruses, EBV is an enveloped virus, which contains a core DNA surrounded by an icosahedral nucleocapsid. The EBV DNA encodes for a series of almost 100 products, resulting in the production of a wide variety of viral particles (EBNA-1, EBNA-2, EBNA-3 family, EBNA-LP, LMP-1, LMP-2 family), two small RNA known as EBV-encoded RNAs (EBER1 and EBER2), BamHI-A rightward transcripts (BARTS) and several microRNAs (miRNAs) (Robertson 2010), triggering the activation of many different anti-apoptotic molecules, cytokines, and signal transducers, to escape immune response of the host and to promote EBV infection and immortalization (De Falco et al. 2009). The EBV life cycle includes a lytic phase, resulting in the production of the new viral particle, and a latent phase, during which the virus remains silent for the lifetime of the host in memory B-cells (Robertson 2010). First, the virus infects the epithelium of the oropharynx and the naïve B-cells of Waldeyer ring. EBV enters in the epithelial cells in a lytic phase, producing viral progeny and amplifying the cell-to-cell spreading. Then, EBV establishes persistent infection expressing a series of latency proteins that allow the virus to drive the infected naïve B-cell into the memory B-cell compartment (Thorley-Lawson 2001; Laichalk and Thorley-Lawson 2005; Thorley-Lawson and Allday 2009; Roughan and Thorley-Lawson 2009). Three different viral latency programs correlated with the phases of maturation of the B-cells have been described. It is thought that this mechanism serves to hide from the immune system of the host. After receiving certain activation signals (BCR stimulation, hypoxia, TGF-β, DNA damage, or chemical agents) EBV can exit the latency phase switching to the lytic one, with subsequent lysis/death of infected cells and proliferation of virions that infect other cells (Kenney and Mertz 2014). However, the aim of EBV is to persist in the chronic infected cells, avoiding killing them and eluding the cytotoxic immune defense (Hatton et al. 2014). The virus starts a precise and elegant interplay between itself, the infected memory B-cells, and the immune system to continue viral persistence for the entire life cycle of the host. Through the molecules mentioned above, EBV has implemented a large number of mechanisms to deregulate the immune response of the host, interfering with apoptosis, antigen processing and presentation functions and many other immune cell pathways (Thompson and Kurzrock 2004). The potential role of the viral lytic cycle has been also studied, suggesting that EBV lytic reactivation may have great relevance in immune deregulation and lymphomagenesis in the context of both active and depressed immune response (Ma et al. 2011; Ambrosio et al. 2017).

3.3 EBV-LPD: An Overview

As can happen in any coexistence, something wrong could occur also in the chronic interaction between EBV and the immune system of the host. Especially in a setting of immunodeficiency, the infected cells in the peripheral blood or in oropharynx proliferate, express many viral genes, proteins, and miRNAs and dysregulate different cellular pathways of the host (Holmes and Sokol 2002). EBV has been correlated with some non-hematological neoplasms and with many LPDs (Rezk et al. 2018):

1. EBV-related reactive lymphoid proliferations
   - EBV-positive reactive lymphoid hyperplasia
   - Acute infectious mononucleosis
   - EBV-positive hemophagocytic lymphohistiocytosis
   - Chronic active EBV disease (CAED)
   - Hydroa vacciniforme
   - Severe mosquito bite allergy
   - EBV-positive mucocutaneous ulcer
2. Hodgkin lymphoma
3. EBV-related B-cell LPDs
   - Burkitt lymphoma
   - Lymphomatoid granulomatosis
   - Plasmablastic lymphoma
   - Primary effusion lymphoma
   - Diffuse large B-cell lymphoma associated with chronic inflammation
   - EBV-positive diffuse large B-cell lymphoma
   - HHV8-positive germinotropic lymphoproliferative disorder
   - Plasma cell neoplasms.
4. EBV-related T/NK-cell LPDs
   - Angioimmunoblastic T-cell lymphoma (AITL)
   - Follicular T-cell lymphoma
   - Extranodal T/NK-cell lymphoma, nasal type
   - Aggressive NK-cell lymphoma
   - Systemic EBV-positive T-cell lymphoma of childhood
   - Peripheral T-cell lymphoma, not otherwise specified (NOS)
5. EBV-associated immunodeficiency-related LPDs
   - HIV-related lymphomas
   - Post-transplant LPDs (PTLDs)
   - Other iatrogenic immunodeficiency-related LPDs
   - LPDs associated with a primary immune disorder
6. EBV-related histiocytic/dendritic disorders
   - Inflammatory pseudotumor-like follicular/fibroblastic dendritic cell sarcoma
7. Non-hematological neoplasms
   - Nasopharyngeal carcinoma
   - Gastric carcinoma
   - Lymphoepithelioma-like carcinoma

According to the purpose of this review and for the potential link to IgG4-RD that we are going to highlight, we want to focus mostly on the histologic characteristics of CAED.

3.4 Clinical and Histopathological Features of CAED

CAED is a rare systemic disease related to persistent EBV infection in patients without known immunodeficiency. It is more common in Asian people and the usual clinical manifestations are fever, lymphadenopathy, and hepatosplenomegaly and the patients have high EBV DNA copies in the peripheral blood. Hemophagocytic syndrome and hypogammaglobulinemia can occur in a subset of patients (Cohen et al. 2009; Pittaluga 2013). Likely, this disorder is due to a cellular immune defect, which failed to control the interaction between EBV and the chronically infected lymphocytes with subsequent reactivation of the lytic cycle of the virus and expression of lytic proteins (Cohen et al. 2011). Although EBV usually resides in memory B-cells, in the majority of CAED cases EBV is detected in T and NK cells (the so-called "T/NK-cell type" of CAED). However, there is also a so-called "B-cell type of CAED" with EBV detected in B cells (Rezk et al. 2018). Morphologically, CAED of T/NK-cell type show a T/NK-cell infiltrate, that is usually sinusoidal in the liver and spleen, sinusoidal or interstitial in the bone marrow and interstitial or diffuse in the paracortical areas of lymph nodes. The architecture of the affected tissue is usually preserved and cytologic atypia is absent within the EBV-positive cells. In situ hybridization reveals the expression of EBV in many cytotoxic T lymphocytes (Rezk et al. 2018). Instead, CAED of B-cell type shows a morphology similar to cases of EBV-positive polymorphic PTLD with an increased number of immunoblasts, plasma cells, occasional Reed-Sternberg-like cells, and paracortical hyperplasia in affected lymph nodes (Fig. 2). In situ hybridization reveals the expression of EBV in small and large scattered B-cells. In some cases, there is a progression to a more monomorphic PTLD-like morphology (Pittaluga 2013). Molecular biology studies have shown that CAED can be polyclonal, oligoclonal, or monoclonal and maybe it could have a different prognostic impact on the evolution of the disease (Ohshima et al. 2008). Monoclonal cases of CAED can be difficult to differentiate from an EBV-positive T-cell or B-cell lymphoma. Some authors believe that monoclonal T-cell types of CAED should be considered

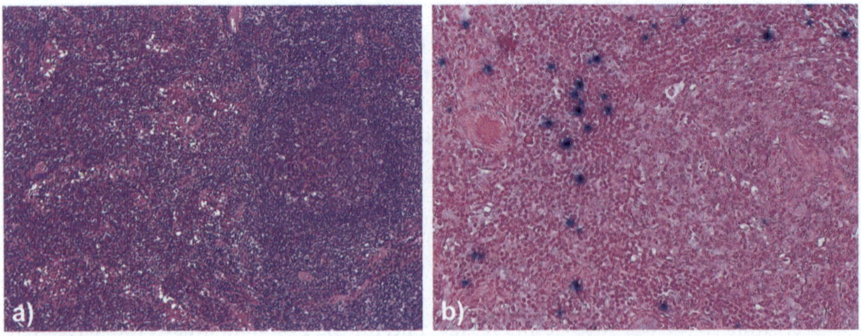

Fig. 2 CAED of B-cell type in a lymph node showing an increased number of plasma cells, immunoblasts, and paracortical hyperplasia (**a**, hematoxylin and eosin stain, 100×) with expression of EBV in small and large scattered B-cells (**b**, EBER in situ hybridization, 200×)

as T-cell lymphomas (Quintanilla-Martinez et al. 2000, 2008), while others indicate that the absence of cytologic atypia and the preservation of tissue architecture are more consistent with a monoclonal lymphoproliferative disorder rather than a true lymphoma (Ghazale et al. 2007). It is our opinion that more studies are needed to better understand this rare and interesting EBV-related disease.

4 Association Between IgG4-RD and EBV: A Systematic and Chronological Review of the Literature

Few reports have suggested a possible relationship between IgG4-RD and EBV. Those papers are summarized in Table 2. Shacks et al. (1985) were the first to report this link, indicating that some cases of acute EBV mononucleosis were associated with a transient 75% increase in mean serum of IgG4 levels. Mean IgG4 levels reached a peak from 2 to 3 weeks post-onset and declined to baseline by 60 days. IgG4 peaked independently of other immunoglobulin classes.

In 2010, Cheuk et al. described four cases of idiopathic cervical fibrosis, a rare tumefactive inflammatory-sclerosing disease involving the soft tissues of the head and neck region and known also as sclerosing cervicitis or cervical fibrosclerosis (Rice et al. 1975). The cause of these lesions has been attributed to infections, autoimmune diseases, traumas, or prior surgery, but most cases are considered idiopathic (Dehner and Coffin 1998). A proportion of the patient also has other synchronous or metachronous inflammatory fibrosclerosing lesions in other anatomic sites, so a possible association with a systemic condition has been suggested (Amorosa et al. 1976). Recently, most cases of these fibro-inflammatory lesions have been considered to be part of the spectrum of IgG4-RD. In the series reported by Cheuk et al. (2010), all cases showed a mass of the cervical soft tissues characterized

Table 2 Papers suggesting a possible relationship between EBV and IgG4-RD

Paper	Manifestation/disease	Number of reported cases
Shacks et al. (1985)	Acute mononucleosis	N.D.
Cheuk et al. (2010)	Idiopathic cervical fibrosis associated with EBV+ classical Hodgkin lymphoma	1
Takahashi et al. (2012)	Lymphadenopathy with prominent granulomatous inflammation	1
Wada et al. (2013)	Lymphadenopathy	1
Choe et al. (2013)	Inflammatory pseudotumor-like follicular dendritic cell sarcoma of the spleen	6
Takeuchi et al. (2014)	Lymphadenopathy	18
Furukawa et al. (2015)	Salivary and lacrimal glands	8
Kotetsu et al. (2017)	Lungs	1
Szczawinska-Poplonyk et al. (2016)	Lungs	1
Nagata et al. (2018)	Graves' disease of thyroid	7

by bands of collagen with few fibroblasts and myofibroblasts and by a dense lymphoplasmacytic infiltrate forming lymphoid aggregates with follicles, admixed with a variable amount of eosinophils. The plasma cells were polytypic. The IgG4+ elements were the majority of the plasma cells, with a high IgG4+/IgG+ ratio (from 63% to 98%). In one of those cases, an 89-year-old man presented also with an enlarged cervical lymph node showing areas of effaced architecture due to the presence of multi-nucleated and nucleolated large cells, scattered in a background of lymphoplasmacytic, eosinophilic, and histiocytic inflammatory infiltrate. The atypical cells showed the immunophenotype of classical Hodgkin lymphoma and were also EBV+. Moreover, the preserved areas of the same lymph node and the other cervical lymph nodes had the features of type II IgG4-related lymphadenopathy, with many IgG4+ polytypic plasma cells between reactive lymphoid follicles. The patient had a long clinical history of chronic sclerosing sialadenitis of left submandibular gland, left parotid mass, and left orbital mass, with high levels of serum IgG4. Moreover, the imaging showed diffuse lymphadenopathy in cervical, mediastinal, axillary, and paraaortic regions. Of interest, this patient was the only one with a systemic disease in this series.

Other two cases involving lymph nodes were described by Takahashi et al. in 2012 and by Wada et al. in 2013. The first was the case of a 60-year-old woman with generalized lymphadenopathy (cervical, supraclavicular, axillary, mediastinal, abdomen, pelvic, paraaortic, and inguinal), elevated levels of serum IgG and IgG4 and positive tests for anti-DNA and antinuclear antibodies (ANA). The biopsy specimen of a cervical lymph node showed an expansion of interfollicular zone characterized by a prominent granulomatous inflammation, with a predominance of epithelioid macrophages and some Langhans multinucleated giant cells and dendritic cells, extending into the adjacent fat tissue. Bundles of spindle cells were also seen. Intermingled with the granulomas there were many plasma cells, small lymphocytes, eosinophils, and neutrophils immunoblasts, plasmacytoid cells, and few Hodgkin-like cells. The plasma cells were polytypic and in the majority were IgG4+, with a IgG4+/IgG+ ratio of 70%. The absolute number of IgG4+ cells exceeded 50/high power field, fulfilling the histological criteria for type V IgG4-related lymphadenopathy. Moreover, EBER in situ hybridization highlighted many EBV+ lymphoid cells scattered throughout the paracortical area. These cells were negative for LMP1 and EBNA2. Instead, Wada et al. (2013) described an interesting case of a 72-year-old woman with generalized lymphadenopathy due to reactivation of EBV and mimicking the clinical features of IgG4-RD. The patient had elevated serum levels of IgG4, hypergammaglobulinemia, proteinuria, and positivity for ANA. Cervical lymph node biopsy specimens showed reactive follicular hyperplasia with enlarged germinal centers and prominent lymphoplasmacytic infiltration in the interfollicular areas. There were many polytypic IgG+ plasma cells, but only few IgG4+ cells. An in situ hybridization to detect EBER of EBV demonstrated many EBV+ lymphocytes in the germinal centers as well as in the interfollicular areas. Tests for EBNA2 and LMP1 were negative. Although this case had many clinical features in common with IgG4-RD, the lymph node histological findings were quite different. In our opinion, this case is very useful in this review to remind the importance of a multi-disciplinary approach to the diagnosis of IgG4-RD. It also highlights the interactions between EBV, immune system, and IgG4+ plasma cells, that in some cases may induce a "false" IgG4-RD as described also by Shacks et al. in 1985 and as we mentioned above. We will return to this topic later, hypothesizing a possible mechanism about the relationship between EBV and IgG4+ plasma cells.

In 2013, Choe et al. described six cases of inflammatory pseudotumor (IPT)-like follicular dendritic cell (FDC) sarcoma of the spleen, a rare variant of FDC sarcoma arising almost invariably in the spleen or liver and associated with clonal expansion of EBV in the neoplastic cells (Brittig et al. 2004; Rosenbaum et al. 2009). The most prominent histologic features of these cases were fibroblastic stromal proliferation and heavy background inflammatory cell infiltration constituted mostly by lymphocytes, plasma cells, and occasional eosinophils and neutrophils. Lymphoid follicles were present in one case. Epithelioid histiocytes were seen in four cases and were prominent in one. Obliterative vasculitis was found in three cases. The neoplastic cells were uniformly positive for EBV by EBER in situ hybridization and at least focally for LMP1, with no EBV-positive cells in the remaining splenic tissue. There was also a large amount of IgG4+ plasma cells accounting from 27 to 128 cells/high-power field (HPF), with a IgG4/IgG ratio ranging from 25% to 75%. The reported splenic IPT-like FDC sarcomas were solitary lesions and the prognosis was good. Because it was a retrospective report, serum IgG4 levels were not available. The authors suggested a possible relationship between IPT-like FDC sarcoma and IgG4-RD maybe mediated by EBV.

The first article trying to understand the possible role of EBV in IgG4-RD was in 2014 by Takeuchi et al. They analyzed 31 cases of IgG4-related lymphadenopathy making a comparison with 22 lymph nodes with lymphoid hyperplasia and with 10 AITL cases. All the IgG4-related lymphadenopathy cases had elevated serum levels of IgG4 and at immunohistochemistry many IgG4+ plasma cells were present ranging from 122 to 477/HPF with a IgG4/IgG ratio always >40%. In 18 cases, EBER in situ hybridization revealed many EBV-positive cells expressing antigens compatible with a type I viral latency. In 15 of those 18 cases, scattered EBV+ lymphocytes were seen in interfollicular areas. Two type III cases showed numerous diffusely distributed EBV+ small lymphocytes and immunoblasts. In one case the EBV+ cells were localized in germinal centers and associated with epithelioid granulomas. There were no significant differences in age, serum IgG4 levels, and numbers of FOXP3+ cells between EBV+ and EBV- cases. Of interest, 9 EBV+ cases that showed features of type IV IgG4-related lymphadenopathy had systemic lymphadenopathy and/or extranodal involvement. Type IV cases are usually localized in submandibular lymph nodes, and less than 10% develop systemic lymphadenopathy (Sato et al. 2012). This is in contrast with the data seen in this study in the group of EBV+ type IV cases, suggesting a possible impact of EBV reactivation in the development of a systemic IgG4-RD. Another interesting point concern the absence of a difference in age between EBV+ and EBV- patients, because the clonal expansion of EBV-infected cells in many cases of IgG4-RD could have been only an age-related phenomenon. There were no differences in age also between IgG4-RD group and lymphoid hyperplasia group, but the latter had significantly fewer cases with increased EBV-infected cells. So, age did not seem to be a factor in the association of IgG4-RD with increased numbers of EBV-infected cells. Finally, only few scattered IgG4+ plasma cells were detected by authors in AITL cases, which suggested that IgG4+ cell proliferation was not simply induced by EBV reactivation or deregulation of helper T-cells and that both IgG4-RD and EBV reactivation may be induced by other factors.

The 8 patients with IgG4-RD collected by Furukawa et al. in 2015 clinically had bilateral hard masses in the salivary and/or lacrimal glands. Serum concentrations of IgG4 were high in all cases. Four patients developed systemic involvement, including pancreatitis, retroperitoneal fibrosis, or inflammatory pseudotumor of the lung, after the initial enlargement of the salivary glands. The histology of all surgical specimens showed high-grade infiltration of IgG4-positive plasma cells and fibrosclerosis, with atrophy and loss of acinic glands. At immunohistochemistry, expression of EBNA2 and EBER by in situ hybridization was seen, while LMP1 expression was not observed. The copy number of EBV DNA in affected salivary glands from patients with IgG4-RD was significantly higher than that in the glands of 12 patients with sialolithiasis used as controls. The copy number of EBV DNA of IgG4-RD patients was also strongly correlated with the serum IgG4 concentration and with the IgG4/IgG plasma cells ratio in the analyzed tissues and was significantly greater in patients with systemic involvement than in those without systemic disease. The authors suggested that EBV-DNA copy number could be used as a diagnostic finding to assess IgG4-RD progression, indicating that careful management and a long period of follow-up should be considered for those patients due to their high risk of developing potentially life-threatening systemic involvement. This study not only demonstrated like other papers the histological association between EBV and IgG4-RD but also assessed EBV-DNA copy number in IgG4-RD patients, trying to correlate this with the prognosis of the disease.

IgG4-RD with lung involvement is rare, with only few clinicopathologic studies (Zen et al. 2009; Matsui et al. 2013). Two rare cases of IgG4-RD localized to lung and associated with EBV infection were reported by Szczawinska-Poplonyk et al. in 2016 and by Kotetsu et al. in 2017. In the first one, Szczawinska-Poplonyk et al. (2016) described the case of a 7-year-old boy with chest pain, fever, dyspnea, submandibular and cervical lymphadenopathy and purulent secretions on the posterior pharyngeal wall. A chest CT scan revealed a solid, well-defined mass in the right lower lobe, surrounded by ground-glass haze. Suspecting a neoplastic lesion, a lung biopsy was performed which resulted unconclusive. So, the mass was surgically removed and the pathologic examination of the lung revealed a sclerotic lesion infiltrating the alveolar septa and surrounding the bronchovascular walls. The fibrosclerosis was accompanied by a lymphoplasmacytic inflammatory infiltrate with rare eosinophils. The IgG4+ plasma cells were $\geq$10/HPF and the IgG4/IgG ratio was $\geq$40. Obliterative phlebitis was also present. The serologic analyses showed normal pre and postoperative IgG4 serum levels (42,16 mg/dl vs 42,59 mg/dl, respectively) and elevated EBV DNA copy number (1,400 copies/ml). The authors classified this case as a rare pediatric manifestation of an IgG4-RD complicated by acute EBV infection. They did not clarify if EBV was present or not in lung tissue. However, the case is another clue of the possible affinity between EBV and IgG4+ plasma cells. Instead, in the paper of Kotetsu et al. (2017) a 70-year-old male presents with a dense consolidation in the right middle lobe and slight mediastinal lymph node swelling in the subcarinal region at computed tomography (CT) scan. There was no other organ involved. Malignant lymphoma was suspected, so surgical biopsy of the right hilar lymph node was performed under thoracoscopy. A lung biopsy was also made. Histologic examination of the resected lymph node revealed diffuse infiltration of plasma cells, largely IgG4+ (ratio IgG4/IgG: 86%). Scattered EBV-infected cells were also observed. The lung specimen was characterized by slight interstitial thickening and IgG4+ plasma cells infiltration into the peripheral bronchiole and alveoli (ratio IgG4/IgG: 70%). Serum IgG4 was increased. Due to the presence of EBV-infected cells, the authors performed an EBV test and found a chronic subclinical asymptomatic EBV viremia. Based on this fact, the authors suggested that many IgG4-RD associated with EBV infection may be overlooked, since EBV infection is not routinely checked when approaching a suspicious IgG4-RD case.

The paper of Nagata et al. in 2018 is the most recent one about the relationship between IgG4-RD and EBV. In their study, the authors investigated if EBV reactivation induced IgG4 production and if EBV-positive B cells or IgG4-positive plasma cells are present in the thyroid tissues of Graves' disease patients with lymphoplasmacytic infiltration. EBER in situ hybridization and immunohistochemistry for IgG and IgG4 performed on seven resected thyroid tissues affected by Graves' disease and characterized by diffuse lymphoplasmacytic demonstrated EBV+ cells and IgG4+ plasma cells in the same areas of thyroid tissues. The analyzed specimens did not show evidence of tumefactive lesions, storiform fibrosis, or obliterative phlebitis. Then they cultured the lymphocytes of 13 Graves' disease patients and 14 controls and induced EBV reactivation to measure IgG4 levels in culture fluids, detecting EBV-reactivated cells with IgG4 on their surface in culture cells and IgG4 production in culture fluids. In all those patients the serum levels of IgG4 were elevated. The authors concluded that a subset of Graves' disease maybe is not an IgG4-RD, but more probably is an IgG4-RD-like condition. Moreover, EBV reactivation seems to stimulate the IgG4 production, which may induce high serum IgG4 levels and promote IgG4-positive plasma cell infiltration of tissues. This study is the first one that try to clarify the cellular pathogenetic mechanism that ties EBV reactivation and IgG4+ plasma cells production, with the possibility of a consequential development of an IgG4-RD or an IgG4-RD-like condition.

In our Unit we have recently diagnosed a case of an IgG4-RD of the salivary glands in a 61-year-old man with a long-standing history of lymphadenopathies due to CAED. We have demonstrated the coexistence of EBV and IgG4+ plasma cells both in salivary gland and lymph nodes by the patient. To the best of our knowledge, this is the first case of a well-known systemic lymphadenopathy due to chronic EBV infection associated with a coexistent IgG4-RD. We are studying the interaction between the two diseases and the clinical implications of this potentially life-threatening association (data to be published).

5 IgG4-RD and EBV: Pathogenetic Hypotheses and Prognostic Impact of This Association

The papers published to date share explanations/hypotheses about the possible pathogenetic relationship between EBV and IgG4-RD. In particular, they focus on the role of T-cells and autoimmunity or immune deregulation. Some important implications about the prognostic impact of this association also emerge.

5.1 Autoimmunity and Immune Deregulation

We know that autoimmunity is thought to play a role in the pathogenesis of IgG4-RD (Wada et al. 2013). In patients with autoimmune pancreatitis (AIP), for example, autoantibodies against lactoferrin and carbonic anhydrase II may be detected together with high serum levels of IgG4 (Okazaki et al. 2008). Moreover, other studies have also demonstrated that exists a molecular mimicry between *Helicobacter pylori* and constituents of pancreatic epithelial cells, indicating that infections may induce the development of AIP (Kountouras et al. 2005). EBV is also well recognized as a possible trigger of autoimmune disorders, such as systemic lupus erythematosus, rheumatoid arthritis, and Sjogren's syndrome (Poole et al. 2006; Toussirot and Roudier 2008; Niller et al. 2008). Usually, the clinical setting in which EBV may interact with autoimmune disorders is characterized by advanced age, transient hypergammaglobulinemia with positivity for autoantibodies, and no significant degree of immunodeficiency (Wada et al. 2013). Patients with IgG4-RD and EBV reactivation often have those clinical features, in particular the presence of ANA, suggesting an immunological dysfunction. As mentioned above, Takeuchi et al. (2014) tend to exclude that EBV infection in IgG4-RD could be due to the advanced age of the patients. We also know that EBV escapes from host defenses through many mechanisms of immune deregulation (Thompson and Kurzrock 2004). So, it seems more likely that also in IgG4-RD setting EBV reactivation induces immune deregulation and not vice versa.

5.2 T-Cells

Upregulation of Treg and Th2 activity has been reported to have a key role to increase the production of IgG4+ plasma cells in IgG4-RD (Zen et al. 2007). Other authors suggested that T-cells, along with IL-17 (a profibrotic cytokine), TGF-β and periostin, contribute to the pathogenic mechanism of IgG4-RD through the synthesis and assembly of extracellular matrix (Ohta et al. 2012, 2014; Furukawa et al. 2015). The relationship between Treg and EBV has not been fully investigated, but it is possible that upregulation of Treg activity might contribute to a disruption of immune balance (Takeuchi et al. 2014). Although the precise mechanism of immune suppression induced by Treg is not fully understood, secretion of IL-10 and TGF-β seems to be involved (Wingate et al. 2009). Previous studies have demonstrated that

EBERs of EBV are responsible for the production of several cytokines that can promote the proliferation of EBV-infected cells. EBERs bind to retinoic acid-inducible gene I (RIG-I), activating its downstream signaling and inducing the production of IL-10 (Pan et al. 2009; Ohta et al. 2012; Furukawa et al. 2015). However, in the histologic sections of the study of Takeuchi et al. (2014) the number of FOXP3+ Tregs was similar between EBER+ and EBER- groups. Although the T-cell balance in EBV infection has not yet been fully investigated (Ohshima et al. 2003), it seems that reactivation or chronic active EBV infection may alter the functions of T-cells producing specific cytokines and leading to increased production of IgG4. Shedding new lights to this pathogenetic pathway, Nagata et al. (2018) analyzed culture fluid of peripheral blood mononuclear cells (PBMCs) after the induction of EBV reactivation and they observed many IgG4+ PBMCs. In fact, activation-induced cytidine deaminase (AICD), necessary for class-switch recombination (CSR), is supplied during EBV reactivation because EBV promotes the expression of AICD mRNA through nuclear factor-kappa B (NF-KB) in newly infected B-cells (Muramatsu et al. 2000; He et al. 2003; Xu et al. 2007; Nagata et al. 2017). The CSR to IgG4 requires Th2 cytokines and in particular IL-10 from Tregs (Aalberse et al. 2009; Nirula et al. 2011). As said before, EBERs stimulate the production of human IL-10. Therefore, this fact can clarify the data observed by Shacks et al. in 1985 in patients with acute mononucleosis and EBV may contribute to IgG4 production even in systemic IgG4-RD (Nagata et al. 2018). However, further investigations are required.

5.3 Prognosis

IgG4-RD with increased numbers of EBER+ cells may have some clinical implications. First of all, many of the cases cited above have a systemic disease (Cheuk et al. 2010; Wada et al. 2013; Takeuchi et al. 2014; Furukawa et al. 2015) or have unusual manifestations (Takahashi et al. 2012; Choe et al. 2013; Szczawinska-Poplonyk et al. 2016; Kotetsu et al. 2017). Takeuchi et al. (2014) highlighted that, in contrast to usual expectation, all the patients with EBV+ type IV IgG4-related lymphadenopathy developed systemic disease and/or extranodal involvement. Furukawa et al. (2015) reported that the copy number of EBV DNA in the affected salivary glands of IgG4-RD patients was strongly correlated with the serum IgG4 concentration and was significantly higher in patients with potentially life-threatening systemic involvement, including pancreatitis, than in those without such involvement. The standard therapy of IgG4-RD with prednisolone seems to sort some effect also in cases with EBV reactivation (Cheuk et al. 2010; Takahashi et al. 2012; Wada et al. 2013; Furukawa et al. 2015; Kamisawa et al. 2015; Kotetsu et al. 2017). However, searching EBV DNA copy number or testing EBER in situ hybridization to detect EBV in tissues could be prognostically relevant in patients with features of IgG4-RD, especially in presence of systemic or unusual manifestations.

6 Conclusion

Reactivation of EBV seems to play a role in IgG4-RD through many pathways of immune deregulation and inducing directly CSR and production of IgG4+ plasma cells. EBV seems also to affect the prognosis of the disease. In particular, EBV DNA copy number seems to be strongly correlated with the risk to develop systemic or unusual manifestations of IgG4-RD. However, further studies are needed to better clarify the pathogenetic link between EBV and IgG4-RD.

Compliance with Ethical Standards The authors declare that they have no conflict of interest. Neither human participants nor animals have been involved in research studies for the writing of this paper.

References

Aalberse RC, Stapel SO, Schuurman J, Rispens T (2009) Immunoglobulin G4: an odd antibody. Clin Exp Allergy 39(4):469–477

Ambrosio MR, Rocca BJ, Ginori A, Mourmouras V, Amato T, Vindigni C, Lazzi S, Leoncini L (2015) A look into the evolution of Epstein-Barr virus-induced lymphoproliferative disorders: a case study. Am J Clin Pathol 144(5):817–822

Ambrosio MR, Mundo L, Gazaneo S, Picciolini M, Vara PS, Sayed S, Ginori A, Lo Bello G, Del Porro L, Navari M, Ascani S, Yonis A, Leoncini L, Piccaluga PP, Lazzi S (2017) MicroRNAs sequencing unveils distinct molecular subgroups of plasmablastic lymphoma. Oncotarget 8(64): 107356–107373

Amorosa LF, Shear MK, Spiera H (1976) Multifocal fibrosis involving the thyroid, face, and orbits. Arch Intern Med 136(2):221–223

Brittig F, Ajtay E, Jaksó P, Kelényi G (2004) Follicular dendritic reticulum cell tumor mimicking inflammatory pseudotumor of the spleen. Pathol Oncol Res 10(1):57–60

Burkitt D (1962) A "tumour safari" in East and Central Africa. Br J Cancer 16(3):379–386

Carruthers MN, Khosroshahi A, Augustin T, Deshpande V, Stone JH (2015) The diagnostic utility of serum IgG4 concentrations in IgG4-related disease. Ann Rheum Dis 74(1):14–18. https://doi.org/10.1136/annrheumdis-2013-204907

Cheuk W, Tam FK, Chan AN, Luk IS, Yuen AP, Chan WK, Hung TC, Chan JK (2010) Idiopathic cervical fibrosis–a new member of IgG4-related sclerosing diseases: report of 4 cases, 1 complicated by composite lymphoma. Am J Surg Pathol 34(11):1678–1685

Choe JY, Go H, Jeon YK, Yun JY, Kim YA, Kim HJ, Huh J, Lee H, Shin DH, Kim JE (2013) Inflammatory pseudotumor-like follicular dendritic cell sarcoma of the spleen: a report of six cases with increased IgG4-positive plasma cells. Pathol Int 63(5):245–251

Cohen JI, Kimura H, Nakamura S, Ko YH, Jaffe ES (2009) Epstein-Barr virus-associated lymphoproliferative disease in non-immunocompromised hosts: a status report and summary of an international meeting, 8-9 September 2008. Ann Oncol 20(9):1472–1482

Cohen JI, Jaffe ES, Dale JK, Pittaluga S, Heslop HE, Rooney CM, Gottschalk S, Bollard CM, Rao VK, Marques A, Burbelo PD, Turk SP, Fulton R, Wayne AS, Little RF, Cairo MS, El-Mallawany NK, Fowler D, Sportes C, Bishop MR, Wilson W, Straus SE (2011) Characterization and treatment of chronic active Epstein-Barr virus disease: a 28-year experience in the United States. Blood 117(22):5835–5849

Crawford DH, Macsween KF, Higgins CD, Thomas R, McAulay K, Williams H, Harrison N, Reid S, Conacher M, Douglas J, Swerdlow AJ (2006) A cohort study among university students: identification of risk factors for Epstein-Barr virus seroconversion and infectious mononucleosis. Clin Infect Dis 43(3):276–282

De Falco G, Antonicelli G, Onnis A, Lazzi S, Bellan C, Leoncini L (2009) Role of EBV in microRNA dysregulation in Burkitt lymphoma. Semin Cancer Biol 19(6):401–406
de la Hera Magallanes AI, Montes-Moreno S, Hernández SG, Hernández-León CN, Lopez M, Pajares R, Pinilla SM, Piris MA (2011) Early phase of Epstein-Barr virus (EBV)-positive diffuse large B cell lymphoma of the elderly mimicking EBV-positive reactive follicular hyperplasia. Histopathology 59(3):571–575
Dehner LP, Coffin CM (1998) Idiopathic fibrosclerotic disorders and other inflammatory pseudotumors. Semin Diagn Pathol 15(2):161–173
Delecluse HJ, Feederle R, O'Sullivan B, Taniere P (2007) Epstein Barr virus-associated tumours: an update for the attention of the working pathologist. J Clin Pathol 60(12):1358–1364
Delgado-Eckert E, Shapiro M (2011) A model of host response to a multi-stage pathogen. J Math Biol 63(2):201–227
Deshpande V, Zen Y, Chan JK, Yi EE, Sato Y, Yoshino T, Klöppel G, Heathcote JG, Khosroshahi A, Ferry JA, Aalberse RC, Bloch DB, Brugge WR, Bateman AC, Carruthers MN, Chari ST, Cheuk W, Cornell LD, Fernandez-Del Castillo C, Forcione DG, Hamilos DL, Kamisawa T, Kasashima S, Kawa S, Kawano M, Lauwers GY, Masaki Y, Nakanuma Y, Notohara K, Okazaki K, Ryu JK, Saeki T, Sahani DV, Smyrk TC, Stone JR, Takahira M, Webster GJ, Yamamoto M, Zamboni G, Umehara H, Stone JH (2012) Consensus statement on the pathology of IgG4-related disease. Mod Pathol 25(9):1181–1192
Dojcinov SD, Venkataraman G, Raffeld M, Pittaluga S, Jaffe ES (2010) EBV positive mucocutaneous ulcer–a study of 26 cases associated with various sources of immunosuppression. Am J Surg Pathol 34(3):405–417
Dojcinov SD, Venkataraman G, Pittaluga S, Wlodarska I, Schrager JA, Raffeld M, Hills RK, Jaffe ES (2011) Age-related EBV-associated lymphoproliferative disorders in the Western population: a spectrum of reactive lymphoid hyperplasia and lymphoma. Blood 117(18):4726–4735
Ebbo M, Grados A, Guedj E, Gobert D, Colavolpe C, Zaidan M, Masseau A, Bernard F, Berthelot JM, Morel N, Lifermann F, Palat S, Haroche J, Mariette X, Godeau B, Bernit E, Costedoat-Chalumeau N, Papo T, Hamidou M, Harlé JR, Schleinitz N (2014) Usefulness of 2-[18F]-fluoro-2-deoxy-D-glucose-positron emission tomography/computed tomography for staging and evaluation of treatment response in IgG4-related disease: a retrospective multicenter study. Arthritis Care Res (Hoboken) 66(1):86–96
Epstein MA, Achong BG, Barr YM (1964) Virus particles in cultured lymphoblasts from burkitt's lymphoma. Lancet 1(7335):702–703
Furukawa T, Shimotai Y, Ohta N, Ishida A, Kurakami K, Suzuki H, Yamakawa M, Hongo S, Kakehata S (2015) Epstein-Barr virus in the enlarged salivary tissues of patients with IgG4-related disease. Laryngoscope 125(9):2079–2084
Ghazale A, Chari ST, Smyrk TC, Levy MJ, Topazian MD, Takahashi N, Clain JE, Pearson RK, Pelaez-Luna M, Petersen BT, Vege SS, Farnell MB (2007) Value of serum IgG4 in the diagnosis of autoimmune pancreatitis and in distinguishing it from pancreatic cancer. Am J Gastroenterol 102(8):1646–1653
Hamano H, Kawa S, Horiuchi A, Unno H, Furuya N, Akamatsu T, Fukushima M, Nikaido T, Nakayama K, Usuda N, Kiyosawa K (2001) High serum IgG4 concentrations in patients with sclerosing pancreatitis. N Engl J Med 344(10):732–738
Hatton OL, Harris-Arnold A, Schaffert S, Krams SM, Martinez OM (2014) The interplay between Epstein-Barr virus and B lymphocytes: implications for infection, immunity, and disease. Immunol Res 58(2–3):268–276
He B, Raab-Traub N, Casali P, Cerutti A (2003) EBV-encoded latent membrane protein 1 cooperates with BAFF/BLyS and APRIL to induce T cell-independent Ig heavy chain class switching. J Immunol 171(10):5215–5224
Hjalgrim H, Friborg J, Melbye M (2007) Chapter 53: The epidemiology of EBV and its association with malignant disease. In: Arvin A, Campadelli-Fiume G, Mocarski E, Moore PS, Roizman B, Whitley R, Yamanishi K (eds) Human herpesviruses: biology, therapy, and immunoprophylaxis. Cambridge University Press, Cambridge

Holmes RD, Sokol RJ (2002) Epstein-Barr virus and post-transplant lymphoproliferative disease. Pediatr Transplant 6(6):456–464

Kamisawa T, Egawa N, Nakajima H (2003a) Autoimmune pancreatitis is a systemic autoimmune disease. Am J Gastroenterol 98(12):2811–2812

Kamisawa T, Funata N, Hayashi Y, Eishi Y, Koike M, Tsuruta K, Okamoto A, Egawa N, Nakajima H (2003b) A new clinicopathological entity of IgG4-related autoimmune disease. J Gastroenterol 38(10):982–984

Kamisawa T, Zen Y, Pillai S, Stone JH (2015) IgG4-related disease. Lancet 385(9976):1460–1471

Kanno A, Nishimori I, Masamune A, Kikuta K, Hirota M, Kuriyama S, Tsuji I, Shimosegawa T (2012) Research committee on intractable diseases of pancreas. Nationwide epidemiological survey of autoimmune pancreatitis in Japan. Pancreas 41(6):835–839

Kenney SC, Mertz JE (2014) Regulation of the latent-lytic switch in Epstein-Barr virus. Semin Cancer Biol 26:60–68. https://doi.org/10.1016/j.semcancer.2014.01.002

Khosroshahi A, Bloch DB, Deshpande V, Stone JH (2010) Rituximab therapy leads to rapid decline of serum IgG4 levels and prompt clinical improvement in IgG4-related systemic disease. Arthritis Rheum 62(6):1755–1762

Khosroshahi A, Carruthers MN, Deshpande V, Unizony S, Bloch DB, Stone JH (2012) Rituximab for the treatment of IgG4-related disease: lessons from 10 consecutive patients. Medicine (Baltimore) 91(1):57–66

Kotetsu Y, Ikegame S, Takebe-Akazawa K, Koga T, Okabayashi K, Takata S (2017) A case of IgG4-related lung disease complicated by asymptomatic chronic Epstein-Barr virus infection. Clin Respir J 11(6):1012–1017

Kountouras J, Zavos C, Chatzopoulos D (2005) A concept on the role of Helicobacter pylori infection in autoimmune pancreatitis. J Cell Mol Med 9(1):196–207

Laichalk LL, Thorley-Lawson DA (2005) Terminal differentiation into plasma cells initiates the replicative cycle of Epstein-Barr virus in vivo. J Virol 79(2):1296–1307

Ma SD, Hegde S, Young KH, Sullivan R, Rajesh D, Zhou Y, Jankowska-Gan E, Burlingham WJ, Sun X, Gulley ML (2011) A new model of Epstein-Barr virus infection reveals an important role for early lytic viral protein expression in the development of lymphomas. J Virol 85:165–177

Matsui S, Hebisawa A, Sakai F, Yamamoto H, Terasaki Y, Kurihara Y, Waseda Y, Kawamura T, Miyashita T, Inoue H, Hata N, Masubuchi H, Sugino K, Kishi J, Kobayashi H, Usui Y, Komazaki Y, Kawabata Y, Ogura T (2013) Immunoglobulin G4-related lung disease: clinicoradiological and pathological features. Respirology 18(3):480–487

Muramatsu M, Kinoshita K, Fagarasan S, Yamada S, Shinkai Y, Honjo T (2000) Class switch recombination and hypermutation require activation-induced cytidine deaminase (AID), a potential RNA editing enzyme. Cell 102(5):553–563

Nagata K, Kumata K, Nakayama Y, Satoh Y, Sugihara H, Hara S, Matsushita M, Kuwamoto S, Kato M, Murakami I, Hayashi K (2017) Epstein-Barr virus lytic reactivation activates B cells Polyclonally and induces activation-induced cytidine deaminase expression: a mechanism underlying autoimmunity and its contribution to Graves' disease. Viral Immunol 30(3):240–249

Nagata K, Hara S, Nakayama Y, Higaki K, Sugihara H, Kuwamoto S, Matsushita M, Kato M, Tanio S, Ishiguro K, Hayashi K (2018) Epstein-Barr virus lytic reactivation induces IgG4 production by host B lymphocytes in Graves' disease patients and controls: a subset of Graves' disease is an IgG4-related disease-like condition. Viral Immunol 31(8):540–547

Niller HH, Wolf H, Minarovits J (2008) Regulation and dysregulation of Epstein-Barr virus latency: implications for the development of autoimmune diseases. Autoimmunity 41(4):298–328

Nirula A, Glaser SM, Kalled SL, Taylor FR (2011) What is IgG4? A review of the biology of a unique immunoglobulin subtype. Curr Opin Rheumatol 23(1):119–124

Ohshima K, Karube K, Hamasaki M, Tutiya T, Yamaguchi T, Suefuji H, Suzuki K, Suzumiya J, Ohga S, Kikuchi M (2003) Differential chemokine, chemokine receptor and cytokine expression in Epstein-Barr virus-associated lymphoproliferative diseases. Leuk Lymphoma 44(8): 1367–1378

Ohshima K, Kimura H, Yoshino T, Kim CW, Ko YH, Lee SS, Peh SC, Chan JK (2008) CAEBV Study Group. Proposed categorization of pathological states of EBV-associated T/natural killer-cell lymphoproliferative disorder (LPD) in children and young adults: overlap with chronic active EBV infection and infantile fulminant EBV T-LPD. Pathol Int 58(4):209–217

Ohta N, Makihara S, Okano M, Kurakami K, Ishida A, Furukawa T, Suzuki Y, Watanabe T, Kakehata S, Aoyagi M (2012) Roles of IL-17, Th1, and Tc1 cells in patients with IgG4-related sclerosing sialadenitis. Laryngoscope 122(10):2169–2174

Ohta N, Ishida A, Kurakami K, Suzuki Y, Kakehata S, Ono J, Ikeda H, Okubo K, Izuhara K (2014) Expressions and roles of periostin in otolaryngological diseases. Allergol Int 63(2):171–180

Okazaki K, Uchida K, Ohana M, Nakase H, Uose S, Inai M, Matsushima Y, Katamura K, Ohmori K, Chiba T (2000) Autoimmune-related pancreatitis is associated with autoantibodies and a Th1/Th2-type cellular immune response. Gastroenterology 118(3):573–581

Okazaki K, Uchida K, Fukui T (2008) Recent advances in autoimmune pancreatitis: concept, diagnosis, and pathogenesis. J Gastroenterol 43(6):409–418

Ota M, Katsuyama Y, Hamano H, Umemura T, Kimura A, Yoshizawa K, Kiyosawa K, Fukushima H, Bahram S, Inoko H, Kawa S (2007) Two critical genes (HLA-DRB1 and ABCF1)in the HLA region are associated with the susceptibility to autoimmune pancreatitis. Immunogenetics 59(1):45–52

Pan SH, Tai CC, Lin CS, Hsu WB, Chou SF, Lai CC, Chen JY, Tien HF, Lee FY, Wang WB (2009) Epstein-Barr virus nuclear antigen 2 disrupts mitotic checkpoint and causes chromosomal instability. Carcinogenesis 30(2):366–375

Pattle SB, Farrell PJ (2006) The role of Epstein-Barr virus in cancer. Expert Opin Biol Ther 6(11): 1193–1205

Pittaluga S (2013) Viral-associated lymphoid proliferations. Semin Diagn Pathol 30(2):130–136

Poole BD, Scofield RH, Harley JB, James JA (2006) Epstein-Barr virus and molecular mimicry in systemic lupus erythematosus. Autoimmunity 39(1):63–70

Quintanilla-Martinez L, Kumar S, Fend F, Reyes E, Teruya-Feldstein J, Kingma DW, Sorbara L, Raffeld M, Straus SE, Jaffe ES (2000) Fulminant EBV(+) T-cell lymphoproliferative disorder following acute/chronic EBV infection: a distinct clinicopathologic syndrome. Blood 96(2): 443–451

Quintanilla-Martinez L, Kimura H, Jaffe ES (2008) EBV-positive T-cell and NK-cell lymphoproliferative disease of childhood. In: Swerdlow SH, Campo E, Harris NL et al (eds) WHO classification of hematopoietic and lymphoid tissues. IARC, Lyon, pp 278–280

Rezk SA, Zhao X, Weiss LM (2018) Epstein-Barr virus (EBV)-associated lymphoid proliferations, a 2018 update. Hum Pathol 79:18–41

Rice DH, Batsakis JG, Coulthard SW (1975) Sclerosing cervicitis: homologue of sclerosing retroperitonitis and mediastinitis. Arch Surg 110(1):120–122

Robertson ES (2010) Epstein-Barr virus: latency and transformation. Caiser Academic Press

Rosenbaum L, Fekrazad MH, Rabinowitz I, Vasef MA (2009) Epstein-Barr virus-associated inflammatory pseudotumor of the spleen: report of two cases and review of the literature. J Hematop 2(2):127–131

Roughan JE, Thorley-Lawson DA (2009) The intersection of Epstein-Barr virus with the germinal center. J Virol 83(8):3968–3976

Sadler R, Chapman RW, Simpson D, Soonawalla ZF, Waldegrave EL, Burden JM, Misbah SA, Ferry BL (2011) The diagnostic significance of serum IgG4 levels in patients with autoimmune pancreatitis: a UK study. Eur J Gastroenterol Hepatol 23(2):139–145

Sato Y, Inoue D, Asano N, Takata K, Asaoku H, Maeda Y, Morito T, Okumura H, Ishizawa S, Matsui S, Miyazono T, Takeuchi T, Kuroda N, Orita Y, Takagawa K, Kojima M, Yoshino T (2012) Association between IgG4-related disease and progressively transformed germinal centers of lymph nodes. Mod Pathol 25(7):956–967

Shacks SJ, Heiner DC, Bahna SL, Horwitz CA (1985) Increased serum IgG4 levels in acute Epstein-Barr viral mononucleosis. Ann Allergy 54(4):284–288

Stone JH, Zen Y, Deshpande V (2012) IgG4-related disease. N Engl J Med 366(6):539–551

Szczawinska-Poplonyk A, Wojsyk-Banaszak I, Jonczyk-Potoczna K, Breborowicz A (2016) Pulmonary manifestation of immunoglobulin G4-related disease in a 7-year-old immunodeficient boy with Epstein-Barr virus infection: a case report. Ital J Pediatr 42(1):58

Tabata T, Kamisawa T, Takuma K, Anjiki H, Egawa N, Kurata M, Honda G, Tsuruta K, Setoguchi K, Obayashi T, Sasaki T (2009) Serum IgG4 concentrations and IgG4-related sclerosing disease. Clin Chim Acta 408(1–2):25–28

Takahashi E, Kojima M, Kobayashi M, Kitamura A, Yokoi T, Hara K, Nakamura S (2012) Primary IgG4-related lymphadenopathy with prominent granulomatous inflammation and reactivation of Epstein-Barr virus. Virchows Arch 460(2):225–229

Takeuchi M, Sato Y, Yasui H, Ozawa H, Ohno K, Takata K, Gion Y, Orita Y, Tachibana T, Itoh T, Asano N, Nakamura S, Swerdlow SH, Yoshino T (2014) Epstein-Barr virus-infected cells in IgG4-related lymphadenopathy with comparison with extranodal IgG4-related disease. Am J Surg Pathol 38(7):946–955

Thompson MP, Kurzrock R (2004) Epstein-Barr virus and cancer. Clin Cancer Res 10(3):803–821

Thorley-Lawson DA (2001) Epstein-Barr virus: exploiting the immune system. Nat Rev Immunol 1(1):75–82

Thorley-Lawson DA, Allday MJ (2009) The curious case of the tumour virus: 50 years of Burkitt's lymphoma. Nat Rev Microbiol 6(12):913–924

Toussirot E, Roudier J (2008) Epstein-Barr virus in autoimmune diseases. Best Pract Res Clin Rheumatol 22(5):883–896

Umehara H, Okazaki K, Masaki Y, Kawano M, Yamamoto M, Saeki T, Matsui S, Sumida T, Mimori T, Tanaka Y, Tsubota K, Yoshino T, Kawa S, Suzuki R, Takegami T, Tomosugi N, Kurose N, Ishigaki Y, Azumi A, Kojima M, Nakamura S, Inoue D (2012) Research Program for Intractable Disease by Ministry of Health, Labor and Welfare (MHLW) Japan G4 team. A novel clinical entity, IgG4-related disease (IgG4RD): general concept and details. Mod Rheumatol 22(1):1–14

Wada Y, Kojima M, Yoshita K, Yamazaki M, Kobayashi D, Murakami S, Nishi S, Nakano M, Narita I (2013) A case of Epstein-Barr virus-related lymphadenopathy mimicking the clinical features of IgG4-related disease. Mod Rheumatol 23(3):597–603

Wingate PJ, McAulay KA, Anthony IC, Crawford DH (2009) Regulatory T cell activity in primary and persistent Epstein-Barr virus infection. J Med Virol 81(5):870–877

Xu Z, Pone EJ, Al-Qahtani A, Park SR, Zan H, Casali P (2007) Regulation of AICDA expression and AID activity: relevance to somatic hypermutation and class switch DNA recombination. Crit Rev Immunol 27(4):367–397

Zen Y, Nakanuma Y (2010) IgG4-related disease: a cross-sectional study of 114 cases. Am J Surg Pathol 34(12):1812–1819

Zen Y, Fujii T, Harada K, Kawano M, Yamada K, Takahira M, Nakanuma Y (2007) Th2 and regulatory immune reactions are increased in immunoglobin G4-related sclerosing pancreatitis and cholangitis. Hepatology 45(6):1538–1546

Zen Y, Inoue D, Kitao A, Onodera M, Abo H, Miyayama S, Gabata T, Matsui O, Nakanuma Y (2009) IgG4-related lung and pleural disease: a clinicopathologic study of 21 cases. Am J Surg Pathol 33(12):1886–1893

Index

© The Editor(s) (if applicable) and The Author(s), under exclusive license to Springer Nature Switzerland AG 2024
N. Rezaei (ed.), *Hematological Cancer Pathophysiology: An Interdisciplinary Approach*, Interdisciplinary Cancer Research 10,
https://doi.org/10.1007/978-3-031-75445-6

SPRINGER NATURE

GPSR Compliance

The European Union's (EU) General Product Safety Regulation (GPSR) is a set of rules that requires consumer products to be safe and our obligations to ensure this.

If you have any concerns about our products, you can contact us on ProductSafety@springernature.com

In case Publisher is established outside the EU, the EU authorized representative is:

Springer Nature Customer Service Center GmbH
Europaplatz 3
69115 Heidelberg, Germany

The manufacturer's authorised representative in the EU is Springer Nature Customer Service Centre GmbH, Europaplatz 3, 69115 Heidelberg, Germany. If you have any concerns regarding our products, please contact ProductSafety@springernature.com

Printed and bound by CPI Group (UK) Ltd, Croydon, CR0 4YY
30/09/2026
02225598-0006